REDUCING REGIONAL INEQUALITIES

Edited by

Alex Bowen
Head of Policy Analysis and Statistics
National Economic Development Office

and

Ken Mayhew
Economic Director (until December 1990)
National Economic Development Office

Foreword by Walter Eltis, D. Litt
Director General
National Economic Development Office

KOGAN PAGE

First published in 1991

Apart from any fair dealing for the purposes of research or private study, or criticism or review, as permitted under the Copyright, Designs and Patents Act, 1988, this publication may only be reproduced, stored or transmitted, in any form or by any means, with the prior permission in writing of the publishers, or in the case of reprographic reproduction in accordance with the terms of licences issued by the Copyright Licensing Agency. Enquiries concerning reproduction outside those terms should be sent to the publishers at the undermentioned address:

Kogan Page Limited
120 Pentonville Road
London N1 9JN

in association with the

National Economic Development Office
Millbank Tower
Millbank
London SW1P 4QX

© Crown Copyright 1991

British Library Cataloguing in Publication Data

A CIP record for this book is available from the British Library.

ISBN 0 7494 0444 2

Typeset by DP Photosetting, Aylesbury, Bucks
Printed and bound in Great Britain by
Clays Ltd, St. Ives PLC

Contents

List of Figures	7
List of Tables	11
About NEDO	17
Notes on the Contributors	18
Foreword by Walter Eltis	23

1. Regional Issues in Economics: Setting the Scene   27
   Alex Bowen and Ken Mayhew

2. Regional Economic Disparities: Causes and Consequences   70
   Jim Taylor

3. Regional Policy and Market Forces: A Model and an Assessment   109
   Patrick Minford and Peter Stoney

4. Regional Earnings and Pay Flexibility   185
   Janet Walsh and William Brown

5. Migration and Regional Policy   216
   Barry McCormick

6. Regional Economic Disparities: The Role of Housing   246
   John Muellbauer and Anthony Murphy

7. Regional Economic Development in the UK: The Case of the Northern Region of England   272
   John B Goddard and Alfred T Thwaites

Reducing Regional Inequalities

8	Transport Infrastructure and Regional Policy	321
	Alan Armitage and Derek Palmer of the Confederation of British Industry	
9	Regional Development in the 1990s: A Trade Union Perspective	333
	Trades Union Congress	
10	Regional Economic Disparities: Some Public Policy Issues *Alex Bowen and Ken Mayhew*	355

Index   367

List of Figures

2.1	Regional unemployment disparities in selected UK regions, 1974–89	71
2.2	Inter-county differences in unemployment rates by region, April 1989	72
2.3	Inter-county disparities in average gross weekly earnings by region: full-time males, April 1989	73
2.4	Inter-county disparities in average gross weekly earnings by region: full-time females, April 1989	74
2.5	Percentage unemployed in selected UK regions, 1927–39	75
2.6	Regional unemployed disparities and percentage unemployment in the UK, 1951–89	76
2.7	Scatter plot of average earnings versus unemployment by UK county, 1988	80
2.8	House price disparities: regional variations in the UK, 1983–89	82
2.9	Estimated efficiency wage in the UK counties by region, 1988	83
2.10	Employment in manufacturing and non-manufacturing in the UK, 1948–88	86
2.11	Regional disparities in unemployment in the EC, 1987	98
2.12	Regional disparities in GDP per capita in the EC, 1986	99
2.13	Percentage unemployment versus capacity utilisation in the UK, 1974–89	102
2.14	CBI capacity utilisation index, 1972–89	103

2.15	House price inflation in the UK: a regional analysis, 1984-9	104
2.16	House prices versus unemployment rate by UK county, April 1988	105
3.1	Typical regions' manual labour market	112
3.2	Regional land market (industrial and housing land)	113
3.3	Index of regional GDP	126
3.4	Terms of trade	128
3.5	The effect of falling manufacturing/services terms of trade	129
3.6	The effect of the northern transport cost disadvantage	132
3.7	Effect of manual labour migration	140
3.8	The union mark-up	142
3.App 1.1	The structure of the traded goods sector in a region	166
3.App 1.2	Equilibrium size (and structure between protected and open sectors) of typical region	167
3.App 3.1	Static predictions over the past: full model	176
3.App 3.1a	Actual and predicted consumption expenditure plus manufacturing investment	176
3.App 3.1b	Actual and predicted average gross weekly wage	177
3.App 3.1c	Actual and predicted manufacturing production	177
3.App 3.1d	Actual and predicted non-manufacturing GDP	178
3.App 3.1e	Actual and predicted non-manufacturing employment	178
3.App 3.1f	Actual and predicted manufacturing employment	179
3.App 3.1g	Actual and predicted unemployment	179

List of Figures

3.App 3.1h	Actual and predicted working population	180
4.1	Regional convergence in earnings, 1960–86	186
4.2	Coefficients of variation in earnings	188
4.3	Decomposition of variability in relative earnings across regions	189
5.1a	Scatter diagram of migration and relative unemployment for ten selected counties, July 1931 – July 1936	224
5.1b	Scatter diagram of migration and relative unemployment by divisions	224
6.1	The ratio of South East to UK mix-adjusted house prices (all dwellings excluding sales not at market prices)	254
6.2	The ratio of South East to UK male earnings (full-time adults whose pay was not affected by absence)	255
6.3	The ratio of HP/W in the South East to HP/W in the UK (HP = mix-adjusted house prices; W = non-manual earnings)	255
6.4	Net migration from South East and 'Greater' South East regions	261
7.1	Change in market access (regional market assumptions used)	295
9.1	Government spending on regional policy in the 1980s	334
9.2	Regional growth prospects to 2000	336

List of Tables

1.1	Structure of the European Community budget	47
1.2	Shares of Regional Development Fund spending	49
2.1	Regional disparities in output per capita, 1971–87	77
2.2	Regional disparities in employment growth, 1971–88	78
2.3	Regional disparities in income and expenditure	79
2.4	Regression analysis of inter-county disparities in average earnings, 1988	80
2.5	Regional disparities in house prices, 1989	81
2.6	Regional migration flows, 1981–7	82
2.7	The contribution of each region's industry mix to its employment growth, 1981–8	85
2.8	Regional disparities in the educational attainments of school leavers, 1986–7	87
2.9	Regional distribution of graduates in GB	88
2.10	New firm formation rates: the growth in the stock of businesses, 1980–8	89
2.11	Factors likely to affect regional disparities in the birth-rate of new firms: some surrogate variables	90
2.12	Regional distribution of investments by the Business Start-Up Scheme and the Business Expansion Scheme, 1981/2 to 1986/7	91
2.13	Regional distribution of venture capital funds and investments	92

2.14	Local control of manufacturing firms: distance of plant from head office	93
2.15	Employment in Japanese-owned manufacturing companies in the UK regions, 1988	94
2.16	Significant innovations in the manufacturing sector in each UK region, 1945-83	96
3.1	Manufacturing and services factor intensity (%)	114
3.2	National employment by occupation and tenure	116
3.3	Regional wages and unemployment	117
3.4	Regional land price trends (1980 prices, £000 per hectare)	118
3.5	Regional land use	120
3.6a	Inter-regional migration, 1980-6 (thousands)	120
3.6b	Migration, population and labour force trends, 1979-87	121
3.7	Manufacturing intensiveness of regions (ratio of manufacturing to unprotected services)	122
3.8	Regional industrial structure	123
3.9	Regional profits as a share of GDP (%), 1971-87	125
3.10	Semi/unskilled labour intensity – whole economy	130
3.11	Transport cost differentials, non-oil trade, Manchester–London	131
3.12	Regional unionisation rates	134
3.13	Domestic rates as a tax (rates paid per annum as a percentage of house prices)	135
3.14	Non-domestic rates as a percentage of approximate market rental value	136
3.15	National unemployment trends from 1982	141

List of Tables

3.16	Central and pessimistic forecast UK and world economic forecast (% pa)	150
3.17a	Regional profiles: North	153
3.17b	Regional profiles: Yorkshire and Humberside	154
3.17c	Regional profiles: East Midland	155
3.17d	Regional profiles: East Anglia	156
3.17e	Regional profiles: South East	157
3.17f	Regional profiles: South West	158
3.17g	Regional profiles: West Midland	159
3.17h	Regional profiles: North West	160
3.17i	Regional profiles: Wales	161
3.17j	Regional profiles: Scotland	162
3.17k	Regional profiles: Northern Ireland	163
3.App 1.1	The model algebra	168
3.App 2.1	Typical northern freight differential	173
3.App 2.2	Rail freight rates	175
3.App 3.1	The sub-region model (estimated over 1976–87)	181
4.1	Average gross weekly earnings of adult full-time male employees by region expressed as a percentage of the unweighted average of all regional averages	187
4.2	South East regional earnings differentials, 1970–89	188
4.3	Most important levels of bargaining influencing pay increases in the private sector, 1984 (1980)	193
4.4	The fragmentation of bargaining on individual items of pay and conditions: all sectors manufacturing, 1986	194

4.5	Matched sample, 1979-86, changes in the structure of bargaining: rates of pay	195
4.6	Multi-employer negotiating bodies dissolved since 1986	195
4.7	Pay settlement pressures in private manufacturing and services	203
4.8	Regional house price disparities, 1989	204
4.9	Increases in Inner London allowances, 1985-9	206
4.10	Local government labour shortages	210
5.1	Persistence of regional unemployment rates	219
5.2	Average net migration and unemployment rates for heads of household, 1983-6	222
6.1	The regional pattern of housing tenure	250
6.2	Trends in housing tenure structure	251
6.3	Regional house prices differentials	254
7.1	The North in recession, 1979-83	277
7.2	Employment change in the North in recovery	278
7.3	Unemployment change, labour market participation and migration, 1983-8	279
7.4	Indicators of regional prosperity, 1983-9	279
7.5	Changes in regional income distribution and wages	280
7.6	Investment in manufacturing industry, 1981-6	281
7.7	The industrial structure of the North, 1987	283
7.8	Shift-share analysis of manufacturing employment change in the North, 1952-79	284
7.9	Some notable job losses in the North East electronics industry	285

List of Tables

7.10	High technology industries in the North	286
7.11	Shift-share analysis of innovations, 1945–79	287
7.12	Adoption of microprocessors in engineering establishments	288
7.13	Technology transfer into the North, 1973–7	288
7.14	Research and development in the North, 1981–6	290
7.15	Occupational structure, Tyne and Wear and Berkshire, 1981	291
7.16	Net change in business registrations, 1980–6	292
7.17	Factors accounting for inter-county variations in local enterprise activity potential, 1980–5	293
7.18	New firms in Teesside and Reading, 1980–5	294
7.19	Expenditure on transport and telecommunications, 1987	296
7.20	Average components of expenditure (£000s) on transport and communication	297
7.21	Computer facilities, 1984	298
7.22	Computer networks and company performance, 1985–7	298
7.23	Service employment in the North, 1978–89	301
7.24	Residual levels of financial and producer service and employment not 'explained' by local levels of 'consumer demand' and 'business demand', 1984	303
7.25	Employment: main locations of non-industrial civil service staff, 1979–87	304
7.26	Components of change analysis of foreign-owned manufacturing establishments in the North, 1978–89	306
8.1	Land-borne traffic, 1988	323
8.2	Infrastructure investment, 1984–7	325

15

Reducing Regional Inequalities

9.1 Main influences on foreign direct investment decisions in industrialised countries 345

About NEDO

These papers have been prepared for publication by the National Economic Development Office.

The National Economic Development Council (NEDC) brings together representatives of Government, management, the trade unions and other interests to assess economic performance and opportunities for improving it. The NEDC meets quarterly, under the chairmanship of the Chancellor of the Exchequer and other Secretaries of State.

There are 18 sector groups and working parties covering different parts of industry or working on practical industrial issues. Sectors covered include, construction, electronics, engineering, food, textiles, and tourism and leisure. Working parties are studying biotechnology, traffic management systems and the European public sector market.

The National Economic Development Office supports the work of the Council and its sector groups and working parties. It carries out independent research and provides advice on ways of improving economic performance, competitive power and the efficiency of industry, stimulating new ideas and practical action.

Notes on the Contributors

Alan Armitage is an economist from University College London who spent the first three years of his career at the CBI, working on macroeconomic topics and analysis of the Industrial Trends Survey. He moved to the National Economic Development Office, spending ten years in the Industrial Division, examining aspects of the engineering, plastics and electronics industries. He returned to the CBI in 1988 as Deputy Director of Company Affairs, heading the Industrial Policy Group which is responsible for CBI policy formation on energy, transport, local government and public procurement.

Dr Alex Bowen is Head of Policy Analysis and Statistics. He studied economics at Clare College, Cambridge and won a Kennedy Scholarship to the Massachusetts Institute of Technology, where he obtained his PhD. Subsequently he has worked as a Lecturer in Economics at Brunel University, a Research Officer at the London School of Economics, and as a consultant to the World Bank, joining the National Economic Development Office in 1987. He has been Head of Policy Analysis and Statistics at NEDO since April 1989.

Professor William Brown is Professor of Industrial Relations and a Fellow of Wolfson College, Cambridge University. He has worked at the National Board for Prices and Incomes and the Industrial Relations Research Unit at Warwick University. His research has been concerned with workplace bargaining, shop steward organisation, payment systems, arbitration, incomes policies and pay determination. He is currently writing a book on the pay policies of large firms. Publications include *Piecework Bargaining* and *The Changing Contours of British Industrial Relations*.

Dr Walter Eltis became Director General of the National Economic Development Office in November 1988, after two years as Economic Director. He was a Consultant to NEDO from 1963 to 1966. From 1963

Notes on the Contributors

until 1988, he was Official Fellow and Tutor in Economics at Exeter College, Oxford.

Professor John Goddard OBE is Director of the Centre for Urban and Regional Development Studies (CURDS) at the University of Newcastle-upon-Tyne. Prior to joining the University in 1975 he was Head of the Department of Geography at the London School of Economics. Professor Goddard is a consultant to the Organisation for Economic Co-operation and Development (OECD) on urban economic development and on telecommunications and regional development. He is an adviser to the Tyne and Wear Urban Development Corporation and is a Director of the Newcastle Initiative, a private sector-led urban regeneration programme.

Ken Mayhew was Economic Director of the National Economic Development Office during 1989 and 1990 while he was on leave from Oxford University, where he is Fellow and Tutor in Economics at Pembroke College. He has been an Economic Assistant at the Treasury, Consultant at the CBI and Senior Tutor of the Oxford University Business Summer School.

Dr Barry McCormick is currently Reader in Economics at the University of Southampton. He has published widely in the areas of labour economics and labour market policy, with articles in the *Economic Journal*, *European Economic Review* and *Review of Economics and Statistics*, among others. Dr McCormick has a long-standing interest in labour mobility and unemployment.

Professor Patrick Minford has been Professor of Applied Economics, University of Liverpool, since 1976. Formerly Adviser to HM Treasury's External Division, serving with HM Treasury's Delegation to Washington DC, 1973-74. Published works include *Substitution Effects, Speculation and Exchange Rate Stability*; *Unemployment: Cause and Cure* and *The Housing Morass*.

Derek Palmer has degrees in economics from the Universities of Liverpool and Birmingham. After working as a researcher at the Centre for Environmental Studies, he joined the House of Commons Select Committee on the Environment as an adviser. He joined the CBI in 1984 and set up the Employee Relocation Council. Derek Palmer took over responsibility for transport at the CBI in 1988 and is Secretary to the CBI's Transport Policy Committee. In November 1989 the CBI published 'Trade Routes to the Future', of which he was co-author. He is a member of the NEDO Traffic Management Working Party. He is also a Fellow of the Institution of Highways and Transportation, as well as being a Member of the Chartered Institute of Transport, and recently joined the CIT Transport Policy Committee.

Dr Peter Stoney joined the Department of Economics at Liverpool University in 1970. His main research area is the Merseyside economy, a subject on which he has written extensively.

Jim Taylor has been Professor of Economics at Lancaster University since 1983. His publications on regional issues include *Regional Economics and Policy* and *Regional Policy and the North–South Divide*, with Harvey Armstrong, as well as several articles on evaluating the effectiveness of regional policy. He has been editor of *Regional Studies*, the Journal of the Regional Studies Association, since 1986.

Alfred Thwaites is Deputy Director of the Centre for Urban and Regional Development Studies (CURDS), with research interests in technological change, in particular within the engineering and traditional industries. He has wide-ranging experience of designing, directing and carrying out research in industry, both at home and abroad, and particularly across Europe.

Dr Janet Walsh is a Research Officer with the Department of Applied Economics. A Fellow of Girton College, Cambridge, her current research is on private sector pay settlements and the construction of an archive of enterprise-level pay data. Her doctoral research was on structural change in the British textile and clothing industry and her publications in this area focus on the corporate strategies of multinational firms in the textile sector.

Note
The National Economic Development Office organised the policy seminar on *Reducing Regional Inequalities* to improve the level of understanding of the issues involved, and we are very grateful to the authors for the time they have devoted to the project. Their articles are, of course, entirely personal and it should not be assumed that any part of this book reflects the opinions or judgements of the National Economic Development Council.

Foreword

This book is about disparities in economic performance among the different regions of the UK, and about how employment and output in those parts of the country which lag behind can be improved to the benefit of the whole nation. It is an account of the second in a series of policy seminars organised by the National Economic Development Office, which provide an opportunity for the discussion of important policy issues by academic economists, representatives of the CBI and TUC, and the business community at large. Civil servants attend as observers. The papers presented at each seminar, together with a summary of the discussion and the conclusions arrived at are then published.

The issues which it is appropriate for our policy seminars to cover are ones where there is scope not only for official policy action, but also for significant changes of direction by employers, trade unionists and individual workers. Thus, the theme of the first seminar was how to improve work incentives for the low paid, and how to encourage those presently at the margins of economic activity to make a fuller contribution to their own and to the country's well-being. This second seminar rests on the belief that the lower than average levels of economic activity, which some areas of the country have persistently experienced, represent an inefficient use of national resources. We have long been familiar with the coexistence of unemployment in some areas and labour shortages in others. At the same time those who live in the South East are all too aware of transport congestion; they are probably less aware of how freely the traffic moves in much of the North and in Scotland.

Diagnosis must precede cure, and therefore we were anxious to explore why regional economic disparities exist and persist. To what extent do they reflect differing endowments of productive resources? To what extent do sluggish market reactions widen such discrepancies and magnify their consequences? In Chapter 2, Jim Taylor makes it clear that the poorer regions of the country are less well supplied with capital, skilled labour and entrepreneurial talent. At the same time he emphasises that to label this a 'North–South divide' can lead to dangerous oversimplification. There are pockets of considerable economic success in the northern regions and pockets of decline and low activity in the south. Any policies and intitiatives therefore need to be carefully targeted.

A predominant reason for the persistence of regional problems is the failure of the poorer areas to improve their competitive position by developing a cost structure that is sufficiently attractive to bring in the new employment opportunities that are needed. In this context two markets receive particular attention in this book – the labour market and the housing market. Chapters 3, 4, 5 and 6, by Patrick Minford and Peter Stoney, Janet Walsh and William Brown, Barry McCormick, and John Muellbauer and Anthony Murphy, are all concerned with these markets. Minford, among others, demonstrates how pay differentials have been failing to reflect the relative strengths of demand for labour in the various regions. There was general (though not universal) agreement about this, but not about what should be done about it. Some, like Minford, argued that social security benefits had set too high a floor for minimum wages in the depressed areas, while others put more emphasis on the still considerable strength of national, as opposed to local, pay bargaining. Similarly, though most contributors believed that the housing market exacerbated regional problems by discouraging mobility, there was controversy about the reasons for this. Some placed particular stress on the planning procedures and regulations that distorted the price of land and helped to make housing unnecessarily expensive in the South East. Others suggested that a major barrier to mobility was the lack of tax subsidies for private rented accommodation, in contrast to the tax breaks available for owner-occupation.

Much of the discussion in this book focuses on what might be done to make these two markets, for labour and housing, work better. Strikingly, however, relatively little mention is made of what used to be thought of as the 'traditional' instruments of regional policy, namely direct government investments and subsidies to reward the relocation of capital to the relatively depressed areas. Much more emphasis is now placed on getting the market supply side right, which has many facets in addition to making the housing and labour markets more efficient. They are well brought out by John Goddard and Alfred Thwaites in Chapter 7, by the TUC in Chapter 9 and the authors from the CBI in Chapter 8. These contributions all argue that the present entrepreneurial resources in the regions can do much to sustain and nourish themselves with the right encouragement and with some seed money. Important among the areas where help can be provided are the policies of large companies, which have too easily fallen into the habit of regarding the regions as only suitable for 'branch-plants'. Research and development and technical

Foreword

training are activities that they mostly confine to the golden triangle of the South East. Advances in communications and information technology, which render the locational dispersion of critical activities far more simple, are too often ignored. In this spirit the TUC advocates the promotion of regional research and innovation centres, and also suggests steps to ensure that the new Training and Enterprise Councils are successful on a country-wide basis.

Some of the proposals raised at the seminar and in this book are potentially costly to the public purse, including the CBI's suggestions for achieving a high quality integrated transport network. But, encouragingly, many of the proposals depend on the success and initiative of business-led bodies in harnessing and encouraging the entrepreneurial spirit that is already evident. We held the seminar in the North East and the local businesspeople who attended gave ample testimony that the entrepreneurial spirit is flourishing. The participants at the policy seminar are especially indebted to Bill Hay of the Newcastle Initiative and to Alastair Balls of the Tyne and Wear Urban Development Corporation for the insights they provided into the progress that is being made in the North East.

Maggie Hobbs and Marianna Bogarios were responsible for preparing the papers for the publishers and assisting Alex Bowen and Ken Mayhew, the organisers of the conference and the editors of this volume. We are especially grateful to the contributors to the conference and to this book for providing us with such a full and detailed account of the opportunities to achieve a better regional balance in the UK.

Walter Eltis
Director General
National Economic Development Office

1
*Regional Issues in Economics: Setting the Scene**

Alex Bowen and Ken Mayhew

I Introduction

Many economic and social problems have an important regional dimension. For instance, unemployment rates, productivity and standards of living vary across the regions. The latest data (those available for January 1991) reveal that unemployment rates (as a percentage of the total workforce) for travel-to-work areas in the UK vary from 24.2 per cent in Strabane (Northern Ireland) to 2.3 per cent in Crawley (Sussex). At a higher level of aggregation, that of the standard planning regions of the UK, the rate varies from 14.1 per cent in Northern Ireland to 5.1 per cent in East Anglia. In 1988/89, consumers' expenditure per capita ranged from 25 per cent above the UK average in Greater London to 20 per cent below it in Northern Ireland. The reasons these variations excite interest among economists is because they may shed light on inefficiencies in the functioning of the national economy. Where there are inefficiencies, two issues arise. First, why do these inefficiencies occur? Second, can public policy correct them or compensate for their effects, or do they occur as a result of intervention in the economy?

These issues are of concern to the National Economic Development Office (NEDO), because the office has a remit to give independent advice on ways of improving the economic performance, competitive power and efficiency of the nation, and to communicate with employers and trade unions with a view to stimulating new ideas and practical action. The second NEDO policy seminar sought to help to fulfil this remit by

---

* This paper expresses the views and judgements of its authors; it does not represent any official view of the National Economic Development Office. The authors are grateful to Tim Walsh, Marco Pianelli and Andrew Wood for their assistance.

stimulating discussion about the reasons for regional economic diversity and the implications for economic policy. Traditional regional policy is one element, but it is evident that many regional problems are related to the way in which the national economy as a whole functions. Thus, labour markets, the housing system, and the provision of the nation's infrastructure will all be subject to examination. Political allegiances vary geographically, giving an extra edge to discussion of regional problems, as exemplified in debates about the so-called 'north–south divide'.

This chapter first seeks to show how different types of economic analysis can be applied to regional issues to identify the main economic questions. A section is then devoted to a discussion of urban problems which are now seen by many as the most pressing reason for regionally differentiated policy. The paper goes on to outline the development of public policy with respect to regional development, noting the recent shift towards urban policy (with an appendix detailing the chronology). Particular attention is paid to the developing role of the European Community (EC). However, regional policy *per se* is not the only set of interventions in the economy with an important geographical impact. Investment in infrastructure, particularly the transport infrastructure, has important consequences for the optimal location of firms and workers. Labour and housing markets are affected by government legislation, local authority behaviour and the institutions of collective bargaining, among other factors. Some of the more important questions arising from these interventions (considered in more depth in other chapters of this book) are also discussed briefly.

II Economic analysis and spatial economic disparities

Any discussion of policies to moderate regional disparities needs to be based upon some analysis of why these disparities arise in the first place. Diagnosis should precede prescription. It may be helpful, therefore, to outline in general terms how economic analysis can illuminate the sources of spatial economic disparities. Other chapters will take this a step further by applying some aspects of this analysis to particular problems of the British economy.

Expenditure – based models: room for regional fiscal policy?

Much of the discussion of regional disparities in Britain has been based on an essentially Keynesian analysis of the causes of regional unemployment. Regional economies are treated in a similar way to national economies, subject to a fixed exchange rate regime (because of the common currency) and an externally determined monetary policy. Such an approach leads one to ask what scope there is for a regionally differentiated fiscal policy and for a devaluation. If unemployment is high in a particular region because aggregate demand there is inadequate, a fiscal stimulus or a boost to regional net exports via devaluation may be useful. But without exchange rate flexibility, a devaluation requires a lowering of unit costs relative to other regions, either by means of a reduction in factor prices (eg wages) or by means of a subsidy (eg the old regional employment premium). The idea of using a regionally differentiated fiscal policy to help high unemployment areas draws attention to the regional pattern of government taxation and expenditure, particularly capital expenditure (because of its additional supply side-effect of adding to productive capacity). Thus, Armstrong and Taylor (1988), for instance, argue that, 'the regional impact of all government expenditure should be estimated and published' (p46), and 'revitalising the economic base of the depressed areas cannot be done without a substantial increase in capital investment' (p45).

The success of policies designed to manipulate aggregate demand depends upon the size of the Keynesian multiplier effects. If these are very small, because a large proportion of extra income is spent on the output of other regions, the analysis is not very helpful; only the first-round effects of an increase in aggregate demand will be focused on the region in question. The size of these leakages depends on factors such as transport costs and the tradability of services. When it is prohibitively expensive to import certain goods and services into a region, the level of their (local) production is determined by the income generated by the industries in the region which produce tradable goods and services.

A cut in money wages (or other important elements of costs) in these industries would be expected to increase employment by making production more profitable in the region concerned, at the expense of similar industries in other regions (including those abroad). There are theoretical objections to using cuts in money wages as a means of reducing unemployment at the national level, based upon the danger of triggering a deflationary spiral. These objections have less force at regional level because the regional economies are relatively open, and prices of most traded goods are determined at a national level.

Whatever the *theoretical* objections to money wage cuts, a more cogent criticism might be that wages are simply not an exogenous variable which can be varied at will. One needs to understand how wages are *actually* set;

Reducing Regional Inequalities

Chapter 4, by Walsh and Brown, addresses this question. A straightforward answer would be provided if there were a stable, simple relationship between regional wage inflation and regional unemployment, that is, a regional 'Phillips curve'. Such a relationship might be expected to show that, as unemployment fell, every additional fall in the jobless total had an increasing impact on money wage increases. This would provide a strong argument for trying to equalise unemployment rates across regions, because doing so would reduce the national average inflation rate at any given national average unemployment rate. However, it should be noted that attempts to estimate regional Phillips curves have not been very successful.

Pure trade models: can product market competition eliminate regional disparities?

Keynesian analysis tends to focus on multiplier effects, tradability of goods and services, and the scope for reductions in factor prices in depressed regions. Traditional trade theory (of the textbook 'Heckscher-Ohlin' variety) focuses on a different set of concerns deriving from its longer-term perspective. In simple trade models, designed to analyse long-run equilibria, the pattern of trade among countries – or in our case, regions – is determined by their differing endowments of factors of production – land, natural resources, skilled labour and so forth. Under certain relatively stringent assumptions, free trade leads to equalisation of factor prices across regions. The assumptions require that regions have access to the same technologies, that they have equal numbers of goods/services and factors of production, that transport costs are negligible and that there is mobility of factors of production within each region. Factor price equalisation takes place without any mobility of factors across regional boundaries being necessary. This serves as a reminder that migration of labour and mobility of capital between regions is *not* always required in order to obtain convergence of wages and profit rates. Thinking about the validity of the simplifying assumptions may help point to some of the reasons why wages and profit rates do not converge in practice.

Traditional trade theory leads one to investigate the relative factor endowments of different regions and to consider what barriers might exist to the free flow of goods and services. The Keynesian theorist is more concerned with short-run adjustment mechanisms, while the Heckscher-Ohlin trade theory is more appropriate in the long run. In both cases, examination of the assumptions made in order to simplify the theoretical models can cast light on the source of regional problems.

In the short run, trade theory has little to say about the determinants of factor price movements, save that they may be necessary to reallocate factors *within* each region after some change in the pattern of demand or in barriers to trade. More institutional theories can be tacked on to explain these movements (eg rigidity of real or nominal wages due to collective bargaining). Neary (1978) has explored the issue of imperfect factor mobility in the context of the standard trade model. When applied to regions instead of countries, his work shows that a change in the national pattern of demand for goods can produce complex changes in real wages and profits across industrial sectors within a region as workers are reallocated. As Williamson (1983) writes, 'the effect of a tariff (or other price change) on income distribution is not a trivial matter to be settled by appeal to some handed-down formula as an eternal truth. Economists can hope to say useful things but only after facts have been examined *and* theories understood.'

Can we learn from development economics?

There have been some very interesting attempts recently to analyse longer-run trade patterns within a framework which owes more to balance of payments theory and development economics than to traditional Heckscher-Ohlin trade theory. These models try to formalise the work of political economists who have sought to explain the backwardness of developing economies and their vulnerability to macroeconomic shocks affecting the world economy (examples include Taylor (1983), Vines (1984) and McIntosh (1984)). They are often characterised by the shorthand 'North-South models', contrasting the economies of the Northern and Southern hemispheres. Their applicability to the so-called 'north-south divide' in the UK has not yet been explored in much detail, but several of the ailments afflicting the 'north' of the UK economy seem to reflect, in attenuated form, the ailments diagnosed in the 'South' of the world economy.

Taylor (1983) provides a summary of the pessimistic conclusions for the less developed countries/regions that can be derived from this type of thought. The backward countries/regions can find that productivity increases hurt their terms of trade and growth rate, inward financial capital flows do little to stimulate growth, and adverse 'world' macroeconomic shocks cause their growth to lag behind the growth of the more successful regions. The essential assumptions made which lead to these conclusions include a low income elasticity of demand for the export of the backward countries/regions and their dependence on the more advanced countries/regions for crucial capital goods. The supply side constraint affecting the backward countries/regions arises from a lack of capital goods, not labour, which is abundant. The superficial similarities between this story and aspects of the UK 'north–south divide' merit, perhaps, further investigation.

Spatial economics, economies and diseconomies of scale

The strands of economic analysis noted above can help to provide a framework for thinking about regional economic disparities, but they lack any specific spatial element. For instance, transport costs and locational decisions are tangential issues. Here we discuss the field of 'spatial economics', which develops location theory, and has a bearing on questions of regional development.

Among the classic works in the field is Hotelling (1929). Its essence can be illustrated by the famous example of the ice-cream salesmen. Suppose there are two ice-cream salesmen on a beach of a given length and that bathers are spread along it at equal intervals. Each of the bathers wants one ice-cream (regardless of price) and walks to the nearest salesman to purchase it. Where will the salesmen locate? In the simplest model, where prices are fixed, salesmen end up locating in the middle next to each other. Welfare would be improved if the two sellers were located one-quarter and three-quarters of the way down the beach, because customers' walking distance would thus be minimised. However, in more complicated versions of this problem, modelled as dynamic games between the two salesmen with variable prices and transport costs, locational choice can differ substantially, depending on the salesmen's payoffs; frequently no equilibrium can be achieved. Although interesting, the models of locational decisions based on game theoretic considerations are very sensitive to the assumptions made about the strategies of the sellers and therefore they have very little predictive power. They do, however, illustrate how locational decisions made under conditions of imperfect competition need not be welfare maximising and how introducing a spatial dimension inevitably raises the question of the degree of competitiveness among producers.

Many location theory models focus on increasing returns to scale and transport costs, both of which can play havoc with conclusions drawn from an analysis of highly competitive markets. In these models, locational equilibria depend on the precise nature of the trade-off between decentralising forces (eg workers' desire for more space as their incomes increase) and agglomeration forces (eg increasing returns to scale) and occur when these forces balance. These models are useful in organising one's thinking about the spatial dimension of economics, but they are very difficult to test. The magnitudes of the decentralising and agglomeration forces are very difficult to estimate, let alone predict.

One model along these lines was developed by Alonso (1964). In his model it is assumed that there is an initial centre where the buying and selling of goods and services take place. In the case of cities, this centre is urban business land where competitive (price-taking) producers sell their goods. Because there are transport costs and other transaction costs, producers want to locate close to this centre and therefore bid up the price of land near it. Location is then determined by profit maximisation, which requires comparing gross revenue obtained at this centre with the sum of wages, rents and material costs at the chosen location, plus transport costs to the centre. However, it does not explain why there is a transactions centre in the first place.

Economic theories of urban growth derived from this sort of approach suggest several reasons why the population and employment structure of cities should change during the process of economic development. First, economies of scale in production are exhausted and so-called 'agglomeration economies' (for example, benefiting from the nearby presence of specialist suppliers, or from high quality infrastructure) are balanced at some stage by the diseconomies imposed by congestion and overcrowding. The increased vertical integration of firms which has been observed reduces the benefits to be derived from agglomeration economies anyway. Second, many firms find it profitable to work outside the centre of cities as the population (especially the skilled, upper-income population which provides market demand and a labour force) moves out to suburbs in response to rising incomes and improved commuting technology. Third, as technologies become more capital intensive, plants tend to require larger or custom-made sites; a city has a comparative advantage in labour-intensive ones. Fourth, cities have also had a comparative advantage in the past in activities requiring frequent face-to-face encounters, such as public administration, corporate headquarters functions, high-level business services, and the arts. Progress in the fields of telecommunications and information technology reduces the comparative advantage of cities in such activities, for instance by allowing teleconferencing and telecommuting. The major shift of producer services from the London local labour market area to the rest of the metropolitan region and to the freestanding towns of southern England

Reducing Regional Inequalities

illustrates this, as does the growth of the 'M4 corridor', spurred largely by the decentralisation of routine office functions. Fifth, advances in transport technology and the reduction in the importance of transport costs in value added make it less valuable for firms to be located in large population centres. Sixth, the fall in the average size of households counteracts the desirability to multiple-worker households of being located in a large labour market.

Considerations like these point to the reasons why cities, and inner cities in particular, are likely to contract in countries at an advanced level of economic development. They help to explain why changes in industrial structure in the British economy as a whole are insufficient to account for the decline in employment in most larger British cities. They do not explain why some cities can contract without major problems (eg Stuttgart), while others, like Liverpool, suffer.

III The urban dimension

Chapter 2 of this book, by Jim Taylor, looks at the role of the industrial mix within regions in contributing to their relative economic performance, making use of shift-share analysis. This is also important for the economic well-being of urban areas. For instance, the shift of the economy from manufacturing towards services has hurt employment in some northern and Scottish conurbations. Employment in Clydeside has suffered from the area's dependence upon older industries such as iron and steel, heavy engineering and shipbuilding, now subject to much greater international competition. Between 1969 and 1981, the major conurbations lost 1.7 million manufacturing jobs, which constituted 79 per cent of the total fall of 2.1 million such jobs. However, while differences in industrial structure help to account for why some cities do worse than others, they do not account for relative urban decline as a whole. This can be seen by considering how urban employment would have grown if each industrial sector represented in a city had grown at the same rate as it grew in the country as a whole. Most cities show a predicted growth rate *higher* than the national average by this method, partly because services (rapidly growing) are over-represented in cities and primary production (growing slowly) is under-represented. But actual growth has been *lower*. This phenomenon has continued, according to the ESRC Inner Cities Research Programme (Hausner, 1987). Bristol in 1971 had a similar proportion of its employment in manufacturing as did Glasgow and Tyneside; yet, in 1987, Bristol had become one of the 20 cities in Britain with the highest proportion of employment in producer services, which is the most rapidly growing service sector. Hence, industrial structure alone is insufficient to explain even differences among cities. Clydeside and Tyneside actually increased their concentration of employment in older manufacturing industries dependent on particular, relatively immobile, resources found only in a limited number of locations.

The ESRC study identifies various factors which have been characteristic of the three old industrial conurbations it covered (Clydeside, Tyneside and the West Midlands). These are:

- low investment rates;
- poor product innovation;
- older, more labour-intensive capacity;
- poor quality premises;
- high labour costs;
- inefficient work practices;
- poor labour relations;
- routine production of a mature product;
- low value-added activities.

These factors all conspire to generate low rates of profit and of innovation. Industries with high rates of investment and firms which wish to expand are discouraged by them. Investment which is not simply designed to replace worn-out machinery is more mobile and therefore more free to choose between alternative locations; it will therefore respond more readily to disincentives such as those listed above and to the economic forces encouraging decentralisation. The fixity of much capital equipment and the burden of start-up costs on new sites discourage low-growth firms from responding to the economic incentives to decentralise. Firms lacking in entrepreneurship similarly do not respond. These firms are more likely to be closed down. Job losses due to closure have accounted for more urban job losses than have outward movements by firms. In several conurbations, such as Coventry, the closure of several relatively large individual plants has been the major factor in the decline of employment. Thus, a problem of adjustment first reflected in the problems faced by inner city firms and plants becomes a problem of labour market adjustment, as unemployment increases.

Profitability in the conurbations tends to be below the national average in a wide range of industries, although it should be noted that profit rates tend to vary more among urban-size categories than they do among regions. This is confirmed in some of the ESRC case studies (eg

Clydeside). This reflects both self-selection by less successful firms which 'stay put' and external circumstances derived from the urban location, such as a higher rates burden and (in some cities) a higher rental charge for land. Recent research (Tyler, Moore and Rhodes, 1988) has shown that a firm's profit, particularly in a low-profit industry or time period, can be very sensitive to rates and rental charges, even when these account for a small part of total costs. A 1 per cent increase in distance from inner London reduced the rates bill for an industrial company by 0.36 per cent on average. Taking other costs into account, it can be calculated that, for industrial firms, gross profits could have been increased by up to 10 per cent by a move of only 30 miles out of London. For commercial firms, the benefit could have been even greater.

Another of the external circumstances which can reduce the attractiveness of inner cities to businesses is the physical environment. Derelict land imposes costs on nearby firms by making the working environment less attractive to employees and encouraging vandalism. This is a classic 'externality'. It may appear to provide opportunities for expansion in congested areas, but the costs of reclamation are high outside assisted areas, despite the grants available under the Derelict Land Act 1982. The price of the land often does not reflect its value, which may be negative before reclamation. Malfunctioning of the market for land can arise because of the non-negativity constraint on prices, or because land is held for speculative purposes, or because owners do not wish to dispose of it below its book value. The potential role of speculation is increased by the system of planning permission, local authority and central government development aid, and zoning, whereby the development potential of sites depends upon their administrative status, which is uncertain. However, a recent case study of inner Manchester (Adams, Baum and McGregor, 1988) found that the main barrier to redevelopment was not restrictions imposed by land planning policies, but the reluctance of owners to part with their land and the existence of problematic site conditions. There are still widespread difficulties in assembling appropriate sites for redevelopment where ownership is diverse. Over the period 1964 to 1979, industrial floorspace in London and the other major conurbations fell slightly, while elsewhere it grew by 30 per cent.

Thus, economic development has tended to make it more profitable for new firms and industries to locate outside the centres of large cities. Older, less innovative and less competitive firms tend to be left behind. This reduces employment growth, which puts pressure on labour markets. Unemployment differentials are not eliminated by wage differentials, nor by outward migration. However, wage bargaining and migration are sufficiently selective to lead to a relative concentration of disadvantaged people in inner cities. This is a further barrier for firms to cross before they decide to create employment in inner cities. The externalities created by the decay of the physical environment, which the property market is unable to reverse, compound the problem.

IV Policy and the regions

Regional policy

The need to assist regions with severe economic problems has been a recurrent issue in policy-making at least since 1934, when the Special Areas Act first gave official status to the disadvantages suffered by the areas assisted and granted special help to them. For a long time regional policy appears to have been based on two guiding principles. The first was that the major concern should be with helping regions with high levels of unemployment; the problem of regional imbalances was virtually identified with the problem of regional unemployment – hence the choice of some areas which received privileges (in the form of subsidies to investment and/or employment and advantages in planning control). The second principle was that policy should attempt to 'take work to the workers' (although New Towns and limited help to economic migrants helped to take workers to the work). These principles were consistent with the Keynesian consensus which emerged after the Second World War that governments had a responsibility to intervene in the economy in order to ensure full employment. The 1944 White Paper on Employment referred to the desirability of 'balanced industrial development', and this was one of the motives behind the introduction of Industrial Development Certificates in 1948 (following the Town and Country Planning Act 1947), which provided an administrative means of directing new plants to the areas in need. An appendix to this chapter, reproduced from Armstrong and Taylor (1988), provides a brief chronology of the evolution of regional policy.

The post-war importance of regional policy increased after 1958, when the Prime Minister of the day, motivated by memories of the depression of the 1930s, responded to economic slowdown by providing financial assistance to more areas of high unemployment (Distribution of Industry Act 1958) and reactivating the Industrial Development Certificate procedure, dormant from 1952. In 1962, a special minister for the North East was appointed (Lord Hailsham), followed by a minister of regions in 1963 (the Rt Hon Edward Heath). The government's interest

Reducing Regional Inequalities

in regional policy continued during the ensuing upswing, and, although the concern with regional unemployment was no longer the stated rationale for policy, it remained the benchmark for defining and demarcating the 'problem regions'. The new strategy was designed to enhance the nation's prosperity by aiming reflation policies more accurately. By doing this, areas of high unemployment where, by definition, resources were being under-utilised, could expand without threatening the already tight labour markets of the South East and Midlands with overheating. Underlying this analysis was the view that regionally directed demand management policies could shift regions along their respective regional Phillips curves (towards a common position), while shifting the national Phillips curve towards its origin (Bowers *et al*, 1970). In short, regional policy was designed to satisfy the desire to 'boom without busting' (Harold Macmillan quoted in Parsons, 1988, p117). Unfortunately, this strategy was characterised by short-term concerns for levels of growth and employment. As a consequence, the policy instruments used were not designed to achieve any longer-term regional or industrial development aims (whether such aims could have been achieved is another question anyway). Most of the grants and subsidies had blanket coverage of the assisted areas, and although some grants were conditional on proving employment creation, most incentives were automatic. Thus, they can be viewed more as examples of regional fiscal policy than a supply side policy.

This approach continued throughout the 1960s, with some relatively minor alterations to the policy mix. The incoming Labour government of 1964 voiced the intention to integrate regional planning into an overall industrial strategy. Unfortunately for the planners, there was a tension from the very beginning between the inherently centralist nature of national planning and the decentralist approach required for regional planning. The Regional and Economic Planning Councils and Boards which were set up by the new Department of Economic Affairs were always subordinate to central ministries, such as the Board of Trade, which retained control over many instruments of regional policy such as the location of new factories, Industrial Development Certificates, and the award of grants and loans. For this reason it is difficult to detect a radical change in policy, although some ambiguities were cleared up concerning the location and scope of development areas (DAs), and policies were adapted slightly in response to a rise in unemployment. The Control of Office and Industrial Development Act 1965 rescheduled DAs and tightened up the Industrial Development Certificate procedure. New larger DAs were set up under the Industrial Development Act 1966.

These areas were not chosen according to the sole criterion of unemployment levels, but were chosen because they were considered to be regions of potential growth. This brought some stability to the location of DAs; they no longer altered as unemployment rates changed.

In November 1967, several special DAs were designated to provide additional assistance to ex-mining areas with high unemployment. The other major change in 1967 was the proposal for a regional employment premium in order to encourage employment creation. Although these subsidies remained an important part of regional policy in terms of expenditure for 10 years, they remain a very controversial instrument, partly because of the difficulties of calculating net costs per job created, but not least for their uncertain long-term effect. Superficially, they appeared to be an effective method of expanding employment through the substitution effect (of labour for capital), thus having a direct effect on unemployment. It was the long-term consequences of this substitution effect which were the major concern, since reducing the demand for capital can act as a disincentive to innovate by discouraging the adoption of the latest techniques. The long-term effects on competitiveness are obvious, and the same criticism can be made of all policies which make assistance contingent upon employment creation. This dilemma between the need to stimulate the creation of jobs, while not interfering with firms' modernisation programmes, becomes particularly important in times of rapid restructuring and innovation.

The Conservative government of 1970–4 also stressed the growth potential of regions, but took a new tack which was intended to make a break with the interventionist policies of the past. An end to the regional employment premium was proposed, together with a more selective approach to incentives which took into consideration the growth potential of regions. Industrial Development Certificate controls were reduced and there was a shift from grants towards depreciation allowances. However, the recession of the early 1970s, with unemployment increasing rapidly, resulted in a return to the old policies of blanket assistance and support for ailing industries (eg the Industry Act 1972 and the creation of an industrial development executive to identify firms in need of assistance), despite the growing criticism surrounding policies which were seen as exercises in damage limitation.

The period of 1958–76 represents a period of continuity with respect to regional policy, with regional issues and their interpretation remaining much the same, despite changes in government. However, there were changes in the mixture of policy instruments used; they were often combined in what appears to have been a rather *ad hoc* manner (Parsons,

1988). The changes seem to have been associated with the fluctuations of the national economy. The contrast with the macroeconomic policies of the period is striking. The policies of demand management and fine-tuning had the clear objectives of stabilising output trends and ensuring an acceptable level of employment and price stability. The means of attaining these objectives were tried and tested, even though they were not always successful. Regional policy, on the other hand, had neither clear objectives nor standard instruments with which to achieve them. The causes of the regional imbalances were never clearly established, and yet policy responses to 'the regional problem' were numerous and generally accepted by all governments.

This fact explains to a large extent why policies frequently appear to have been *ad hoc* and inconsistent. It also helps to explain how regional policy reflected national policies, both in objectives and in thinking. Part of the problem was the extent of the success of the 'Keynesian Revolution', with the resulting emphasis on national homogeneous aggregates which were simply not conducive to an analysis of regional issues (Chisholm, 1987). This hole in economic theory almost inevitably left regional policy to become an appendage to national concerns.

After 1974, the new Labour government initially strengthened regional policies, for instance by reactivating the regional employment premium and increasing its size. But there was a policy change after two years in office, with reductions in expenditure on regional policy contributing to the 1976 expenditure cuts.

With unemployment rising in all regions, and investment rates falling nationally, it was increasingly difficult to use administrative policies, such as the use of Industrial Development Certificates, which were costless to the Exchequer. A substantial proportion of those investment projects which were refused location in the South East or Midlands either were not completed or were relocated in other countries. Policy innovations were reduced, not only because of the cuts in public expenditure, but also because the economic environment was changing in a way which made traditional regional policies both less attractive (since unemployment was rising in all regions) and less effective (since the supply of mobile new manufacturing investment had diminished). The debate about whether the UK was suffering 'deindustrialisation' took on a regional dimension because traditional manufacturing industry was concentrated in particular parts of the country.

The late 1970s and 1980s represent a transition towards new priorities. There are many elements to this change. There has been a continued effort to reduce taxes and government expenditure as a proportion of national income. The actual effectiveness of regional policies, their 'value for money', was being questioned and with the general collapse of the old Keynesian consensus and its emphasis on aggregates, there was a return to the principle of helping markets function freely rather than manipulating them. Linked with this was a recognition of the changing nature of the British economy, in particular, the declining importance of manufacturing. In this context, the support of ailing manufacturing industries could be seen as standing in the way of progress, which some identified with the service sector.

Although an important aspect of this approach is the identification of the regional problem with one of wage cost rigidities and labour immobilities (Minford, 1988), implying the need to 'free' the labour market, it is important to note that some of the old policy instruments remain, despite their interventionist nature. For example, grants and subsidies continue to be important, although automatic assistance is increasingly rare, being replaced by discretionary assistance aimed specifically at employment creation. However, the scale of regional assistance has decreased markedly, with a reduction in the size of assisted areas from covering 47 per cent of the labour force in 1979 to 15 per cent in 1985; and a reduction in expenditure (at 1985 prices) from £1,011.4 million in 1982/3 to £477.9 million in 1987/8 (according to *Regional Trends*, annual, HMSO).

There are two issues here: first, that of labour versus capital subsidies; and, second, that of automatic versus discretionary incentives.

The arguments in favour of labour subsidies include:

- they appear to be directly related to the level of employment/unemployment;
- given the importance of labour costs in total factor costs, they represent an effective regional real devaluation, thus increasing product demand;
- multiplier effects are more likely to be contained within the region (whereas capital subsidies involve a high leakage of the multiplier effect due to the need to import capital goods);
- the mobility of labour across regions is lower than that of new investment, so local labour may be less competitive than the market for capital.

The arguments against labour incentives include:

- a bias against high-tech 'jobless' growth, which nevertheless may have important beneficial spin-off effects in the region, for

instance, by stimulating technical progress in regional supply chains;
- the disincentives to modernisation or restructuring.

Capital subsidies may also result in employment spin-offs within the region due to forward and backward linkages. Neutral subsidies would subsidise *output*, allowing producers to decide the optimal factor mix to use in producing it. An attempt to come to terms with this dilemma was made after 1984 with regional development grants (subsequently abolished), which combined an element of both capital and labour subsidies. This was achieved by offering a grant of £3,000 per job created, or a capital grant (15 per cent of eligible capital expenditure), depending on which was more favourable to the recipient. This scheme therefore represented an attempt to move in the direction of neutrality between the factors of production (Allen, 1989).

Turning now to the question of automatic versus discretionary incentives, the latter allow savings on projects which would have occurred regardless of availability of grant. They are more appropriate for a flexible system of incentives which is capable of being adjusted (in principle) to the requirements of the recipient. On the other hand, they are less 'visible'. They generate uncertainty. With sometimes lengthy gaps between the application for a grant and notification of eligibility, the firm is likely to omit discretionary schemes from their investment appraisal systems. Complicated and bureaucratic application procedures can result in a bias towards large firms over small firms. Marginal firms which receive assistance may displace more efficient firms which do not receive assistance. Firms may consume assistance in order to receive more assistance (see Swales (1988) for empirical research on these issues). There are ways to counter some of the problems associated with automatic incentives. For example, they can be restricted to small firms and/or they can be limited by means of a ceiling, after which awards are discretionary (note the investment premium scheme in the Netherlands; Allen, 1989).

There have been many efforts to estimate the quantitative impact of various regional policies. One of the more comprehensive was the investigation by Moore, Rhodes and Tyler (1986). This estimated that regional policy over the period 1960-81 created around 600,000 manufacturing jobs in DAs, of which about 150,000 did not survive throughout the period up to 1981 (this ignores subsequent multiplier effects). A majority (56 per cent) were created in the 1960s. The most cost-effective policy for the Exchequer was the use of Industrial

Development Certificates, but these came second to investment incentives in terms of the total number of jobs created. The authors estimated that one-third of the net employment created was in the conurbations of Merseyside, Clydeside and Tyneside. However, of the jobs which were attracted from outside the DAs, these conurbations captured only 20 per cent.

Wren (1988, 1989) has investigated the employment effects of financial assistance in the particular case of Cleveland. He concludes that, in practice, the regional development grant functioned like a labour subsidy in the 1984–8 period (after the 1984 changes), and was an improvement on the old regional employment premium because it was closer to a true *marginal* employment subsidy. It helped small manufacturing companies the most. It is noticeable that his methodology leads to estimates of costs per job substantially lower than those of Moore, Rhodes and Tyler, reinforcing the conclusion that such numbers must be treated with great scepticism.

Urban policy

It has been argued above that the economic role of large cities is changing, and that some decentralisation of employment is to be expected. However, the process of change has casualties, because the reallocation of resources and population does not take place smoothly. The markets for land, housing and labour are imperfect. The costs of urban squalor and deprivation are not borne fully by the economic agents who cause them. Hence, there is a rationale for intervention by public agencies to assist cities, particularly inner cities, adjust to their new role. In this section, the policies which have been pursued to this end are reviewed briefly.

Central government's concern with inner city problems has evolved since the 1960s, when the urban programme was set up under the aegis of the Home Office (1968). The original focus of concern was race relations and urban deprivation, and this led to the defining of areas of 'special social need' – areas which had high proportions of poor housing, large families, the unemployed, children in need of care and ethnic minorities.

During the 1970s, the contribution of economic forces to urban development was increasingly recognised, and the Department of the Environment displaced the Home Office as the lead department (taking over the Urban Programme in 1976). The Inner Urban Areas Act 1978 increased the resources available to the Urban Programme, partly by switching some spending from the new towns programme (as policy-makers began to realise that decentralisation of population and employment did not rely upon government action). It set up a hierarchy of local authorities – partnership authorities, programme authorities and other designated authorities – which were to be entitled to various degrees of financial assistance. This assistance could be used for loans (or, in some special areas, grants) for the acquisition of land or site preparation and towards the cost of setting up co-operative ventures. Inner-city partnership teams had to produce inner-city programmes and programme authorities had to prepare inner-area programmes. Small improvement areas, with a lifespan limited to 10 years, could be declared by the designated local authorities, within which grants and loans were to be available for the improvement of the environment, and grants could be obtained for the improvement or conversion of old industrial or commercial buildings. Spending under the Urban Programme has increased substantially since 1978, with a greater proportion of aid going towards projects concerned with economic regeneration (of the sort envisaged by the 1978 Act), and within this proportion, more emphasis being given to attracting and supporting private sector investment. This has helped to tackle the problem of dereliction and site redevelopment and recognised the need to provide incentives to private investors to develop businesses in the inner cities. Discretionary rather than automatic assistance has been the rule.

Subsequent government initiatives have included the establishment of urban development corporations (UDCs) following the Local Government, Planning and Land Act of 1980 (first in London Docklands and Merseyside). UDCs are single purpose planning authorities designed to bring land and buildings into use by the private sector and to provide infrastructure. Thus, they are aimed at the problem as noted above, that inner city redevelopment is often inhibited by the difficulties of assembling viable sites with the appropriate planning permissions. Unlike projects under the Urban Programme, the work of the UDCs is largely independent of their local authorities. On the one hand, this helps to cut through the red tape, especially where there are several local authorities in the area, but, on the other, the benefits of the local expertise and accountability of local authorities are lost. The 1980 Act also introduced Enterprise Zones, in which developers were to receive 100 per cent capital allowances from corporation and income tax on industrial and commercial property, and various other benefits. Again it can be seen that the rationale behind the introduction of enterprise zones was to reduce bureaucratic barriers to development and to provide financial incentives to private sector entrepreneurs.

Despite their objectives (in a sense creating 'little Hong Kongs'), a major feature of the Enterprise Zones is their financial inducement to inward investment, and consequently they have been described as 'mini assisted areas' (Damesick and Wood, 1987). A feature of the rhetoric which surrounds Enterprise Zones is the increasing faith in the ability of small firms to plug the gap left by declining large employers. Yet there has been some cynicism expressed about the feasibility of small firms creating sufficient long-term jobs (Sayer and Morgan, 1986), as well as some concern about the spatial impact of the government's small firm policies.

The business expansion scheme (BES) has been criticised for being biased towards the South East, not only because the South East is more dynamic and contains more entrepreneurial talent, but also because the BES funds are predominantly located in the South East (56 per cent in Greater London). A direct result of this policy is a net flow of venture investment funds from north to south (Mason and Harrison, 1989). This point has also been made with regard to venture capital in general, with 85 per cent of venture capital funds being located in London, partly explaining why 65 per cent of those funds are invested in the South East (Martin, 1989).

Since the innovations heralded in 1980, further components have been added to the government's inner city policy. The fact that other government programmes – such as training support and housing policy – have an impact on inner cities has been recognised by the publication of *Action for Cities* (for the Cabinet Office, HMSO, 1988), which draws together a wide range of government interventions. These include:

- city action teams in programme areas to bring together officers from all the relevant government departments and managers seconded from private industry;
- inner city task forces at a more local level, acting as brokers between the private and public sectors;
- urban development grants and urban regeneration grants (to be consolidated into city grants) for specific projects to draw in subsequent private investment;
- powers for local authorities to set up simplified planning zones;
- compacts between schools and colleges on the one hand and employers on the other, in which firms offer job prospects in return for the attainment of agreed educational standards;
- various programmes which have particularly important effects in inner cities, eg the Youth Training scheme and Enterprise

Reducing Regional Inequalities

Allowance Scheme (because of the relatively high rates of unemployment in inner cities) and housing action trusts (because of the high proportion of council estates).

The EC's structural policies

The EC's role in regional development is likely to increase in the future, as the 'social dimension' of the '1992' changes is developed. It has not been insignificant in the past; its members have not been happy to leave the regional distribution of income and economic activity to the market alone. In this section, we review the background to the EC's interventions in this field and consider how they are likely to change in the future. The important developments are an effort to focus the various forms of regional aid more narrowly and to ensure the 'additionality' of EC aid to member states.

EC structural policies are implemented under three main funds: the European Social Fund (ESF); the Guidance Section of the European Agricultural Guarantee and Guidance Fund (EAGGF); and the European Regional Development Fund (ERDF).

Article 123 of the Treaty of Rome requires the ESF 'to improve employment opportunities for workers in the common market and to contribute thereby to raising the standard of living'. A directive by the Council of Ministers in 1971 provided that 60 per cent of ESF resources should be devoted to increasing employment in 'problem' regions. The ESF had 3.5 billion ECU in 1989.

Title II of the Treaty of Rome which set out the Common Agriculture Policy (CAP) specifically stated that due account would have to be taken of the 'structural and natural disparities between the various agricultural regions'. In particular, the Guidance Section of the EAGGF was required to pay due regard to the regional consequence of the CAP and had 1.5 billion ECU in the 1989 budget.

The ERDF was established in 1975 with the purpose of correcting 'the principal regional imbalances within the community resulting in particular from agricultural preponderance, industrial change or structural underemployment'. The ERDF had some 4.5 billion ECU at its disposal in the 1989 budget.

In addition, the EC has some financial resources and policy instruments for specific structural purposes including instruments provided under the Coal and Steel Treaty for industrial loans and retraining, but these have been modest in size and limited in scope. The New Community Instrument, established in 1978, enables the Commission

Regional Issues in Economics: Setting the Scene

to borrow funds on the capital markets for on-lending through the European Investment Bank. The bank's first task is to contribute to the 'balanced and steady development' of the EC by granting loans financing 'projects for developing less developed regions' and it has, in practice, operated as a Europe-wide regional development bank.

The three funds are not large relative to the total community budget, relative to total EC GDP or relative to the scale of inequalities and structural problems within and between member states. Since 1975, the

Table 1.1 *Structure of the European Community budget*

		\multicolumn{3}{c}{Shares in total expenditure %}	\multicolumn{3}{c}{Shares in Community GDP %}
		1972	1980	1986	1972	1980	1986
1	Agriculture and fishing of which	76.2	73.6	65.5	0.42	0.61	0.64
1.1	price guarantee	75.0	69.7	62.4	0.42	0.58	0.61
1.2	fishing	–	0.3	0.5	–	0.00	0.01
1.3	structural guidance	1.2	3.6	2.4	0.00	0.03	0.02
2	Other sectoral policies (R&D) transport, energy and other industries	3.6	1.9	2.3	0.02	0.02	0.02
3	Social Fund	2.9	4.7	7.2	0.02	0.04	0.07
4	Regional Fund	–	6.7	6.8	–	0.06	0.07
5	Mediterranean Programmes	–	–	0.4	–	0.00	0.04
6	Development and co-operation	6.1	3.1	3.3	0.03	0.03	0.03
7	General administration, etc	5.9	5.0	5.2	0.03	0.04	0.04
8	Repayments	5.3	5.1	9.4	0.03	0.04	0.09
9	Total (1 to 8)	100.0	100.0	100.0	0.56	0.83	0.97
10	Structural Funds (1.3 +3 +4)	4.1	15.0	16.3	0.02	0.12	0.16

Source: The Regions of the Enlarged Community, Commission of the European Communities, 1987

resources made available under the structural funds have increased substantially, with grants rising from 0.35 billion ECU, in 1973 to 0.83 billion ECU in 1975, rising again to 4.25 billion ECU in 1982. Table 1.1 shows that the three funds have increased from 4.1 per cent of EC expenditure in 1972 to 16.3 per cent in 1986 (and are set to rise to 25 per cent in 1992). The structural policies are put in further perspective when measured against total EC GDP. While the share of structural funds in country GDP has increased considerably, by 1986 they still amounted to less than one-fifth of 1 per cent of income generated in the EC.

The UK has been one of the principal beneficiaries of EC structural spending. In 1986 the UK received 21.9, 9.4 and 23.8 per cent of ESF, Guidance Section of EAGGF and ERDF spending respectively. Thus, the UK received a substantial part of the EC regional expenditures, compared with, say, Ireland or Italy. One reason for this is that, under the terms of the ERDF, each country was originally allocated fixed national quotas on the basis of areas eligible to receive member states' *own* regional aid. The role of national quotas has been progressively reduced, but the UK still received a disproportionate share: 14.5 to 19.3 per cent in 1986. Table 1.2 shows that the UK share of ERDF spending in 1987 stood at 20.8 per cent compared with, for example, 5.3 per cent for Ireland and 11.6 per cent for Greece. The decline in the share received by the UK since the mid 1970s reflects the accession of Greece, Spain and Portugal to the EC.

Infrastructure projects are the single biggest recipient of ERDF money – they receive three times as much support as do industry or tourism projects. Major investments have included local authority road schemes, water supply, sewage provision and industrial estate services.

The development of ports, airports and telecommunications has also been important. The future is likely to see an 'increased emphasis on tourist infrastructure facilities', particularly in areas not traditionally recognised as tourist centres. 'Such diversification of the economic base of traditional manufacturing areas is equally as important as the attraction of manufacturing firms to areas traditionally dependent on agriculture and summer tourism.' (Commission of the European Communities, 1987: para 5-9). Inner urban areas will also obtain increasing support from infrastructure investments to renew and increase the capacity of obsolescent services (the European Commission recently announced an urban development programme).

The ESF concentrates its support in areas of high unemployment, including those covered by regional assistance. It made a major contribution to training schemes operated by the Training Agency in

Regional Issues in Economics: Setting the Scene

Table 1.2 *Shares of Regional Development Fund Spending*[1]

(%)

	1976	1980	1984	1987
Ireland	6.5	9.6	7.7	5.3
Germany	4.8	6.9	3.3	2.9
France	10.4	13.7	14.9	12.3
Italy	40.7	34.2	32.2	22.2
Netherlands	1.9	1.1	1.1	0.8
Belgium	2.2	0.9	0.4	0.9
Luxembourg	0.1	0.1	0.2	0.1
UK	31.9	32.0	22.0	20.8
Denmark	1.4	1.3	2.1	0.6
Greece	-	-	16.0	11.6
Spain	-	-	-	13.6
Portugal	-	-	-	8.8
Total EC	100	100	100	100

[1] Payment appropriations

Source: Official Journal of the European Communities

priority regions (and by the Department of Economic Development in Northern Ireland) and to employment creating manufacturing projects in assisted areas. The UK has also continued to be a significant beneficiary of loans from the European Investment Bank and grants from the Guidance Section Funds.

The piecemeal development of EC structural policies meant that there were few mechanisms for ensuring that expenditures under different funds were consistent with those pursued under other policies (eg environment) and led to limited success in meeting the EC's regional and structural objectives. In particular, assistance from the funds was closely linked to national policies on regional development and in financial terms represented only a small proportion of these. The EC's ability to ensure that the objectives which it had set were complied with and thus to make its assistance subject to its own criteria – conditionality – was circumscribed. This was partly due to the low degree of EC autonomy in decision-making. These problems were compounded by the considerable

Reducing Regional Inequalities

diversity in measures and types of aid provided by the funds (productive investment, infrastructure categories, wage and income support, funding for vocational training, intangible investments, etc), and the significant differences between the funds (in terms of legal bases, specific objectives, scope, links with EC policies, operating characteristics and rules). Greater coherence in the structural funds was clearly needed and was brought about by a combination of economic recession and internal reforms.

Another source of pressure of change was the enlargement of the EC, which substantially widened the range of under-developed regions. The ratio of Denmark's income per capita to that of Ireland was 1.8 (in purchasing power parity terms) and that of Denmark to Portugal 2.8. The addition of Spain and Portugal to the EC in 1986 was not accompanied by a proportionate increase in the structural funds.

It was, however, only with the Single European Act that structural and regional funds were systematically reviewed and reformed, in the so-called 'accompanying measures', designed to make a success of its implementation. Implicit in the reforms was the principle that EC expenditures should reflect the EC's regional priorities and not simply member states' national priorities. Under the reforms, the activities of the three funds were related to five priority objectives, each one fundamental to EC policy in the areas where the treaties allow the funds to intervene (regional, social and agricultural policy). At the same time, each of the objectives is designed to provide a clear link to one aspect of economic and social cohesion. They are:

1. to promote the development and structural adjustment of the regions whose development is lagging behind;
2. to convert the regions, frontier regions or parts of regions (including employment areas and urban communities) seriously affected by industrial decline;
3. to combat long-term unemployment;
4. to facilitate the occupational integration of young people;
5. with a view to reform of the CAP
   (a) to speed up the adjustment of agricultural structures, and
   (b) to promote the development of rural areas.

Under objective (1), only regions with a per capita GDP less than 75 per cent of the EC average should be eligible and, for objective (2), regions with a previously high level of industrial employment and high unemployment should as a general rule qualify. In this way, regions accounting for about 20 per cent of the EC population would be covered under

Regional Issues in Economics: Setting the Scene

objective (1) and an additional 12-15 per cent under objective (2). These figures give an idea of the sort of concentration of aid sought by the Commission and compares with the present 41 per cent and 32-5 per cent coverage. In the 1987 EC budget the total appropriations for the three funds were about 7.2 billion ECU. This total implies that, compared with the situation before enlargement, there has been a substantial fall in the real value of fund assistance per head in the EC.

The financial and regional implications of the reforms for the UK are far from clear. Large parts of the UK have been included in the Commission's 'priority regions' list under objective (1) and for support under objective (5)(b). Total structural funds to these regions will, in general, be doubled by 1992, provided suitable plans are submitted. All regions covered will not necessarily receive a doubling of funds.

The pattern of general government spending in the UK

As noted above, much of the thinking behind the analysis of regional development issues has fallen within a Keynesian framework, yet policy has fallen short of specifically addressing the fiscal stance within the regions. Attempts to estimate levels of taxation and expenditure on a regional basis are hindered by the lack of comprehensive data.

However, regional data are available for some aspects of UK government expenditure such as defence, and this has been used to show how small explicit regional assistance is in comparison. For example, in 1983/4 regional aid totalled £727 million, whereas defence procurement expenditure totalled £6,939 million (Breheny, 1988). But more importantly, 65 per cent of defence procurement expenditure was received by firms in the South East or South West of England. This is important, not only in that it dwarfs the expenditure on regional assistance, but also because much of this expenditure is on research and development, and hi-tech equipment. This emphasises the importance of the nature of government expenditure as well as the regional pattern of the expenditure. It may be the case that assisted areas receive more from the government when transfer payments are included, but '£100 million of support for R&D in a region has very different long-term implications from £100 million in transfer payments for that region' (Sayer and Morgan, 1986).

It has been argued that the government has played an important role in influencing this spatial concentration of 'leading edge' sectors of the economy (Breheny, 1988; Sayer and Morgan, 1986). The argument is based upon the fact that Ministry of Defence (MoD) procurement decisions are taken by defence research establishments which are located in the south of England. To be located near defence research establishments is important to the British electronics industry because of the importance of regular and detailed negotiation between contractors and defence research establishments.

Clearly, other agglomeration economies are important to the location of the electronics industry, but Breheny points out that the MoD equipment budget accounts for 50 per cent of the output of the UK aerospace industry and 20 per cent of the output of the electronics industry (Breheny, 1988, p190). So it is argued that historically the location of defence research establishments has been crucial to the location of Britain's electronics industry, and there has followed the establishment of an environment, infrastructure, culture and skilled workforce which attracts similar firms.

Whether a policy which relocates defence research establishments to the assisted regions would nurture a similar evolution of the locality is open to doubt. Breheny is cautiously optimistic and sees it as 'the one large, and potentially powerful tool of regional policy as yet untried'. Perhaps the most important point is that there are many government policies which have a very important but almost accidental differential impact on the spatial economy.

V Labour markets

Economists analysing unemployment in the context of traditional trade theory have usually identified downward inflexibility of wages as the primary cause. Applied to regions this suggests that the reason for some suffering higher unemployment rates is that real wages in them are too high. A more Keynesian analysis also points the finger at wage rates, among other factors; they have to be flexible downwards if a depressed region is to recover by means of an effective devaluation. Hence, evidence about regional wage rates is important.

Egginton (1988) has shown how the variation in regional unemployment rates appears to have been much greater than that of regional earnings. This contrast has been persistent over time. As Egginton points out, the more flexible regional wages are, the smaller should be the effects on employment and unemployment of a given shift in regional demand. Variation in earnings is *higher* for non-manual workers than for manual workers, but they have *less* variation in unemployment rates. This is consistent with the proposition that greater inter-regional variation in earnings reduces the disparity in regional unemployment rates.

Regional Issues in Economics: Setting the Scene

However, the rank ordering of regions by unemployment rates tends to be very similar from year to year. Thus, the implied solution to regional diversity in unemployment is to establish lower wages than average for a long time in the high unemployment regions. While this might be a solution to the unemployment problem, it could replace it by a 'standard of living' problem. Therefore, one should consider not just barriers to wage adjustment which stand in the way of this solution, but also *why* full employment in some areas requires lower wages. A variety of explanations can be offered. The disadvantaged areas may have been subjected to a series of adverse shocks relative to other areas, such as the reductions in demand for the products of the shipbuilding, iron and steel, and mining industries over the long term. In principle, even one adverse shock could cause a long-term problem if wages are completely inflexible. However, some flexibility is introduced by labour turnover and the changing industrial mix within a region. Pissarides (1978) has argued that (small) wage differentials do arise in response to demand variations. It seems implausible that regional wage differentials should be much less responsive than the aggregate wage level (which itself adjusts slowly in the face of unemployment, according to econometric evidence about the Phillips curve).

A second explanation would argue that equalisation of efficiency wages across the country requires lower wages in depressed areas because they have access to less advanced techniques of production. This has some plausibility given the concentration of research and development in the south, but many of the fruits of R&D are highly mobile. A third explanation could be developed resting on a larger wedge between consumption wages and the factor cost to the employer in terms of his own produce price in the high unemployment areas. However, it is not clear why this should be the case. A lower cost of living in high unemployment areas might mean that real consumption wages are higher in them, but this will not cause problems for producers whose product prices are determined on national markets. There is little evidence about regional variations in the cost of living, except in housing costs.

One way of analysing the unemployment–vacancy (u/v) relationship when there are a number of different labour markets has been put forward by Layard and Nickell (1985). Here, the real wage in each market is a function of excess demand in that market and a combination of factors such as trade union power and the wedge between producer and consumer prices. One can show that there is a relationship between aggregate excess demand on the one hand and aggregate unemployment and the sectoral dispersion of excess demands on the other. The greater

Reducing Regional Inequalities

the aggregate excess demand, the higher are the number of unfilled vacancies. Also, the higher the sectoral dispersion of excess demands, the further away the u/v curve will be from its asymptotes. Hence, the greater the sectoral dispersion, the higher is unemployment and the more concave the wage response to unemployment. In addition, as Hughes and McCormick (1987) have argued, the higher the regional mobility of labour, the lower is the regional dispersion of excess demands in regional labour markets and so the lower the wage pressure. Labour mobility can act as an equilibrating mechanism. This issue is investigated in some depth by McCormick in Chapter 5. He draws attention to the roles of occupation, education and housing tenure in determining migration propensities. The latter aspect is discussed more fully in the next section.

VI Housing and wages in the UK

One central issue in the recent economic literature on unemployment is the apparent unresponsiveness of wage inflation to high rates of unemployment. In this section we shall only be concerned with explanations of wage behaviour which focus on the interaction of labour and housing markets. In particular, we shall concentrate on the work done in this area by Bover, Muellbauer and Murphy (1989), Hughes and McCormick (1981, 1985); and Minford, Peel and Ashton (1987, 1988). Below, we review the theoretical background on aggregate wage determination in sectoral labour markets and the role of housing. Then we turn to the empirical work done by the authors above in this area. Finally, we discuss the different positions taken by these authors.

Bover *et al* (1989) have argued that regional differences in demand shocks can be best measured by regional differences in house price/wage ratios. Suppose that markets clear and that houses are privately owned. There is a positive demand shock in one region. Because house supply is inelastic in the short run, house prices are driven up, partly by higher wages, but also by the arrival of newcomers to the area. Moreover, higher house prices generate a wealth effect, stimulating local consumer expenditure and a liquidity effect due to an expansion in consumer credit. Hence, local house prices increase *relative* to local earnings, which themselves increase. In the longer run, the ratio decreases due to the construction of new houses and, assuming mobility, both retired workers and firms moving elsewhere. Nevertheless, the house price/earnings differential would show the original region-specific demand shock.

It must be pointed out, however, that there are a number of alternative

Regional Issues in Economics: Setting the Scene

interpretations of the relationship between wage pressure and regional differentials in house prices relative to earnings. One interpretation is based on the migration or mobility hypothesis (see, for example, Clark and Van Liercop, 1986). In the basic model residents face a choice over their location. The probability of moves in each direction depends on utility comparisons in which, for a given location, a trade-off takes place between higher earnings and higher living costs (including house prices). Since house prices are only part of living costs, earnings differentials would play a greater role than differentials in housing costs. There is some evidence that this is a plausible model. Mitchell (1988) has analysed moves between London and each of the nine regions and has found that inward moves are positively correlated to relative wages and negatively correlated to relative house prices. The coefficient for the former is higher.

Muellbauer and Murphy (1988), contend that, in practice, many households are likely to be credit rationed, so that relative house prices take on a greater weight than do relative earnings when living standards in different areas are compared. This is reflected in their study, where a house price/non-manual earnings ratio is utilised as an explanatory variable. They argue that, since their findings point to a *negative* association between net migration and regional house price/earnings differences, the latter cannot just be a proxy for relative labour demand shocks. If it were, the association would have to be positive; that is, the greater the (positive) shock, the greater the differential.

Bover *et al* attempted to assess the importance of various potential determinants of real wages over the period 1958-86. Among their empirical findings were the following:

- Changes in mismatch, or, more precisely, changes in the sectoral dispersion of excess demand, are significant in explaining deviations of the real product wage from the productivity trend.
- There is some evidence that wage pressure is related to potential mobility (measured by an index of housing tenure type), but that is only part of the story. It is probable that regional house price/earnings differences reflect the greater power of workers in areas of higher labour demands to obtain compensation for the higher local costs of living.
- Nevertheless it seems likely that house price/earnings differentials play a substantial role in determining migration.
- Average house prices are a significant factor in the wedge between the cost of living and producer prices.

- Union power has a significant effect on wage pressure (the coverage of collective bargaining is greater in the north than in the south).

The authors also pointed out that migration is small in relation to the mismatch, so that variations in migration need to be extremely large in order to compensate for the regional dispersion in earnings. Consequently, the tenure hypothesis, although important, should not be overstated.

There are important dynamic distortions in the owner-occupied housing market due to mortgage interest tax relief, the absence of capital gains tax on the household's main residence, and other institutional distortions on planning and zoning control. Hence, distortions favouring owner occupation (and the portfolio effects associated with it) and variations in liquidity are likely to be more important than rent and tenure control. This is a fundamental reason why house prices in the South East have tended to overshoot. It follows that a more neutral tax system would reduce economic pressure arising from regional demand differentials.

Let us now turn to look at the mobility-tenure hypothesis. This is associated with the work by Minford *et al* (1987) and Hughes and McCormick (1981, 1985). According to this approach, the greater are the barriers to mobility, the greater the segmentation of labour markets and the greater the wage pressure. These studies suggest that council house tenants are by far the least mobile and tenants in the private furnished sector are the most. Immobility of households yields both unemployment and congestion. The basic argument is that the absence of a rented sector ruled by freely undertaken contracts is a major explanation of the low rates of labour mobility in the UK, especially among manual workers. Such a hypothesis was initially put forward by Hughes and McCormick (1981), who discussed different migration rates for different households' housing tenure.

They found that, whatever the socio-economic background of the head of household, migration rates for council tenants are very low, whereas they are relatively high for rental tenants. Hence, the authors argued that council housing is a major contributor to labour immobility.

Minford *et al* obtained similar results, emphasising the deleterious effects of the Rent Act 1965. The authors utilised a mobility index in order to estimate the gap between actual and free market rents in the different tenure groups. They then constructed regional mobility indices for 1963-79, based on this gap, and on regional variations in council house tenancy. They then tried to explain regional unemployment rates using these indices and other variables for the regional labour and product markets. Results seemed to confirm their hypothesis, although the performance of the mobility index is not impressive.

We have seen that housing seems to be an important factor in explaining regional variations in unemployment. Nevertheless, researchers disagree on why this is so and on the policies that ought to be implemented to alleviate the problem.

Bover *et al* argue that estimates of immobility caused by council housing may be seriously biased in many cases. Council housing is housing of the last resort, that is, designed to accommodate the people within any given group suffering more adverse economic or social circumstances. Hence, the independent role for council housing may have been overestimated. With regard to the latter, Bover *et al* argue that, although it is true that higher house prices in the South East should provide a powerful incentive for net out-migration, in practice this does not take place. Hence, the roots of the problem must be elsewhere, and Bover *et al* argue the case for reform of the institutional distortions in the finance of housing as discussed above. Nevertheless, it is probable that policies designed to reform council housing and private sector rent control would help in reducing regional unemployment, inflationary wage pressure and regional dispersion in unemployment.

McCormick has noted that the crucial mobility index used by Bover *et al* is based on non-manual wages. Now, if it is the case that manual wages do not vary much across regions and that non-manual wages do, then discussions of migration should also be concerned with the skill composition of regional labour forces. The council house system may powerfully reinforce the resulting occupational bias. Hence, McCormick would argue that the policies of mobility he recommends may encourage migration of retired and manual workers out of the South East, and encourage more immigration of non-manual workers. Paradoxically, the results would be further occupational concentration that could hurt the peripheral regions even more.

Hughes and McCormick (1987) considered some of the implications of housing resource for labour mobility and drew comparisons between the UK and US housing markets. First of all some facts. In the UK there is a significant proportion of households living in council houses (about 28 per cent of all households). With respect to the housing stock, the latest figures (December 1988) are that 24.9 per cent of all houses in the UK consist of council houses (this proportion has decreased over the 1980s), whereas the proportion of owner-occupier houses in the UK is

Reducing Regional Inequalities

64.9 per cent. The remaining stock of houses is made up by unfurnished and furnished rentals. The importance of stocks of council houses varies by region. For example, in 1988 the stock of council houses was 46.4 per cent of all houses in Scotland, and only 22.6 per cent in England. As council stocks are concentrated in high unemployment areas, the asymmetry in the housing markets prevents migration.

There is a substantial difference between the ratios of US to UK migration rates for non-manual and manual heads of households. For non-manual workers, the US migration rate is about twice the UK rate, although job-related migration is higher in the UK. But for manual workers, the US migration rate is over 5.7 times the UK rate, and for job-related movements the ratio is 18 to 1. But can differences in migration rates between the US and the UK be explained by the UK tenure system? In order to answer this question, Hughes and McCormick (1987) examined what UK migration rates would be if the aggregate tenure pattern in the UK were the same as that in the US (while leaving UK migration propensities within tenure types constant). The increase in migration rates that would be brought about by such a theoretical redistribution of households out of council housing is only limited. The main reason for this is that in the UK council house tenants tend to be older than average manual workers, with few educational qualifications. Hence, their migration rates would be low regardless of tenure. It follows that although a case still exists for reform of the council system in the UK, mobility would still not match US mobility should reforms be implemented. Other socio-economic factors may be important.

Hughes and McCormick (1981, 1985) have argued that:

- living in a high unemployment rate region does not exercise a positive effect on migration;
- an unemployed individual is more likely to plan to migrate if he or she is a non-manual, instead of a manual worker.

The authors argue that the second point is particularly important for the UK case. The manual and non-manual labour markets are strikingly different. First of all, unemployment rates are low and vary little across regions for non-manual workers, whereas they are much higher and show a high degree of regional variation for manual workers.

Nevertheless, the differing migratory propensities of the two groups are not sufficient to explain such differentials. One finds that manual workers move out of the South East where the unemployment rate for their group is the lowest in the UK. Since council house dwellers are mostly manual workers, this casts serious doubts on whether freeing up

Regional Issues in Economics: Setting the Scene

the council house system would improve very much the performance of the manual labour market in terms of reduced unemployment differentials.

In the Hughes and McCormick model, immobility among dwellers in council houses occurs because stocks of council houses are greater in those regions where unemployment is higher, and since council tenants would have to move to smaller council houses, movement from high unemployment regions to low unemployment regions is deterred. Nevertheless, data on local authority dwellings vacant for 1988 show that vacancies are more numerous in those areas where unemployment is actually lower. This is very much in line with the finding that manual workers are actually moving out of the South East.

In view of the evidence that white collar workers migrate from depressed areas and that blue collar workers migrate from the South East, one could then argue that, although reform of the council sector would reduce price distortions in the housing markets and improve labour force mobility, it is unclear whether labour force mobility would benefit the development of high unemployment regions. The compositional effects in the labour market, that is, an increasingly non-manual labour force in the South East and an increasingly manual labour force in the other regions, should be analysed in depth when considering regional issues.

VII Concluding remarks

Regional economic disparities persist in the UK. The regional variation in unemployment, in particular, indicates that these disparities are a problem, because of the waste of resources entailed. The infrastructure resources available in these areas may also be under-utilised. Aggregate output and employment could be increased if the less favoured regions were able to reach the low unemployment rates of the more successful regions without triggering greater inflation. The potential exists for the nation as a whole to move to a more favourable short-run inflation/unemployment trade-off and a lower long-run unemployment rate. What actions should be taken to bring this about?

Regional policy *per se* has tended to be implemented in a rather *ad hoc* way, and attempts to evaluate its impact suggest there has only been a modest one, given its cost. Policy-makers have tended to conclude from this that policy should be more narrowly focused (eg on deprived urban areas) and that aid should be subject to more stringent conditionality (which helps if the conditions are sensible and can be monitored easily). But should more attention be paid to the *differences* among the older policy instruments, so that the more successful ones could be refurbished? For instance, the pricing and auctioning of Industrial Development Certificates merit some consideration. Regional policy spending has always been dwarfed by regional spending on other government programmes. Should there be more investigation of their impact? Defence spending has been examined by some economists, but other programmes - notably the National Health Service - are also important.

The functioning of the labour market is of great importance in discussing regional disparities. There are strong arguments for encouraging wage differentials in favour of the regions with relatively high labour demand. However, wages are not an exogenous policy instrument. How can wage determination be related more closely to local employment conditions? Relying on decentralisation of collective bargaining is insufficient, because there is no guarantee that it will take on board the interests of local 'outsiders' - the unemployed. Efficiency considerations may clash with those of equity if regional relativities are regarded as unfair. This certainly seems to be the case with respect to government transfer payments.

The housing market is another area which receives a lot of attention in the contributions to this book. Although the equilibrating role of migration can be exaggerated, it is clear that the British council housing system inhibits it greatly. Owner-occupation also inhibits migration, relative to a rental system, and some economists believe house price differentials actually generate perverse signals to potential movers. What housing reforms could encourage labour mobility? Should the emphasis be on the development of more flexible types of tenure or should more attention be paid to reducing moving costs of all types?

Infrastructure investment decisions are very difficult to evaluate, because of the difficulty of estimating true social costs and benefits. A greater degree of pricing of infrastructure services - perhaps through more use of 'shadow prices' - is desirable. Only then will it be clear what is the appropriate response to congestion arising from infrastructure inadequacies. It may well be appropriate for development to be directed to regions where the existing infrastructure is under-utilised.

Appendix

A chronology of British regional policy (from Armstrong and Taylor, 1988)

1928 **Industrial Transference Scheme and Juvenile Transfer Scheme.**
Grants and loans to unemployed migrants.

1934 **Special Areas (Development and Improvement) Act.**
Four special areas designated in South Wales, Scotland, North East and West Cumberland. Small loans available to industry. Trading estates established. Loan powers strengthened by the Special Areas Reconstruction (Amendment) Act 1936.

1937 **Special Areas (Amendment) Act.**
Extension of loan powers in special areas. Tax, rent and rate subsidies in special areas. Extension of trading estate policy. Initial funding of £2 million.

1940 **General Transfer Scheme.**
Migration assistance considerably increased for displaced workers. Replaced industrial transference scheme.

1945 **Distribution of Industry Act.**
Assisted areas enlarged and designated as development areas (DAs). Basic responsibility for regional policy given to the Board of Trade, which acquired (from the special areas commissioners) powers of factory building and leasing, finance for trading estates, land reclamation, and grants and loans to firms. Wartime system of building licences was retained to restrict expansion of manufacturing in low unemployment regions.

1946 **Resettlement Scheme.**
Gradual replacement of wartime general transfer scheme. Unemployed migrants from all regions eligible. Wide range of migration costs met. Incorporated a facility to assist temporary migrants (the voluntary temporary transfer scheme), as well as provision to assist key workers required by firms moving to assisted areas. Extended to workers sent to other regions for retraining (subsequently renamed the nucleus labour force scheme).

1947 **Town and Country Planning Act.**
Extension of wartime building controls. Industrial Development Certificates (IDCs) introduced. All new manufacturing establishments or extensions of over 5,000 sq ft required an IDC. Exemption limit altered frequently after 1947.

1950 **Distribution of Industry Act.**
A further small extension of Board of Trade loan and grant powers for firms moving to DAs. Regional policy regarded as being 'weak' during the early and mid 1950s, even though powers existed for a much stronger policy.

1951 **Key Workers Scheme.**
Assistance to key workers moving with their firms to DAs. Separated from the resettlement scheme and became a scheme in its own right. As with location of industry policies, the resettlement schemes were less actively pursued in the 1950s. Assistance to temporary migrants was discontinued during 1950-7.

1958 **Distribution of Industry (Industry Finance) Act.**
Extension of loan and grant assistance to a number of high unemployment areas outside the DAs. Gradual revival of regional policy with a tightening of location controls and increased expenditure.

1960 **Local Employment Act.**
Repeal of Distribution of Industry Acts. DAs replaced by a fragmented patchwork of development districts delimited on the basis of unemployment rates exceeding 4.5 per cent. Retention and strengthening of earlier Board of Trade powers. Introduction of new building grants. Reform of industrial (formerly trading) estates policy. Controls over location of industry retained.

1962 **Resettlement Transfer Scheme.**
Resettlement scheme (1946) replaced by resettlement transfer scheme. Key worker scheme and nucleus labour force scheme retained. The beginning of a more active resettlement policy.

1963 **Budget.**
Free depreciation introduced for firms in assisted areas. Allowed firms to write off investment in plant and machinery against corporation tax at any rate they chose. Strengthened the financial advantage in investing in assisted areas.

1964 **Regional Planning.**
Regional economic planning councils and boards set up by the Department of Economic Affairs. Functions mainly advisory.

1965 **Control of Office and Industrial Development Act.**
Control of office development in London and Birmingham in the form of office development permits.

1965 **Highlands and Islands Development (Scotland) Act.**

Regional Issues in Economics: Setting the Scene

Highlands and Islands Development Board established with extensive powers of loans, grants, equity participation and new factory building for a wide range of economic activities.

1966 **Industrial Development Act.**
One hundred and sixty-five development districts replaced by five large DAs. Replacement of free depreciation by a system of 40 per cent investment grants in DAs (20 per cent elsewhere). Remaining policies retained, eg location controls, building grants (25–35 per cent), land reclamation grants of 85 per cent.

1967 **Finance Act.**
Manufacturing firms in DAs could reclaim payments made under selective employment tax (SET) and were entitled to the selective employment premium (SEP) (37.5p per man per week) and the regional employment premium (£1.50 per man per week and lower rates for women and juveniles). SET and SEP withdrawn in 1970.

1967 **Special Development Areas.**
Special development areas established in Scotland, North East, West Cumberland and Wales with additional incentives and rent-free premises, 35 per cent building grants and some operating cost subsidies.

1970 **Local Employment Act.**
Seven intermediate areas established. Government-built factories, building grants and derelict land clearance grants. Financed by withdrawal of SEP from DA.

1970 **October Mini-Budget.**
Investment grants replaced by accelerated depreciation in the DAs. Other existing powers retained and strengthened.

1972 **Industry Act.**
Several major revisions to regional policy including: a return to investment grants for plant, machinery and building (the regional development grant scheme); proposed phasing out of regional employment premium; end of location controls in DAs and SDAs; increase in other forms of existing assistance; selective assistance for industry under section 7 of the Act; grants and low-interest loans available for services.

1972 **Employment Transfer Scheme.**
Resettlement transfer scheme renamed and strengthened. Key workers and nucleus labour force schemes retained.

1973 **Entry into European Economic Community.**
This gave UK assisted areas access to loans, grants and other

Reducing Regional Inequalities

assistance from EEC funds and institutions such as the European Coal and Steel Community, the European Investment Bank, the European Social Fund and the European Agricultural Guidance and Guarantee Fund. UK was required to adhere to EEC competition policy controls. These included limits on investment subsidies in the central regions of the EEC (including the intermediate areas). Disapproval of subsidies such as the regional employment premium.

1973 **Hardman Report.**
Proposal to disperse 31,000 London-based civil service jobs.

1974 **Regional Employment Premium.**
Doubled.

1975 **European Regional Development Fund.**
A major new EEC fund established specifically to assist depressed areas. Investment grants and interest rebates on other EEC loans for industrial, craft, service and infrastructure projects. Each member state was guaranteed a predetermined share (or quota) of the fund. Regional policy committee established to help to co-ordinate member state and EEC regional policies and to stimulate regional research and regional planning.

1976 **Development Agencies.**
Scottish and Welsh Development Agencies set up. Scottish Development Agency given substantial powers to invest in industry, create new companies, provide finance and advice for industry, build and manage industrial estates, lease or sell factories, reclaim derelict land and rehabilitate the environment.

1976 **Service Industry and Office Location Subsidies.**
Strengthening of grants and rent relief for firms in the services and office sector locating in assisted areas. Preferential loans available.

1977 **Regional Employment Premium.**
Abolished.

1979 **Reform of European Community Regional Policy.**
Major reform of the European Regional Development Fund. The fund was divided into a quota section (comprising 95 per cent of the total fund) and a non-quota section (5 per cent of the fund). The quota section to be allocated to member states in the form of predetermined share of expenditure. Non-quota section to be allocated more at the discretion of the EC. A new system

for co-ordinating regional policy within the EC gradually introduced, including a regional impact assessment scheme designed to monitor and respond to all EC policies expected to have adverse regional effects. Series of initiatives introduced: regional development programmes; periodic reports on regional problems; regional priorities; and guidelines.

1979–83 **Phased reform of UK regional policy.**
Announcement of a major package of reforms to be phased in during 1979–83. Reforms included: planned cuts in the regional budgets; retention of the fourfold categorisation of assisted areas (SDAs, DAs, intermediate areas and derelict land clearance areas), but a gradual redrawing of assisted areas boundaries designed to reduce the percentage of working population covered by assisted areas from 47 per cent in 1979 to 28 per cent by 1982; regional development grants retained at 22 per cent in the SDAs, but cut from 20 to 15 per cent in DAs and abolished in intermediate areas; selective financial assistance retained but to be provided only where necessary for a project to proceed; office development permits abolished and Industrial Development Certificates no longer required in intermediate areas. Easing of floorspace limit for IDCs in non-assisted areas.

1980 **Regional Economic Development Councils.**
Abolished.

1980 **Enterprise Zones.**
Nine enterprise zones created. Firms locating in an enterprise zone exempt from local rates and development land tax; planning applications to be dealt with quickly; less information to be provided to government departments. Rapid increase in number of zones to 25. Designated for a ten-year experimental period.

1981 **Industrial Development Certificates.**
Suspended and subsequently abolished.

1984 **Reform of European Community Regional Policy.**
Reforms to EC regional policy. Included the abolition of the distinction between quota and non-quota sections of the European Regional Development Fund. Member states to receive a share of the fund lying within an indicative range. Gradual shift of fund expenditure from one based on projects to one based on programme contracts. These compromise a number of initiatives and projects to meet specified objectives.

Reducing Regional Inequalities

Gradual increase of EC programme contracts in preference to national programme contracts. Improved co-ordination of regional policies within the EC. Greater assistance to small and medium-sized firms and to indigenous firms in depressed areas. Extension of initiatives to areas of urban deprivation and trans-frontier areas.

1984 **Reform of British regional policy.**
Announcement of major set of reforms. Cuts of £300 million in annual expenditure on regional policy by 1987/8. SDAs abolished and downgraded to DAs. Assisted area boundaries redrawn. Regional development grants available only in DAs (now containing 15 per cent of the working population). West Midlands designated as an intermediate area. Regional development grants to be extended to certain service industries. A cost-per-job limit of £10,000 imposed on RDGs (except for firms employing fewer than 200 employees). Alternatively, subsidies of up to £3,000 for each new job created. Firms to receive either investment grant or subsidy per job, whichever is the most favourable. Gradual shift of emphasis from automatic grant to selective assistance.

1984 **Freeports.**
Six freeports designated at various airport and dockland sites. A freeport is enclosed zone regarded as being outside of customs area. Customs duties are only paid when goods leave the freeport for the rest of the UK or for other EC countries.

1988 **Reform of British regional policy.**
Announcement of a major set of reforms. Regional development grants abolished. Savings in expenditure to be switched (initially at least) to the regional selective assistance scheme which is retained, and to new schemes. Selective assistance is given only to those projects which would not otherwise go ahead. Two new types of grant introduced: (i) small firms under 25 employees (15 per cent investment grant up to a maximum of £15,000 and/or 50 per cent innovation grant up to a maximum of £25,000); (ii) consultancy grants for firms with under 500 employees (67 per cent of costs, compared to 50 per cent for non-assisted area firms). Greater emphasis to be placed on providing assistance in inner-city areas.

1988 **European Regional Development Fund.**
To be doubled in size by 1992. This is in response to the Single

European Act which encourages the complete abolition of remaining barriers to internal trade by 1992.

Bibliography

Adams, CD, Baum AG and McGregor, BE (1988) 'The Availability of Land for Inner City Development: A Case Study of Inner Manchester', *Urban Studies*, vol 25/1, February.
Albrechts, L *et al* (1989) *Regional Policy at the Crossroads: European Perspectives*, Jessica Kingsley.
Allen, K (1989) 'Requirements for an Effective Regional Policy' in Albrechts *et al*, op cit.
Alonso, W (1964) *Location and Land Use*, Harvard University Press, Cambridge, Mass.
Amin, A and Goddard, J (1986) *Technological Change, Industrial Restructuring and Regional Development*, Allen and Unwin, London.
Armstrong, H and Taylor, J (1985) *Regional Economics and Policy*, Philip Allen.
— (1988) *Regional Policy and the North-South Divide*, Employment Institute, London.
Ashcroft, B and McGregor, P 'The Demand for Industrial Development Certificates and the Effect of Regional Policy', *Regional Studies*, 23.4, pp301-14.
Blackaby, DH and Manning, DN (1987) *Regional Earnings Revisited*, The Manchester School.
Bover, O, Muellbauer, J and Murphy, A (1989) 'Housing, Wages and UK Labour Markets', *Oxford Bulletin of Economics and Statistics*, 51, pp97-162.
Bowers, J, Cheshire, P and Webb, A (1970) 'The Change in the Relationship Between Unemployment and Earnings: A Review of Some Possible Explanations, *National Institute Economic Review*, vol 4, pp44-63.
Breheny, M (ed) (1988) *Defence Expenditure and Regional Development*, Alexandrine Press.
Chisholm, M (1987) 'Regional Development: The Reagan-Thatcher Legacy', in *Environment and Planning C: Government and Policy*, vol 5, pp197-218.
Clark, W A and Van Liercop W F (1986) 'Residential Mobility and Household Location Modelling' in Nijkamp, P (ed) *Handbook of Regional and Urban Economics*, vol 1, North Holland.
Commission of the European Communities (1987) *European Regional Development Fund: UK Regional Development Programme 1986-90*, Brussels.
Damesick, P and Wood, P (eds) (1987) *Regional Problems, Problem Regions and Public Policy in the United Kingdom*, Clarendon Press, Oxford.
Egginton, DM (1988) 'Regional Labour Markets in Great Britain, *Bank of England Quarterly Bulletin*, August.
Fothergill, S and Gudgin, G, (1982) *Unequal Growth: Urban and Regional Employment Change in the UK*, Gower, Aldershot.

Hall, P et al (1987) *Western Sunrise; The Genesis and Growth of Britain's Major High-Tech Corridor*, Allen and Unwin, London.

Hart, R (1989) 'The Employment and Hours Effects of a Marginal Employment Subsidy', *Scottish Journal of Political Economy*, 36.4, pp385-95.

Hausner, V A (1987) *Economic Change in British Cities*, Clarendon Press, Oxford.

Heim, C (1988) 'Government Research Establishments, State Capacity and Distribution of Industry Policy in Britain'. *Regional Studies*, 22.5, pp375-86.

Henley, D, Nairn, A and Swales, J (1989) 'Shift-Share Analysis of Regional Growth and Policy: A Critique', *Oxford Bulletin of Economics and Statistics*, 51.1, pp15-33.

Hotelling, H (1929) 'Stability in Competition', *Economic Journal*, 39, pp41-57.

Hughes, GA and McCormick, B (1981) 'Do Council Housing Policies Reduce Migration Between Regions?', *Economic Journal*, vol 91, pp919-37.

— (1985) 'Migration Intentions in the UK: Which Households Want to Migrate and Which Succeed?', *Economic Journal*, vol 95, Conference Supplement, pp76-95.

— (1987) 'Housing Markets, Unemployment and Labour Market Flexibility in the UK', *European Economic Review*, pp615-45.

Jackman, R and Roper, S (1987) 'Structural Unemployment', in *Oxford Bulletin of Economics and Statistics*, vol 49, pp9-36.

Kanbur, SMR and Vines, D (1984) *North-South Interaction and Commodity Control*, Centre for Economic Policy Research Discussion Paper No8.

Layard, R and Nickell, S (1985) 'The Causes of British Unemployment', *National Institute Economic Review*, Feb, pp62-85.

McIntosh, J (1984) 'North-South Trade: Export-Led Growth with Abundant Labour', unpublished manuscript, Essex University.

Martin, R (1989) The Growth and Geographical Anatomy of Venture Capitalism in the UK, *Regional Studies*, 23.5, pp389-403.

— and Rowthorn, B (1986) *The Geography of De-Industrialisation*, Macmillan, London.

Martin, S (1989) 'New Jobs in the Inner City: The Employment Impacts of Projects Assisted under the Urban Development Grant Programme', *Urban Studies*, 26, pp627-38.

Mason, C and Harrison, R (1989) 'Small Firms and the 'North-South' Divide in the United Kingdom: The Case of the Business Expansion Scheme', in *Transactions of British Geographers*, 14.

Minford, P, Peel, M and Ashton, P (1987) *The Housing Morass*, Institute for Economic Affairs, London.

Minford, P, Ashton, P and Peel, M (1988) 'The Effects of Housing Distortions on Unemployment', *Oxford Economic Papers*, June 1988, 40(2), pp322-45.

Minford, P and Stoney, P (1988) *Regional Policy and Market Forces: A Model and an Assessment*, Liverpool University Department of Economics and Accounting, Working Paper No 8901.

Moore, B, Rhodes, J and Tyler, P (1986) *The Effects of Government Regional Policy*,

Department of Trade and Industry, HMSO.
Morgan, K (1986) 'Re-Industrialisation in Peripheral Britain: State Policy, the Space Economy and Industrial Innovation', in Martin and Rowthorn (1986) op. cit.
Muellbauer, J and Murphy, A (1988) *Housing Prices and Migration: Economic and Investment Implications*, Shearson Lehman and Hutton Securities Research Report.
Neary, JP (1978) 'Short-Run Capital Specificity and the Pure Theory of International Trade', *Economic Journal*, September.
Parsons, W (1988) *The Political Economy of British Regional Policy*, Routledge, London.
Pissarides, CA (1978) 'The Role of Relative Wages and Excess Demand in the Sectoral Flow of Labour', *Review of Economic Studies*, vol 45, October, pp453-67.
Pissarides, C and Wadsworth, J (1989) 'Unemployment and the Inter-regional Mobility of Labour', *The Economic Journal*, 99, pp739-55.
Sayer, A and Morgan, K (1986) 'The Electronics Industry and Regional Development in Britain', in Amin and Goddard (1986), op. cit.
Stoker, G (1989) 'Urban Development Corporations: A Review'. *Regional Studies*, 23.2, pp159-67.
Swales, K (1988) 'Are Discretionary Regional Subsidies Cost-Effective?', *Regional Studies*, 23.4, pp361-76.
Talbot, J (1988) 'Have Enterprise Zones Encouraged Enterprise? Some Empirical Evidence from Tyneside', *Regional Studies*, 22.6, pp507-14.
Taylor, L (1983) *Structural Macroeconomics*, Basic Books, New York.
Tyler, P, Moore, BC and Rhodes, J (Oct 1988) 'The Rate Burden: An Accounting Framework for Assessing the Effect of Local Taxes on Business', *Regional Studies*, 22(5), pp387-97.
Vickerman, R (1988) *Transport and the Integration of the UK in the European Economy: The Case of the Channel Tunnel*, Channel Tunnel Research Unit.
— (1989) 'Measuring Changes in Regional Competitiveness: The Effects of International Infrastructure Investments, *Ann Reg Sci* 23: 275-86.
— (1989) *Regional Development Implications of the Channel Tunnel*, Channel Tunnel Research Unit.
Vines, D (1984) *A North-South Growth Model Along Kaldorian Lines*, Centre for Economic Policy Research Discussion Paper No 26.
Williamson, J (1983) *The Open Economy and The World Economy*, Basic Books, New York.
Wren, C (1988) *Some Evidence on the Employment effects of the Revised Regional Development Grant Scheme*, University of Reading Economics Department Discussion Paper No 39.
— (1989) 'Factors Underlying the Employment Effects of Financial Assistance Policies', in *Applied Economics*, vol 21, pp497-513.

2
Regional Economic Disparities: Causes and Consequences

Jim Taylor

Regional economic disparities are an inevitable consequence of a changing world. Regions have to create new markets as old ones disappear. Each region's workers and entrepreneurs have to adapt to these changing circumstances and opportunities if they are to survive. Regions which do not change with the times are likely to lose their competitiveness in world markets, the consequences of which include slow growth, high unemployment, low incomes, dilapidated infrastructure, net outward migration and generally poor prospects for their residents, compared to regions which are able to maintain their competitiveness. The key to a region's economic success is therefore whether or not it can compete effectively in markets which are in a constant state of flux. If regions are to survive and prosper, they must be competitive.

This chapter examines the causes and consequences of differences in the competitiveness of regions in the UK economy. This will involve enquiring into why regional economic disparities occur and why they persist. Before undertaking this task, however, it is necessary to establish the meaning of the term 'regional economic disparities' and to describe the underlying trends in some of the main economic variables.

Section I discusses the meaning of regional economic disparities in the context of the familiar notion of the north-south divide. Section II provides some background data which help to paint a picture of the pattern of regional economic disparities in the UK. This is followed in section III by a discussion of the various causes of regional economic disparities and an explanation of why such disparities have persisted for many decades. Finally, section IV examines some of the main consequences of regional economic disparities and argues that a

Regional Economic Disparities: Causes and Consequences

reduction in these disparities would bring about real and substantial gains to the nation as a whole.

I Regional economic disparities and the north-south divide

Interest in regional economic disparities has waxed and waned since the government first designated a set of assisted areas under the Special Areas Act in 1934. The recent surge of interest began, not surprisingly, when unemployment rates in the northern regions began to rise to what had seemed impossibly high levels only a few years before. The economic depression which followed Sir Geoffrey Howe's use of severe deflationary policies during 1979-81 to reduce inflation led to an ever-widening gap in unemployment rates between the north and the south (see Figure 2.1). This widening unemployment gap between the north and the south accurately reflected regional disparities in employment prospects during the early 1980s, especially for unemployed males and young people. The term used to reflect this widening gap in employment opportunities is 'the north-south divide', which has been used in a plethora of books and

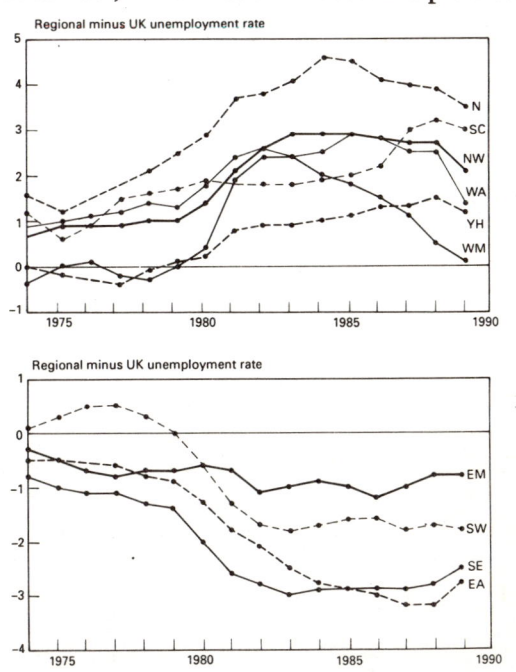

Source: Regional Trends, 24, 1989

Figure 2.1 *Regional unemployment disparities in selected UK regions, 1974-89*

Reducing Regional Inequalities

articles in recent years (eg Armstrong and Taylor, 1988; Balchin, 1989; Lewis and Townsend, 1989; Mason and Harrison, 1989; and Smith, 1989).

The north-south divide is a notion which has caught the imagination of people across the whole of the UK, so much so that there can be few who do not understand its broad meaning. The 'divide' is often represented as an imaginary line running from the Severn to the Humber. The north consists of Scotland, Wales, Northern Ireland, North, North West and Yorkshire and Humberside. The south includes the South East, East Anglia, South West and the East Midlands. The West Midlands has been somewhat of a floating partner of both the north and the south; it belonged to the north in the early 1980s when it became a high unemployment area, but shifted back southwards during the second half of the 1980s.

The stereotype image of the north is a place where unemployment is high, incomes are low, job prospects are poor (particularly for men), the infrastructure is of poor quality, and the most able workers are likely to seek jobs in the south, which has the opposite image. Reality is somewhat more complex, as Joe Rogaly has so vividly pointed out:

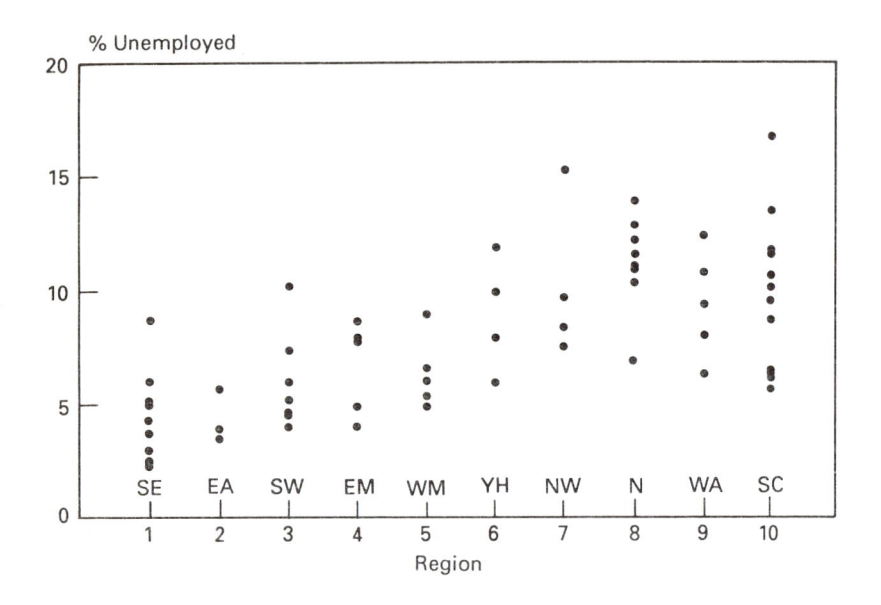

Source: Employment Gazette, June 1989

Figure 2.2 *Inter-county differences in unemployment rates by region, April 1989*

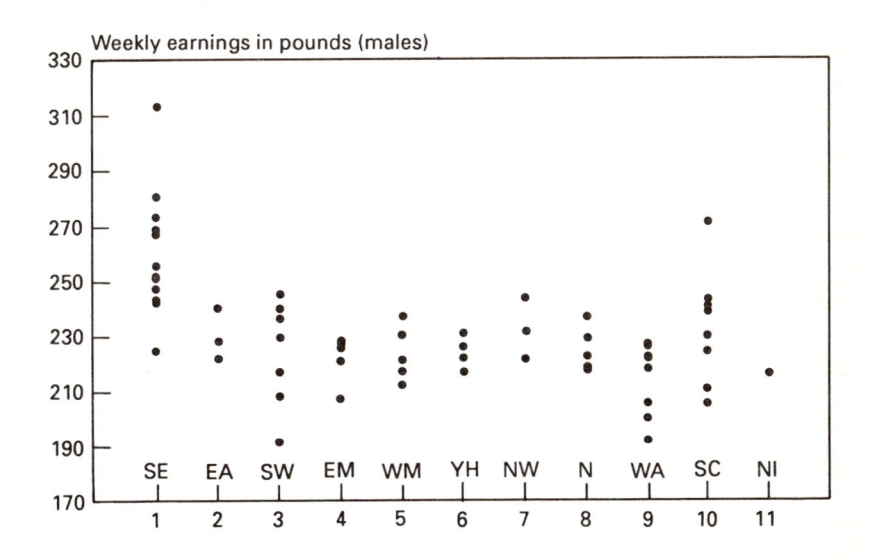

Regional Economic Disparities: Causes and Consequences

[Figure: scatter plot of Weekly earnings in pounds (males) by region, with regions SE(1), EA(2), SW(3), EM(4), WM(5), YH(6), NW(7), N(8), WA(9), SC(10), NI(11), y-axis from 170 to 330]

Source: Regional Trends, 24, 1989

Figure 2.3 *Inter-county disparities in average gross weekly earnings by region: full-time males, April 1989*

One of the most confused notions is that of the north–south divide. It implies that virtually all the wealth is in the south, even the South East, while the rest of the country is a zone of unrelieved devastation.

The truth is that there is an archipelago of wealth in the north, just as there is an archipelago of poverty in the south. For example, parts of Brixton or some of the council estates around Kings Cross in London are as depressing as their counterparts in, say, Manchester. (*Financial Times* 31 March 1987)

There is no shortage of evidence to support the claim that the north–south divide is a grossly simplistic notion which gives a very misleading impression of the true nature of *spatial* economic disparities in the UK. Figures 2.2, 2.3 and 2.4 demonstrate that disparities in variables such as unemployment and earnings are very often greater *between localities within regions* than they are *between regions*. None the less, the simplicity of the north–south divide (as a concept) helps to focus attention on the existence of a fundamental economic imbalance between different geographical areas in the UK. Concentrating upon the

Reducing Regional Inequalities

Weekly earnings in pounds (females)

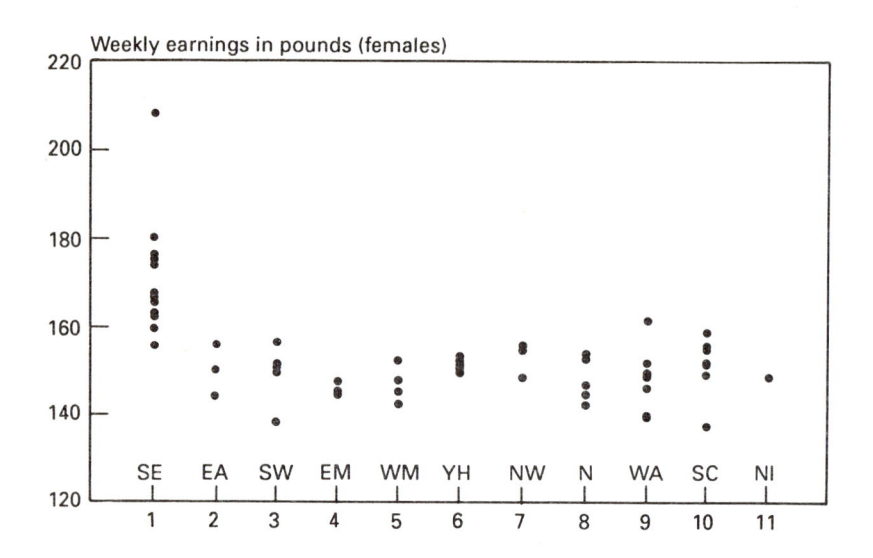

Source: Regional Trends, 24, 1989

Figure 2.4 *Inter-county disparities in average gross weekly earnings by region: full-time females, April 1989*

economic imbalance between a relatively small set of geographical areas (ie the standard economic regions of the UK) also helps to focus on the main issues rather than becoming bogged down in the detail which a more spatially-disaggregated analysis would entail.

II Regional disparities in the UK: some facts

Regional economic disparities in the UK are longstanding and deep-rooted. These disparities are reflected in a wide range of economic variables, all of which support the widely-held view that a significant and substantial economic gulf exists between the northern and southern regions of the UK. The extent of these regional disparities and recent trends in their magnitude are demonstrated in this section by examining six major economic variables:

- unemployment;
- GDP per capita;
- employment growth;
- income (and expenditure);

Regional Economic Disparities: Causes and Consequences

- migration;
- competitiveness.

Unemployment

Regional disparities in unemployment have persisted since the 1920s. For over six decades, the southern regions of Britain have had persistently lower unemployment rates than those prevailing in the northern regions. Figure 2.5 shows the situation during the inter-war years: Wales, Northern Ireland, Scotland, the North East and the North West all experienced substantially higher unemployment rates than the southern regions. The poor performance of the northern regions during the 1920s and 1930s is explained by their heavy reliance on the old staple industries such as cotton, coal, steel and shipbuilding, the decline of which was largely responsible for the far poorer employment prospects of those living in the northern regions compared to those living in the south.

After remaining fairly constant during the 1950s, 1960s and 1970s, regional unemployment disparities widened suddenly and dramatically in the early 1980s (see Figure 2.1 on p71). This widening of regional

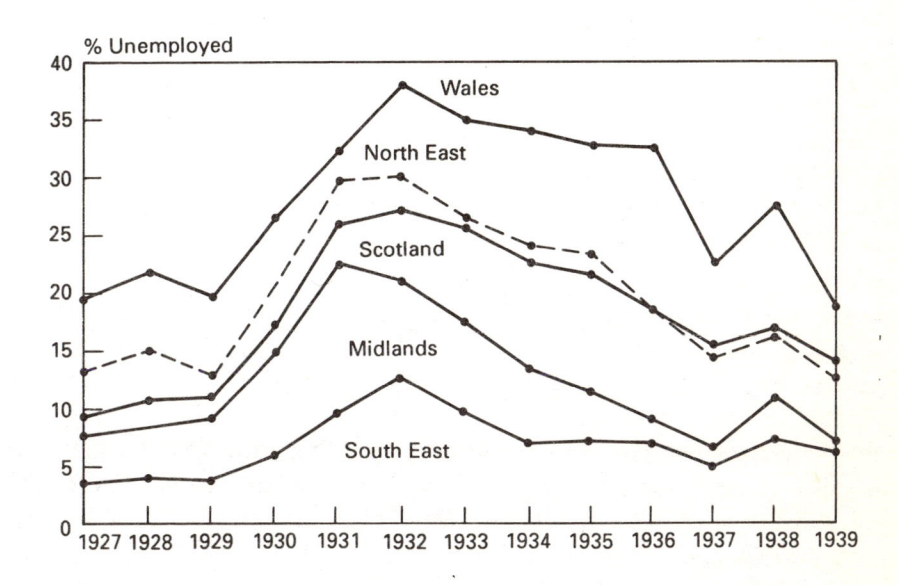

Source: Historical abstract of Labour Statistics

Figure 2.5 *Percentage unemployed in selected UK regions, 1927-39*

Reducing Regional Inequalities

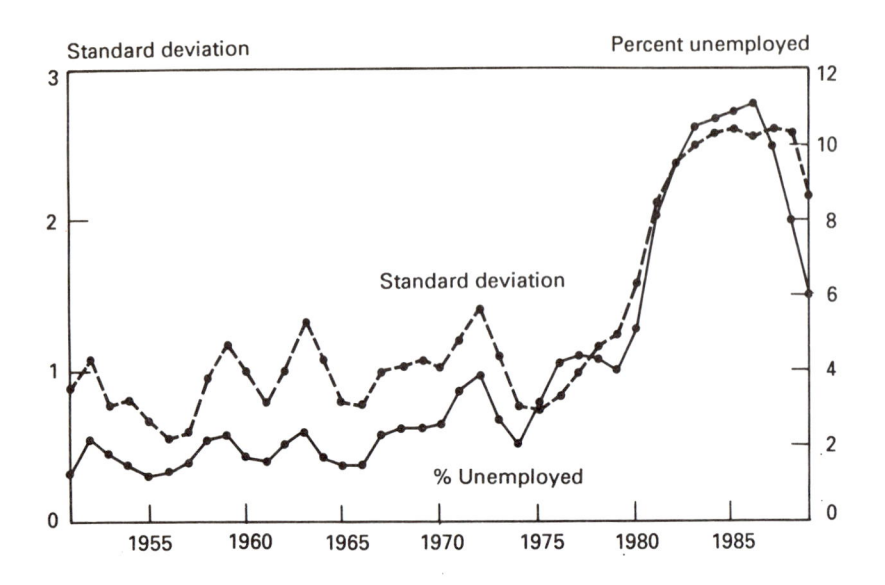

Source: Unemployment data for 1951-73 based on old method of measuring unemployment; 1974-89 data based on new method

Figure 2.6 *Regional unemployed disparities and percentage unemployment in the UK, 1951-89*

unemployment disparities was caused by the steep increase in the national unemployment rate from 4 per cent to over 10 per cent between 1979 and 1983. As the national economy fell into severe recession, the impact was felt far more severely in the northern regions than in the southern regions of the UK. This is simply a reflection of the economic weakness of the north compared with the south. Exactly the reverse happened during 1987-9: the rapid fall in the national unemployment rate has been followed by a corresponding fall in regional unemployment disparities (see Figure 2.6). The obvious conclusion to be drawn from this very close positive relationship over the long run between the UK unemployment rate and regional disparities in unemployment is that a low national rate of unemployment is necessary if regional unemployment disparities are to be at an acceptable level.

GDP per capita

Regional disparities in GDP per capita have remained fairly stable during the 1970s and 1980s (see Table 2.1). The most notable exception to this

Regional Economic Disparities: Causes and Consequences

Table 2.1 *Regional disparities in output per capita, 1971-87*

Region	Gross domestic product per capita (UK = 100) 1971	1979	1987	% change in GDP per capita 1979-87
South East	113.7	116.2	118.5	2.7
East Anglia	93.6	94.4	99.8	4.1
South West	94.8	91.3	94.0	3.4
East Midlands	96.6	96.6	95.1	2.3
West Midlands	102.8	96.2	91.6	1.4
Yorkshire and Humberside	93.3	92.7	92.7	2.0
North West	96.2	96.1	92.8	1.3
North	86.9	90.7	88.9	1.5
Wales	88.3	85.2	82.4	1.7
Scotland	93.0	94.6	94.5	1.8
Northern Ireland	74.3	78.2	77.4	2.3
UK	100	100	100	2.1

Source: Regional Trends 24, 1989

stability is the significant decline of the West Midlands, which fell from a comfortable second place in 1971 to eighth place in 1987. The South East and East Anglia both managed to increase their GDP per capita relative to other regions during the period, with East Anglia doing particularly well during the 1980s.

Employment growth

The widening gulf between the north and the south is most vividly demonstrated by regional disparities in employment growth. During the 1970s, the North West and the West Midlands both experienced negligible employment growth, whereas employment in Scotland, Wales and Northern Ireland grew at a faster rate than in the South East (see Table 2.2).

The picture changed dramatically in the 1980s. Employment growth was negative in all northern regions during 1979-88, reaching -10 per cent in the North West and Wales. By contrast, employment growth in the south was greater in the 1980s than it had been in the 1970s: employment expanded by 8 per cent in the South East, 10 per cent in the

Reducing Regional Inequalities

South West and an incredible 26 per cent in East Anglia. These regional disparities in employment growth leave absolutely no doubt about the growing gulf between the north and the south during the 1980s.

Table 2.2 *Regional disparities in employment growth, 1971-88*

Region	% change in employment 1971-9	1979-88
South East	3.5	8.2
East Anglia	15.2	26.2
South West	7.2	10.6
East Midlands	8.9	7.3
West Midlands	0.9	0.0
Yorkshire and Humberside	3.7	-2.6
North West	0.8	-10.0
North	3.0	-6.3
Wales	7.8	-10.2
Scotland	5.6	-5.2
Northern Ireland	10.9	-7.0
UK	4.4	1.9

Note: Employment includes employees in employment plus the self-employed

Source: Regional Trends, 24, 1989

Income, expenditure and the standard of living

Income data does not lend much support to the notion of a north-south divide - with the notable and very substantial exception of the South East (see Table 2.3). No matter how income is measured, the South East does considerably better than any other region. There is more evidence of a north-south divide, however, in expenditure per person: the residents of East Anglia and the South West spend considerably more than all other regions (except the South East).

The huge income differential between the South East and other UK regions can be explained by two main factors. First, the South East has a disproportionate share of high wage occupations compared to other regions (eg the South East had 16.5 per cent in managerial and professional occupations in 1987, compared to 12 per cent on average in

Regional Economic Disparities: Causes and Consequences

Table 2.3 *Regional disparities in income and expenditure*

Region	Household weekly income (£) 1986–7	Weekly earnings (£) Full-time men 1988	Weekly earnings (£) Full-time women 1988	Weekly disposable income per head (£) 1987	expenditure per person (weekly) 1986–7
South East	302	283	188	116	87
East Anglia	221	230	150	99	76
South West	251	228	152	98	76
East Midlands	227	223	145	94	66
West Midlands	225	226	149	90	62
Yorkshire and Humberside	210	224	151	94	64
North West	232	231	153	93	68
North	198	224	149	92	62
Wales	207	218	150	86	61
Scotland	213	233	152	96	65
Northern Ireland	208	215	148	84	61
UK	245	246	164	100	72

Note: all data at current prices

Source: Regional Trends, 24, 1989

other regions). Second, labour shortages in the South East can be expected to raise wage levels relative to other regions. This is supported by an inter-county analysis of the statistical relationship between average earnings and the unemployment rate (see Table 2.4): low unemployment counties tend to have higher earnings than high unemployment counties. Approximately one-third of the inter-county variation in earnings is explained (statistically) by corresponding differences in the unemployment rate (see Figure 2.7). Part of the remaining variation is likely to be explained by differences in the occupational mix of counties.

Regional disparities in income levels only provide information about one side of the standard of living equation. To assess regional disparities in *real* incomes, it is also necessary to take account of regional disparities in the cost of living. Unfortunately, regional price indices which cover all items of expenditure are not published. It is, however, possible to investigate the effect of one major item of expenditure on regional living costs, namely the cost of housing. Table 2.5 shows that house prices (for the same type of house) have been considerably and consistently higher in the South East than in any other region. Regional house price

Reducing Regional Inequalities

Table 2.4 *Regression analysis of inter-county disparities in average earnings, 1988*

Dependent variable = average gross weekly earnings for each UK county, October 1988

Explanatory variable	Estimated coefficients			
	Males		Females	
	(i)	(ii)	(iii)	(iv)
Constant	5.638	5.626	5.188	5.177
Unemployment rate	-0.098	-0.095	-0.078	-0.075
	(-4.72)	(-5.32)	(-4.61)	(-5.86)
$\bar{R}^2$	0.26	0.32	0.26	0.37

Notes:
1. Both variables were logged before running the regression equations
2. Equations (ii) and (iv) omit Greater London from the analysis
3. Figures in parentheses are t-ratios

Source: Regional Trends, 24, 1989

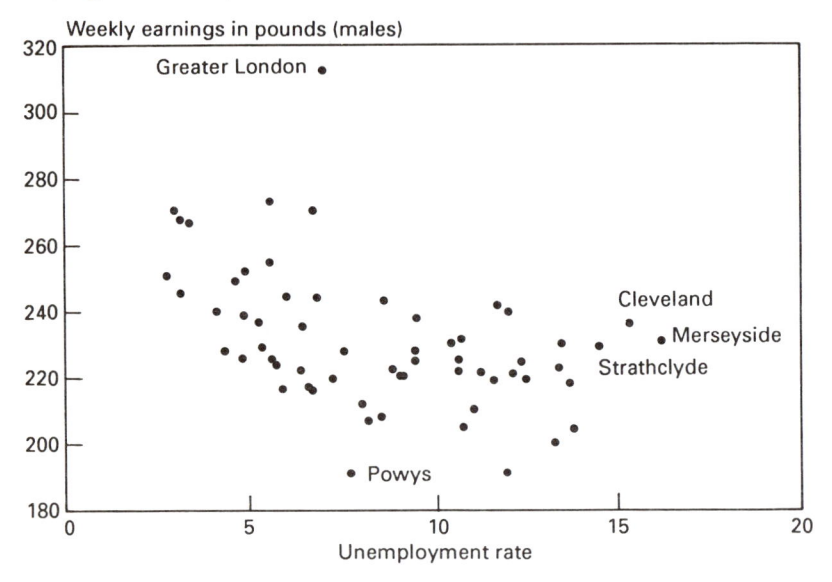

Source: Regional Trends, 24, 1989

Figure 2.7 *Scatter plot of average earnings versus unemployment by UK county, 1988*

Regional Economic Disparities: Causes and Consequences

differentials are not, however, constant over time with the result that regional disparities in the standard of living can fluctuate considerably over short periods. Figure 2.8 shows a strong increase in regional price disparities during 1983–8, a period of rapid GDP growth. This was followed by a very sudden and very sharp downward adjustment during 1988–9 as house price increases in the south came to an abrupt halt, while the north continued to catch up on lost ground.

Migration

Perhaps the most outstanding feature of inter-regional migration in the UK during the 1980s is the very substantial *gross* flows of people into and out of *all* regions (see Table 2.6). These gross inflows and outflows tend to swamp the *net* migration into or out of each region. A considerable number of migrants therefore move *into* as well as *out of* the northern regions. Many of those migrating to northern regions, however, are

Table 2.5 *Regional disparities in house prices, 1989*

Region	Average price of a semi-detached house[1] in the 3rd quarter of each year				Estimated[2] annual difference in gross mortgage repayments between each region and the SE in 1989
	1983	1985	1987	1989	
South East	34,230	40,770	61,490	76,750	–
East Anglia	25,200	29,200	44,020	57,550	2,590
South West	28,360	32,300	44,110	63,240	1,830
East Midlands	21,350	24,320	32,700	51,140	3,460
West Midlands	22,550	23,630	30,740	53,950	3,080
Yorkshire and Humberside	21,580	23,970	28,260	49,020	3,740
North West	22,670	24,370	29,090	48,820	3,770
North	23,950	25,870	29,250	47,420	3,960
Wales	22,220	23,460	28,630	48,770	3,780
Scotland	27,080	30,660	33,060	44,510	4,350
Northern Ireland	24,260	25,330	27,440	27,940	6,590

Notes:
1. Not including new houses but only those built after 1960
2. These estimates assume a mortgage interest rate of 13.5 per cent. The difference in the price of a semi-detached house is used as an estimate of the difference in the amount of the mortgage between the South East and other regions

Source: *The Halifax House Price Index, Regional Bulletin*

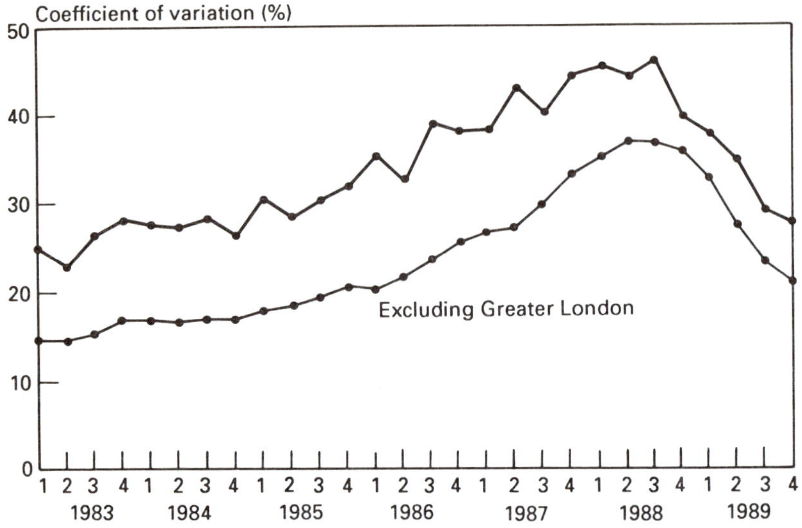

Source: Halifax Building Society, Regional House Price Index, Regional Bulletin

Figure 2.8 *House price disparities: regional variations in the UK, 1983–9*

Table 2.6 *Regional migration flows, 1981–7*

Region	Gross in-migration in thousands	Gross out-migration in thousands	Net migration in thousands	Migration as a % of 1981 population: Gross in-migration	Gross out-migration	Net migration
South East	1,657	1,709	-52	9.7	10.0	-0.3
East Anglia	432	319	113	22.9	16.9	6.0
South West	896	648	248	20.5	14.8	5.7
East Midlands	606	546	60	15.7	14.2	1.6
West Midlands	540	610	-70	10.4	11.8	-1.3
Yorkshire and Humberside	522	568	-46	10.6	11.5	-0.9
North West	551	706	-155	8.5	10.9	-2.4
North	298	346	-48	9.6	11.1	-1.5
Wales	352	317	35	12.5	11.3	1.2
Scotland	318	376	-58	6.1	7.3	-1.1
Northern Ireland	55	82	-27	3.6	5.3	-1.8

Note: Errors in net migration column due to rounding

Source: Population Trends, 55, Spring 1989, Office of Population Censuses and Surveys. p58

Regional Economic Disparities: Causes and Consequences

'return migrants'; they have sampled life elsewhere and have decided to return home.

As might be expected, net migration flows in the UK have been dominated by a north-south drift. There are, however, two exceptions to this general rule: the South East has been a net exporter of people (though only marginally so) and Wales has been a net recipient. The net outmigration from the South East is probably a consequence of the increasing attractiveness of its immediate neighbours (East Anglia, South West and East Midlands), particularly the substantially lower housing costs in those regions. A closer look at the migration data reveals, in fact, that it is Greater London which has been losing people while the rest of the South East has been a net gainer. The net inward migration into Wales during the 1980s is partly explained by its attractiveness for retired people. This is supported by data showing the age distribution of migrants.

Competitiveness

Since the economic future of each region depends crucially upon its competitiveness, it would be useful if its competitiveness could actually

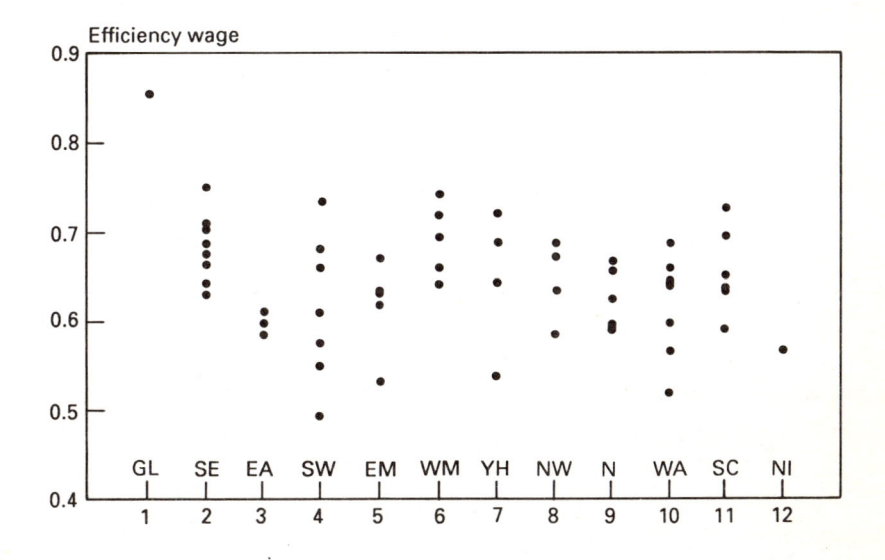

Note: Efficiency wage = real wage/output per worker, which is approximated by labour's share in the value of output

Figure 2.9 *Estimated efficiency wage in UK counties by region, 1988*

be measured. It is tempting to use regional wage levels as a measure of competitiveness, but this could give a totally misleading picture. The reason for this is simple: competitiveness does not depend only on costs, but also on productivity. A measure based upon *unit* labour costs is therefore more appropriate. One such measure is the efficiency wage, which is the real wage divided by labour productivity. Estimates of the efficiency wage are given for UK counties and regions in Figure 2.9, which indicates very substantial variations between UK counties. It is also clear from Figure 2.9 that many of the counties with a low efficiency wage (and thus a high level of competitiveness) are in the south. The counties of East Anglia, for example, have relatively low efficiency wages – a fact consistent with the high growth of employment in this region during the 1980s.

III The causes of regional economic disparities

The previous section established that the economic performance of UK regions has not only varied substantially, but that these regional disparities in performance have persisted for many decades. In general, the southern regions have consistently out-performed the northern regions. This section explores some of the possible causes of these persistent disparities in the economic performance of UK regions. In particular, answers are sought to the following questions.

- Why do regional economic disparities occur?
- Why do these disparities persist?
- Why are these disparities not automatically eliminated by market forces?

Industry mix

Perhaps the most commonly cited reason for the existence of disparities in the economic performance of regions is that some regions possess a more favourable industry mix than others. Regions which are endowed with a high proportion of nationally fast-growing industries can expect to have faster output and employment growth than regions which have a high proportion of nationally slow-growing industries. On the other hand, the same industry may grow at different rates in different regions and it is therefore possible for regions with a 'poor' industry mix to grow quickly. This will happen if nationally slow-growing industries happen to grow quickly in a region which has a high dependence on such industries.

Regional Economic Disparities: Causes and Consequences

The extent to which each region's industry mix has affected its growth performance can be estimated by using shift-share analysis (see Table 2.7). The main result is that regional disparities in employment growth are predominantly accounted for not by regional differences in industry mix but by 'other factors'. Only in the case of the South East did the industry mix have a substantial beneficial effect on its growth performance. The two main 'losers' were the West Midlands and the East Midlands, both of which had a more unfavourable industry mix than any other region during the 1980s. This was due to their higher dependence on manufacturing industries than other regions. The long-term decline of the manufacturing sector as a generator of jobs is vividly demonstrated in Figure 2.10. Regions relying heavily on the manufacturing sector, particularly in the 1980s, were obviously more likely to suffer from employment decline.

Table 2.7 *The contribution of each region's industry mix to its employment growth, 1981–8*

Region	Employment growth in region *minus* employment growth in the UK	Components of change: Industry mix effect	Components of change: Other factors	% employed in manufacturing in 1981
South East	3.4	+3.2	+0.2	23.2
East Anglia	21.9	−0.5	+22.4	27.2
South West	2.8	+0.6	+2.2	25.7
East Midlands	5.4	−4.0	+9.4	36.3
West Midlands	1.3	−4.2	+5.5	39.9
Yorkshire and Humberside	−2.9	−2.7	−0.2	31.4
North West	−9.3	−1.0	−8.3	32.8
North	−2.6	−3.0	+0.4	30.2
Wales	−8.1	−2.0	−6.1	25.6
Scotland	−6.7	−0.3	−6.4	25.4
Northern Ireland	−4.9	+1.0	−5.8	24.3

Note: See Armstrong and Taylor (1985) for an explanation of the shift-share method

Although regional disparities in industry mix have undoubtedly contributed (in some cases substantially) to regional differences in employment growth, Table 2.7 indicates that 'other factors' have played a more significant role. In other words, individual industries have been growing considerably faster in some regions than they have in others. Shift-share analysis indicates that the main losers have been the northern

Reducing Regional Inequalities

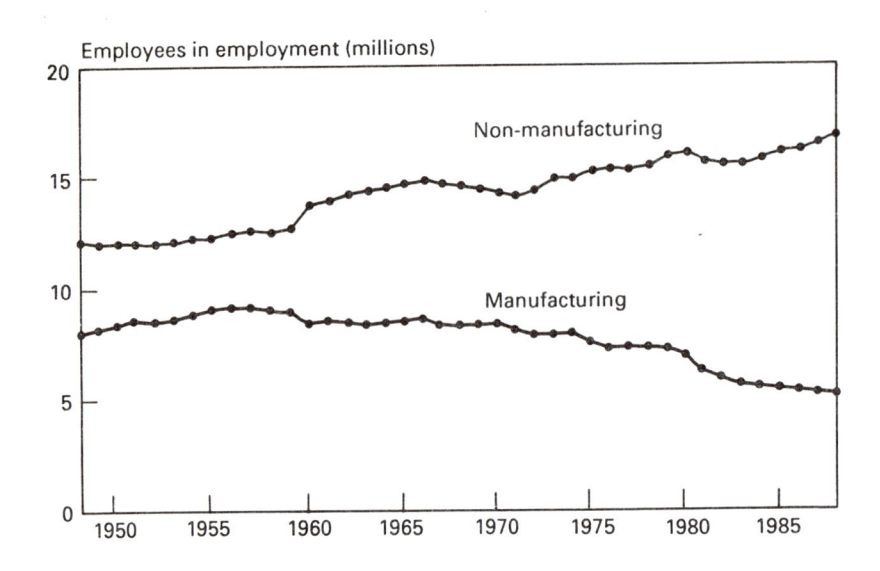

Sources: Annual Abstract of Statistics; British Labour Statistics; Employment Gazette

Figure 2.10 *Employment in manufacturing and non-manufacturing in the UK, 1948-88*

regions. The negative employment growth in the North West, Scotland, Wales and Northern Ireland is explained largely by a poor growth performance of individual industries and not by a poor industry mix.

Shift-share analysis therefore indicates not only that the northern regions generally have a poorer industry mix than the southern regions, but also that individual industries tend to grow faster in the south than they do in the north. Some of the reasons why individual industries tend to grow faster in the south than in the north are discussed in the remainder of this section.

Characteristics of the workforce

Since the characteristics of a region's workforce can be expected to have an effect on its growth prospects, it is pertinent to ask whether regional disparities in economic performance are correlated with corresponding disparities in the quality of each region's workforce. The workforce characteristics which may be expected to have some effect on its labour

Regional Economic Disparities: Causes and Consequences

productivity are likely to include such variables as educational attainment.

Measures of educational attainment generally confirm the notion of a north–south divide. The highest percentage of school leavers with no graded results, for example, occur in Scotland, Wales and Northern Ireland, and the lowest percentages occur in the southern regions (see Table 2.8). There is less evidence of a north–south divide in the case of the percentage of school leavers with three or more A-levels. The south gains substantially, however, by attracting graduates from other regions (see Table 2.9). Taken together, the South East and East Anglia attract about 20 per cent more graduates into jobs than could be expected from their share of GB employees.

Table 2.8 *Regional disparities in the educational attainments of school leavers, 1986–7*

Region	% of school leavers with no graded results Males	Females	% of school leavers with 3 or more A-levels[1] Males	Females
South East	10.6	7.1	12.5	10.7
East Anglia	11.5	7.9	8.8	9.0
South West	6.9	4.9	10.4	8.5
East Midlands	9.5	7.4	10.0	7.6
West Midlands	12.0	8.6	9.2	8.2
Yorkshire and Humberside	11.4	9.4	10.2	7.6
North West	12.7	9.9	10.7	9.5
North	12.3	9.2	8.9	8.0
Wales	19.6	12.7	8.4	8.3
Scotland	21.9	16.5	10.8	11.2
Northern Ireland	26.5	14.2	12.3	14.6
UK	12.8	9.1	10.7	9.5

Note:
1. For Scotland, five or more Scottish highers are assumed to be equivalent to three or more A-levels

Source: Regional Trends, 24, 1989

Reducing Regional Inequalities

Table 2.9 *Regional distribution of graduates in GB*

Region	% of all those graduating in 1980 located in each region in 1986/7	% of total GB employees in 1984	Ratio of column 1 to column 2
South	50.3	42.1	1.19
East Anglia	4.8	3.4	1.41
Midlands	13.1	16.4	0.80
North	19.0	24.6	0.77
Wales	3.7	4.3	0.86
Scotland	9.1	9.1	1.00
GB	100	100	1.00

Notes:
1. The data refer to graduates of UK universities, polytechnics and colleges who graduated in 1980 and who were in full-time employment in 1986/7
2. The regions are defined as follows: South = SE and SW; Midlands = West Midlands and East Midlands; North = N, NW and Yorkshire/Humberside

Source: *National Survey of 1980 Graduates and Diplomates*, Department of Employment, London; *Employment Gazette*, January 1987

Entrepreneurial activity

The importance of new and small firms in the growth process has become a topic of major interest to both academics and politicians in the 1980s. Research sparked off by the original work of Birch (1979) in the US has indicated that small firms have been a major and very neglected source of employment growth in the past. (See Storey and Johnson, 1988, for a review and critique of this research.) This research into the significance of small firms in the job creation process has led to the view that there are significant regional differences in the birth rate of entrepreneurs and that this is likely to be a major factor in explaining regional growth disparities. Some regions may well be a more fertile breeding ground for new firms than other regions.

That new firm formation rates differ substantially between regions is supported by information obtained from VAT data. Table 2.10 shows that the rate of increase in the stock of businesses registered for VAT during 1980-8 was around twice as high in the South East as in the

Table 2.10 *New firm formation rates: the growth in the stock of businesses, 1980-8*

Region	% increase in the stock of businesses registered for VAT 1980-8
South East	31
East Anglia	24
South West	24
East Midlands	22
West Midlands	19
Yorkshire and Humberside	18
North West	17
North	16
Wales	15
Scotland	14
Northern Ireland	10
UK	22

Source: *Labour Market Review: North West*, Department of Employment, Autumn 1989, Table 16, p17

northern regions of the UK. Indeed, the new firm formation rate is inversely related to a region's distance from the South East.

Spatial disparities in the new firm formation rate are investigated in recent research by Moyes and Westhead (1990). They begin by identifying the factors which have been cited in the new firms' literature as being influential in determining the extent to which any specific area is likely to provide a favourable environment for new firm formation. These factors include: the presence of incubator firms; occupational structure; educational qualifications; access to capital; industry mix; market demand; and 'push' factors such as the redundancy rate (see Table 2.11). A wide range of surrogate variables was then constructed by Moyes and Westhead in order to examine the extent to which each county provides a favourable environment for the birth of new firms. Table 2.11 shows the main results of correlating the new firm formation rate (across counties) with each of the surrogate variables. Using cluster analysis, Moyes and Westhead go on to show that the conditions for new firm

Reducing Regional Inequalities

formation are far more favourable in some groups of counties than they are in others.

A further twist is added to regional disparities in the new firm

Table 2.11 *Factors likely to affect regional disparities in the birth-rate of new firms: some surrogate variables*

Factor	Surrogate variables	Direction of relationship	Level of significance
Incubator firms	% of mfg employees in small firms (11-19), 1979	+	***
	% of mfg employees in large firms (500+), 1979	−	***
Occupational structure	% of workforce self-employed, 1981	+	***
Educational qualifications	% of school leavers with no graded results, 1981	−	**
	% of employees with higher educational qualifications 1981	+	*
Access to capital	% owner occupiers, 1979	+	***
	% of dwellings rented from local authority, 1979	−	***
Structure of industry	% of total employees in heavy industries, 1981	−	***
	% of total employees in manufacturing, 1981	−	***
	% of total employees in services, 1981	+	***
Local market demand	% change in employment, 1975-81	+	***
	% change in population, 1971-81	+	***
Push factors	% unemployed, 1983	−	**
	Change in % unemployed, 1979-83	+	***
	Closure rate, 1981	+	***

Note:
*** = significant at 0.1%
** = significant at 1%
* = significant at 5%

Source: Moyes and Westhead, 1990

formation rate by the government's small firms policy. According to Storey and Johnson (1988), this policy unintentionally favours the southern regions. They argue that small firms policy favours areas of low unemployment rather than areas of high unemployment, consequently leading to a widening of regional disparities rather than a narrowing. An example of this bias towards the south is provided by Mason and Harrison's (1989) research into the regional impact of the Business Expansion Scheme (BES). The purpose of this scheme is to increase the flow of equity capital into small firms (by allowing private income earners to claim tax relief at the top marginal rate on investments in new equity in unquoted companies up to a maximum investment of £40,000 per year). Over 60 per cent of the investments under the BES (and its predecessor, the Business Start-Up Scheme) during 1981/2 to 1986/7 were made in the South East and East Anglia while the northern regions were able to acquire less than 20 per cent (see Table 2.12).

Table 2.12 *Regional distribution of investments by the Business Start-Up Scheme and the Business Expansion Scheme, 1981/2 to 1986/7*

Region	Amount invested £m	% of UK total	Amount invested per worker (civilian workforce 1987) £
South East	94.8	54.1	10.3
East Anglia	11.6	6.6	11.5
South West	10.6	6.1	5.1
East Midlands	8.5	4.9	4.4
West Midlands	12.2	7.0	4.7
Yorkshire and Humberside	9.6	5.5	4.1
North West	10.8	6.2	3.6
North	3.8	2.2	2.6
Wales	5.0	2.9	4.2
Scotland	7.0	4.0	2.8
Northern Ireland	1.3	0.7	1.9
UK	175.2	100	6.3

Source: Mason and Harrison (1989), p42

According to Mason and Harrison, the under-representation of investments in the peripheral regions of the UK during 1981-7 was due to a lack of demand for equity capital in the north because of a dearth of expanding new and small businesses. By contrast, the London-based

Reducing Regional Inequalities

fund managers have been able to choose between a wide range of investment opportunities close at hand. They have therefore had little incentive to search for investment opportunities in the north. This is true not only of the BES, but for venture capital funds more generally. Martin (1989) shows that 85 per cent of the UK's venture capital funds (ie funds committed to investing in high risk unquoted companies) were located in the London capital market in 1987. London-based venture capital firms therefore dominate the supply of venture capital (Mason, 1987). Moreover, about 60 per cent of the total venture capital funds have been invested in the South East compared to under 20 per cent in the north (see Table 2.13).

Table 2.13 *Regional distribution of venture capital funds and investments*

Region	% of venture capital investments financed from London (1986)	% of total venture capital investments invested in each region (1987)
Greater London	94.0	31.3
Rest of South East	92.0	30.6
East Anglia	77.0	2.4
South West	62.0	6.9
East Midlands	85.0	6.5
West Midlands	54.0	3.6
Yorkshire and Humberside	63.0	5.8
North West	66.0	3.0
North	44.0	0.4
Wales	20.0	4.3
Scotland	25.0	5.1
UK	–	100

Source: Martin (1989)

It therefore appears that the north is at a distinct disadvantage compared to the south in being able to generate new firms. Not surprisingly, the more favourable environment for new firm formation in the south means that more capital is available for financing start-ups and the expansion of small businesses. Virtuous circle effects are consequently at work in the south since growth tends to feed on itself once it gets underway.

External control of firms: the regional consequences of acquisitions and mergers

Several researchers have argued that the peripheral areas of the UK have suffered as a result of the loss of local control of manufacturing firms. This loss of local control has happened because of the post-war trend towards greater industrial concentration and towards more oligopolistic market structures (Harris, 1989). Government policy has reinforced this trend by inducing large manufacturing companies to locate new branch plants in designated assisted areas. Although no direct evidence is available of the extent to which manufacturing firms are controlled from outside of each region, some idea can be obtained from table 2.14. This shows the proportion of workers employed in plants located over 100 miles from their head office. Apart from Yorkshire and Humberside, workers in the northern regions of Britain are far more likely to be employed in plants which are not located in the same region as their head office.

Table 2.14 *Local control of manufacturing firms: distance of plant from head office*

Region	% of employees in plants located over 100 miles from head office (1980)
South East	12.7
East Anglia	33.2
South West	45.2
East Midlands	17.1
West Midlands	26.7
Yorkshire and Humberside	31.0
North West	55.5
North	53.7
Wales	68.1
Scotland	65.9

Source: Harris (1988a)

Evidence that regional policy has led to greater dependence on decision-makers located in other regions is provided by the increasing

Reducing Regional Inequalities

presence of Japanese branch plants in the main assisted area regions during the mid 1980s. These peripheral regions have nearly 60 per cent of the manufacturing jobs created by Japanese inward investment into the UK. Table 2.15 shows that the regional distribution of employment in plants owned by Japanese companies has been heavily weighted towards the main assisted areas. Although the total number of jobs in Japanese-owned factories is still relatively small (about 1 in 200 of total manufacturing jobs in 1989), the sharp increase in the new Japanese plants established since 1984 indicates an ever-increasing dependence on external decision-making in the assisted areas.

Table 2.15 *Employment in Japanese-owned manufacturing companies in the UK regions, 1988*

Region	Employees in Japanese owned companies Number	% of total	% of total UK manufacturing employees (1988)
South East	1,630	6.5	26.0
East Anglia	600	2.4	4.3
South West	1,130	4.5	7.1
East Midlands	1,400	5.6	9.7
West Midlands	4,720	18.9	13.7
Yorkshire and Humberside	400	1.6	8.7
North West	460	1.8	11.8
North	4,280	17.1	5.1
Wales	6,690	26.7	4.2
Scotland	2,540	10.2	7.5
Northern Ireland	1,160	4.6	1.9
UK	25,010	100	100

Source: Invest in Britain Bureau, Department of Trade and Industry, unpublished data. Aggregated up to regional level by author.

It has been argued that a loss of local control of companies can have several harmful consequences. Harris, for example, argues that takeovers of local firms by national and multinational companies can harm the local economy in three main ways (Harris, 1988a, 1989).

1. Branch plants are more likely to close down during recessions (or

reduce output and employment significantly) than if the operations were locally-owned independent companies. Branch plants are also susceptible to the adverse effects of company reorganisations (eg due to the acquisition of the parent company).
2. Branch plants tend to have fewer local linkages than locally-owned firms. The growth of branch plants therefore has less impact on other firms in the region than would occur if plants were locally-owned.
3. Branch plants concentrate more heavily on processing activities and are likely to lose some of their high level functions (such as R&D, financial control, sales and marketing) to the parent company after acquisition.

These claims receive varying amounts of statistical support. In a survey of manufacturing firms in Northern Ireland, Harris (1988a) shows that externally controlled firms have a significantly higher ratio of unskilled to skilled manual workers and that externally controlled firms are less well integrated into the local economy. They tend to have strong production linkages with plants in other regions within the same company. Using the 1984 Workplace Industrial Relations Survey, Harris (1989) also shows that as the distance between a plant and its head office increases, the proportion of senior and technical staff falls.

Similar results were obtained in a detailed survey of 54 Scottish manufacturing companies which had been acquired by externally based companies during 1965–80. Ashcroft *et al* (1987) provide evidence that acquisition was unfavourable to the Scottish economy on three counts. First, demand for services provided by Scottish firms (eg banking and commerce) declined after acquisition relative to what would otherwise have been expected. Second, key management functions and key posts were lost. Third, R&D and marketing functions were transferred to the parent company. On the other hand, the effects on the acquired companies themselves were found on balance to be favourable: sales increased while employment was unaffected. There is therefore some evidence from recent research that the acquisition of a region's companies through mergers and takeovers may actually have beneficial effects on the growth prospects of the firms which are acquired. On balance, though, the available evidence appears to point in the direction of unfavourable effects overall (Love, 1990).

The propensity to innovate

Technological change has figured prominently in growth theory and hence in explaining why different economies have developed at different

Reducing Regional Inequalities

rates. In the early neo-classical growth models, technological change was treated as an *exogenous* variable 'falling upon an unsuspecting economy like manna from heaven'. More recent work has argued that technological change is *endogenously* determined; that is, it is a consequence of growth as well as a determinant of it (Dixon and Thirlwall, 1975). In addition, the appearance of new technologies based on the microchip has undermined the simple neo-classical view that technological change is a steady continuous function of time and greater emphasis is now placed on theories which stress the discontinuity of technical change. These discontinuities may also have a spatial dimension in so far as some regions may be in the forefront of developing and adopting new technologies while other regions lag behind (and in so doing lose competitiveness).

The significance of technological change to economic development is

Table 2.16 *Significant innovations in the manufacturing sector in each UK region, 1945-83*

Region	Innovations per unit of output 1945-63	1964-79	1945-79	% of UK innovations 1945-83	% of UK manufacturing output (1984)
South East	6.7	6.8	6.7	38.7	28.8
East Anglia	8.7	5.2	7.1	2.7	3.5
South West	7.8	5.0	6.5	5.8	7.1
East Midlands	6.6	5.5	6.1	7.7	8.2
West Midlands	5.3	4.0	4.7	10.4	11.3
Yorkshire and Humberside	3.7	5.7	4.6	8.4	8.1
North West	4.2	5.6	4.8	12.9	13.0
North	7.4	3.2	5.5	5.6	6.0
Scotland	2.9	2.9	2.9	5.0	8.1
Wales	2.2	2.5	2.3	1.9	4.3
Northern Ireland	3.4	1.2	2.4	0.7	1.7
UK	–	–	–	100	100

Note: The SPRU at the University of Sussex compiled the significant innovations database by consulting nearly 400 experts in each major sector of industry and asked them to identify (and rank) significant technical innovations (leading to new products or new processes) that had been successfully commercialised in the UK since 1945. The location of the plant at which each innovation occurred was obtained from the firm responsible for the innovation.

Source: Harris (1988b)

obviously critical. *Product* innovations help regions to improve their product range and hence their product mix, while *process* innovations help regions to improve their efficiency and produce products of higher quality. Both types of innovations are necessary for maintaining competitiveness. It is therefore important to examine the extent to which the innovation rate varies between regions. This has been done by Harris (1988b), who uses the database of significant innovations constructed by the Science Policy Research Unit (SPRU) at Sussex University.

The regional distribution of significant innovations in the UK is given in Table 2.16. This shows the dominating role of the South East, which accounted for nearly 39 per cent of the UK's manufacturing innovations during 1945–83, even though it produced less than 30 per cent of the UK's manufacturing output. The opposite is the case for Scotland, Wales and Northern Ireland, which together accounted for under 8 per cent of the UK's manufacturing innovations, even though they produced over 14 per cent of manufacturing output. A very similar result is obtained when the number of innovations per unit of output is used as the measure of the propensity to innovate (see Table 2.16).

The northern regions of the UK therefore appear to be at a significant disadvantage in the adoption of innovations compared with the southern regions.

Market failure

If labour markets were perfectly competitive, regional unemployment disparities would not exist since all markets would clear. In practice, wage flexibility is generally weak (Blackaby and Manning, 1990). Labour markets do not clear due to a lack of wage flexibility induced in part by national wage bargaining and strong trade unions. Recent research by Walsh and Brown (1990) suggests that the national wage bargaining explanation of regional wage inflexibility is too simplistic. They show that wage bargaining has become increasingly decentralised in recent decades. This decentralisation of wage bargaining has not been undertaken, however, so that firms can take advantage of local labour market conditions in determining wage levels. In large multi-plant firms, the decentralisation of wage bargaining has resulted from the restructuring of these enterprises, which have been split into profit centres and product divisions in order to increase their overall profitability. Wage bargaining has therefore been increasingly decentralised to profit centres and product divisions within companies. According to Walsh and Brown, firms which have been freed from the constraints of national pay

Reducing Regional Inequalities

agreements have not exhibited any great desire to take local unemployment levels into account in determining wage levels. In other words, the decentralisation of wage bargaining has not been accompanied by the adoption of geographically differential payment systems. The decentralisation of wage bargaining has not therefore helped to make wages more responsive to local unemployment levels. Indeed, regional earnings differentials have narrowed substantially since the late 1960s.

In the absence of wage flexibility, the main response mechanism to regional unemployment disparities is outward migration from areas of high unemployment and inward migration into areas of low unemployment. Research on inter-regional migration indicates that although the migratory flows in the UK are equilibrating (Pissarides and Wadsworth, 1989), there is an insufficient volume of migration to make a significant impact on regional unemployment disparities.

There are several reasons why net migration flows are not large enough to remove regional unemployment disparities. First, the costs of migrating from high unemployment regions to low unemployment regions may involve very high initial costs, especially if the migrant has to sell low valued property and buy high valued property (Bover, Muel-

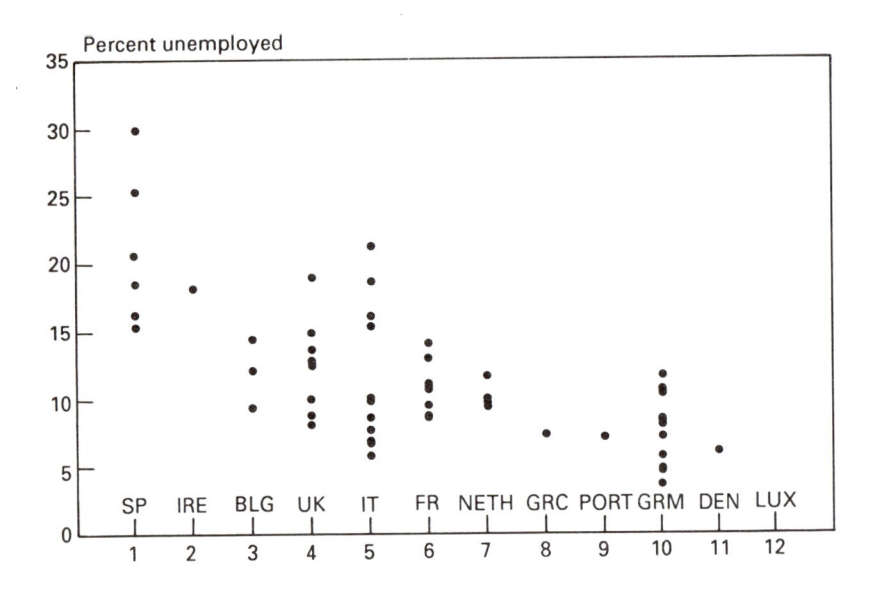

Source: Regional Trends, 24, 1989

Figure 2.11 *Regional disparities in unemployment in the EC, 1987*

Regional Economic Disparities: Causes and Consequences

Figure 2.12 *Regional disparities in GDP per capita in the EC, 1986 (EC=100)*

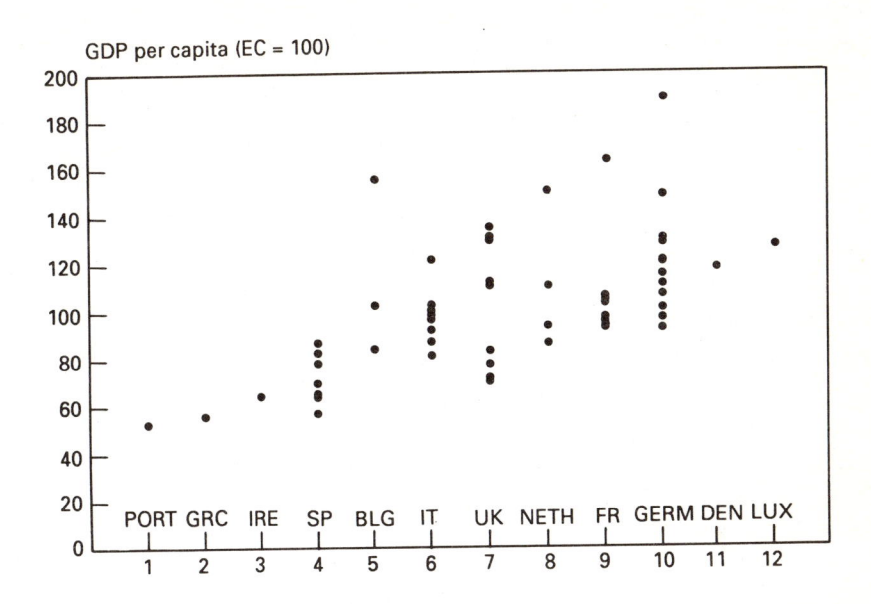

Source: Regional Trends, 24, 1989

lbauer and Murphy, 1989). The capital losses from migrating may therefore be too great for migration to be worth while. Second, even if the lifetime benefits outweigh the lifetime costs, lack of access to capital may prevent potential migrants from moving. Third, migrants suffer from psychic costs due to the loss of social and family ties. The very high rate of return-migration suggests that these psychic costs may weigh heavily in the migration decision. Fourth, the existence of income support (including housing subsidies as well as unemployment and related benefits) acts as a disincentive for unemployed persons to seek out employment opportunities in other regions (Minford, 1985). This effect is reinforced by a dearth of local authority housing in areas of low unemployment (Hughes and McCormick, 1987).

The reinforcement of regional disparities: the Single European Market

Regional economic disparities in Europe are already immense. This can be seen from the unemployment rate and GDP per capita given for all EC regions in Figures 2.11 and 2.12. To what extent are these disparities likely to increase as a result of the completion of the internal market?

Under the Single Europe Act, all remaining barriers to the free

Reducing Regional Inequalities

movement of goods, services, capital and people within the EC are to be removed by 1992. This will have far-reaching effects on the member states since the removal of tariff and non-tariff barriers to trade will result in greater competition within the EC. The economic costs and benefits of this change will be widely distributed both within and between the member states. The consequences for regional disparities could be particularly severe and it is because of the potentially harmful consequences on peripheral areas that the Single Europe Act contains a specific commitment to reduce regional disparities and to increase support for the less-favoured regions on the EC periphery. The main beneficiaries of the increase in regional aid (through the Structural Funds of the EC) are likely to be Portugal, Greece, Spain and Ireland, however, and not the less prosperous regions of the UK (apart perhaps from Northern Ireland).

One of the primary effects of 1992 on industry will be to extend opportunities for reducing unit costs through scale economies. Individual regions will therefore gain or lose depending upon their ability to respond to the new opportunities and challenges presented by 1992. Past and recent experience suggests that the southern regions of the UK are in a better position than the northern regions to gain from the increase in trade between European regions. For many of the reasons suggested above, the southern regions already have a competitive advantage over the northern regions and this is unlikely to diminish after 1992. In addition to having a more favourable industry mix, a better quality workforce, more innovation potential, a higher entrepreneurial birth rate and more control over their own industries, the southern regions are nearer to the centre of economic gravity of the European market. This locational advantage should not be over-stressed, however, since the Far Eastern industrial economies appear to overcome their locational disadvantage with considerable ease.

IV The consequences of regional economic disparities

The consequences of regional economic imbalance have been accurately summarised by the previous Secretary of State for Scotland:

> At a time when costs in the South East of England are escalating, when congestion is increasing, it surely makes sense to move to Scotland. Relocation does not only make sound sense in economic terms but also helps to balance employment opportunities throughout the country and beyond that to demonstrate the political unity of the kingdom. (Malcolm Rifkind quoted by the *Financial Times*, 27 January 1989)

Regional Economic Disparities: Causes and Consequences

This statement argues that the nation as a whole loses out as a result of regional economic disparities. National output is below the level that *could* be achieved if there was a greater geographical balance between the demand for and supply of labour. This is supported by the previous Secretary of State for Wales who has argued that: 'With the inflationary over-heating of parts of England it is vital that we procure greater economic activity in those areas seeking employment' (Peter Walker in a lecture to the Tory Reform Group on 10 April 1989). In addition, excessive growth in the South East has resulted in congestion costs, which amount to several billion pounds per annum in lost production according to the CBI (1989).

There are therefore three main economic consequences of regional economic disparities, all of which adversely affect the efficiency of the economy. These are:

- regional economic disparities lead to a higher national unemployment rate;
- regional economic disparities lead to more severe inflationary pressures;
- regional economic disparities lead to a sub-optimal use of the nation's social overhead capital.

In addition to these consequences on the economic efficiency of the economy, regional economic disparities may have adverse effects on the political unity of the nation. The sense of unfairness and injustice felt by many in the North causes dissatisfaction with the control over the economy exercised by the politically powerful southern half of the country. The result is a call for devolution and self-determination by significant and vociferous minorities, especially in Scotland, Wales and Northern Ireland.

The remainder of this section concentrates on the economic rather than the political consequences of regional economic disparities. We turn first to the effect of regional economic disparities on national employment. It has been argued (Armstrong and Taylor, 1990) that national employment and output could be significantly higher if regional unemployment disparities were reduced. This is because sustained disparities in the unemployment rate between regions indicate the existence of a fundamental geographical mismatch between labour demand and labour supply. Examination of the relationship between the unemployment rate and capacity utilisation during 1974–89 (see Figure 2.13) suggests that there was a substantial increase in structural unemployment in the UK during the 1980s. Between 1980 and 1983, for

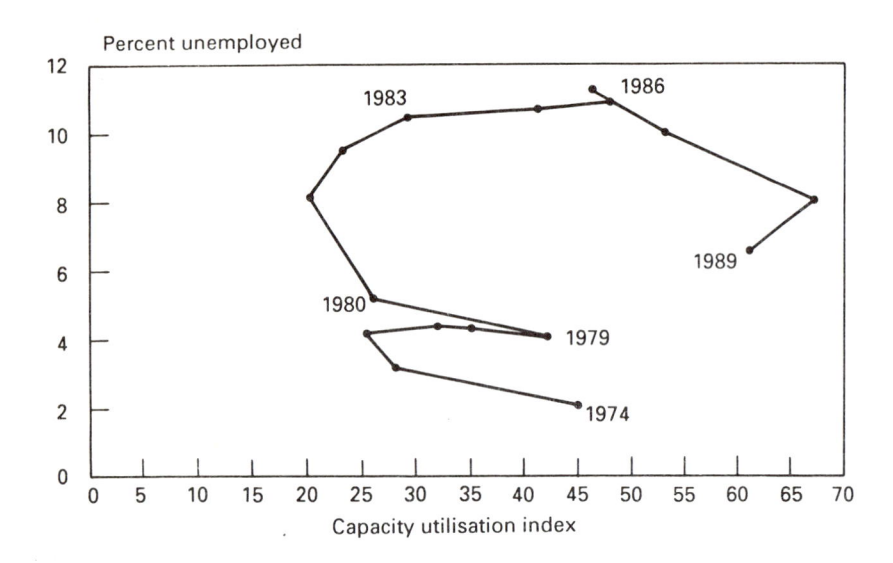

Reducing Regional Inequalities

Figure 2.13 *Percentage unemployed versus capacity utilisation in the UK, 1974-89*

Note: Capacity utilisation index was obtained from the CBI Industrial Trends survey

example, the unemployment rate increased from 5.1 per cent to 10.5 per cent, while capacity utilisation was at about the same level in both years. Similarly, capacity utilisation was at about the same level in 1979 and 1986 even though unemployment was 4.0 per cent and 11.1 per cent respectively in these two years. This implies that structural unemployment increased considerably during 1979-86; perhaps by as much as seven percentage points. This conclusion is consistent with the fact that the UK economy was running at a very high level of capacity utilisation throughout 1988, while unemployment was 8 per cent for the UK as a whole. Regional unemployment disparities were also falling very slowly during 1986-9, even though the UK rate was falling quickly. This suggests that ways need to be found for tackling the fundamental structural problems caused by the geographical mismatch between the demand for and supply of labour in the UK.

The second main economic benefit of reducing regional unemployment disparities is that inflationary pressures would be less severe at any *given* national unemployment rate if there were a more even geographical distribution in the demand for labour. This would allow the national economy to be run at a higher level of the pressure of demand without

Regional Economic Disparities: Causes and Consequences

leading to an unacceptably high rate of inflation. Experience at the end of the 1980s clearly indicated a build up of inflationary pressures as the national unemployment rate fell below the 8 per cent mark. Recent experience supports this view. Unemployment rates in many parts of the South East fell below 3 per cent in 1989, thus leading to an increase in wage pressures due to a shortage of labour in these areas. This labour shortage in parts of the South East has been accompanied by a very high level of capacity utilisation during 1988-9 (see Figure 2.14). Output growth has outstripped the underlying growth in productive capacity and has caused potentially serious wage pressures to arise in labour-scarce areas; hence the imposition of increasingly restrictive monetary policy (in the form of higher interest rates) during 1988-9.

Evidence that the inflationary pressures which emerged during 1988 had their origins in the south rather than in the north is provided by data on house price inflation. Figure 2.15 shows that house price inflation began to rise steeply at the beginning of 1986 in the south and the Midlands, but did not do so in the north until two years later. Not only did the south's house price inflation take off sooner than in the north, it also peaked sooner, falling from over 40 per cent per annum in 1988 (3rd quarter) to -10 per cent one year later. The very rapid fall in the inflation

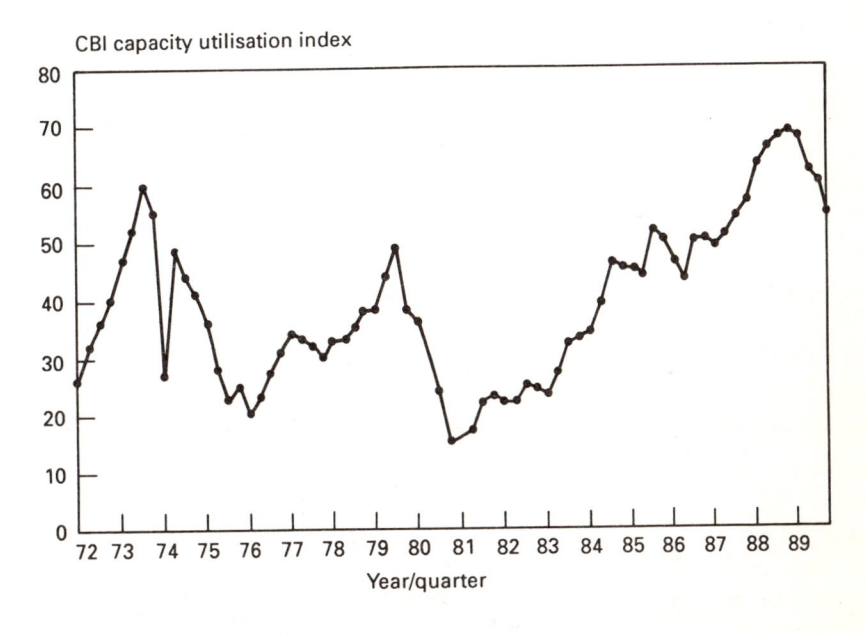

Source: Confederation of British Industry

Figure 2.14 *CBI capacity utilisation index, 1972-89*

103

Reducing Regional Inequalities

Annual house price inflation (%)

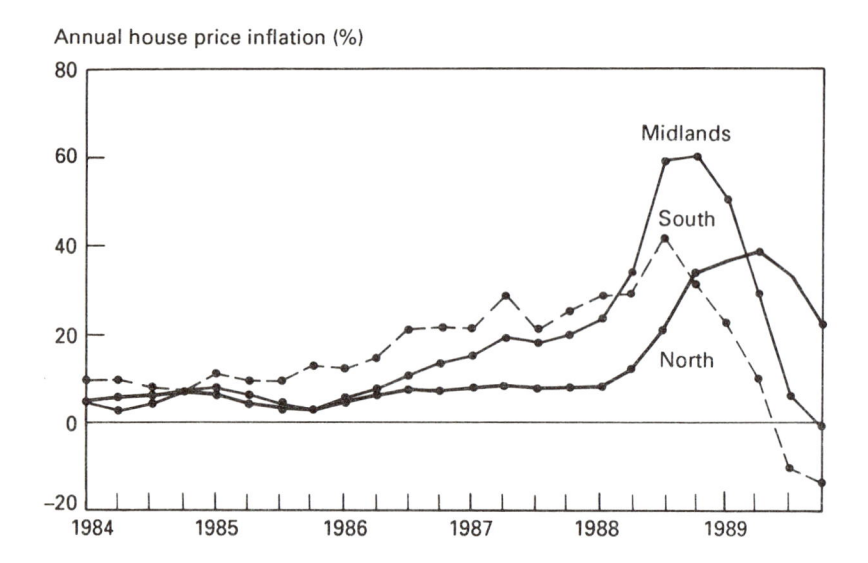

Source: Halifax Building Society, Regional House Price Index

Figure 2.15 *House price inflation in the UK: a regional analysis, 1984-9*

of house prices in the south (and the Midlands) during 1989 indicates that inflationary pressures in the housing market responded very quickly to the doubling of interest rates from 7.5 to 15 per cent between July 1988 and April 1989.

Some idea of the influence of the labour market on inflation can be seen from the high correlation between house prices and the unemployment rate across GB counties (see Figure 2.16). Excluding Greater London, the correlation between these two variables (taking logarithms to allow for the non-linearity in the relationship) is 0.85 for April 1988. This means that nearly three-quarters of the inter-county variation in house prices is explained (statistically) by corresponding variations in the unemployment rate. This result is consistent with the view that inflationary pressures emanate primarily from the tight labour markets of the South East.

The third consequence of regional economic disparities is the suboptimal utilisation rate of social overhead capital. The clearest indicator of excessive use of social overhead capital is the massive congestion which new transport facilities (such as the M25) manage to relieve for only very brief spells. Road, rail and air transport all suffer from over-utilisation.

Regional Economic Disparities: Causes and Consequences

Congestion is a problem faced by all major cities, of course, no matter where they are located. But the problems facing Greater London are in a class of their own and are not made easier by the persistently better growth performance of the south as compared to the north. The problems of congestion in Greater London could therefore be tackled by encouraging a more even regional distribution of employment and population growth – that is, inducing a geographical relocation of demand, rather than constantly trying to relieve the consequences of rapid growth in the South East.

A further aspect of the increasing concentration of economic activity in the South East is the consequences for the environment (Smith, 1989). The London Green Belt is under intense pressure and it is seen as the source of a potential bonanza by many developers. Similar attempts are being made to develop other greenfield sites in the South East. The government's increasing interest in preserving the countryside, however, is reflected by the decision of the Department of Environment to scrap the controversial 'presumption in favour of housing development' introduced in the early 1980s (as reported in the *Financial Times* editorial of 5 October 1989). Christopher Patten's reversal of the provisional approval given by Nicholas Ridley to the Foxley Wood New Town

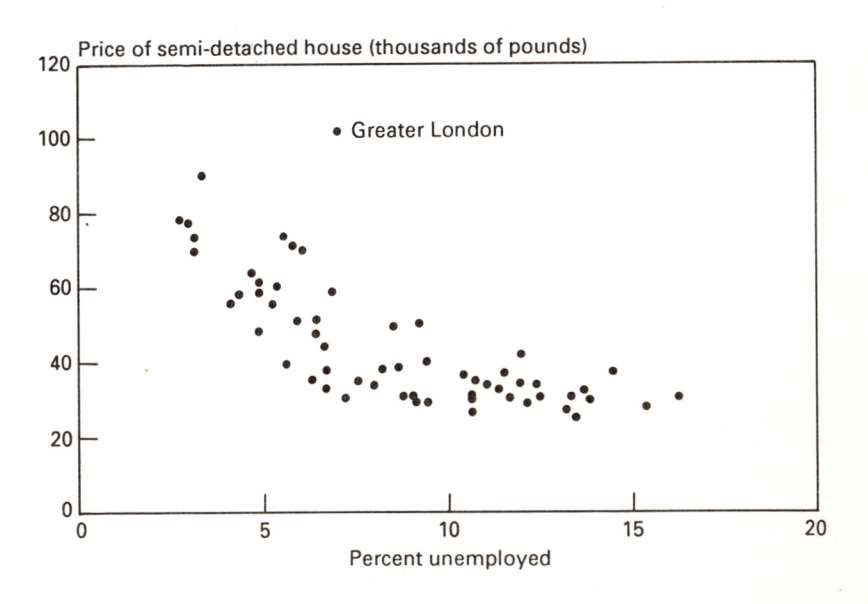

Source: Halifax Building Society and Regional Trends, 24, 1983

Figure 2.16 *House prices versus unemployment rate by UK county, April 1988*

105

development in Hampshire is a further example of a sea change in government policy towards land use planning in the South East.

Conclusion

This chapter has attempted to identify the causes and consequences of the 'north–south divide' in the hope that a better understanding of regional economic disparities will provide useful input into the process of policy formulation.

Several reasons why the southern regions of the UK have outperformed the northern regions during recent decades have been identified. The most significant of these are as follows.

1. The northern regions have had a less favourable mix of industries. They have concentrated more heavily on industries in the manufacturing sector which has become increasingly less important as a job creator in recent decades. Of more significance, however, has been the fact that individual industries have performed better in the south than in the north.
2. The southern regions have benefited from a far greater level of entrepreneurial activity – as reflected by the birth rate of new firms. This appears to have been aided by a more readily available supply of venture capital, as well as by a set of socio-economic characteristics which inherently favour the creation of new start-ups in the south.
3. Manufacturing industries in southern regions have a higher propensity to innovate compared to northern regions.
4. The increasing dependence of northern production units on decision-makers located outside of the region (predominantly in the South East) may also have had harmful consequences on the north. Investigations of the effect of external take-overs on firms in the peripheral areas have indicated that branch plants are more vulnerable to economic downturns and are more likely to concentrate the activities requiring their most skilled personnel at the parent plant. The south consequently benefits from a more highly skilled workforce, which increases its attractiveness to expanding firms.
5. Perhaps of greatest significance, however, in explaining differences in performance between the north and the south is the apparent inability of northern labour markets to improve their competitive position through lower unit labour costs. If efficiency wages in northern regions were more responsive to their relatively high unemployment rates, this would help the regions to overcome the inherent economic advantages possessed by southern regions.

The main consequences of regional economic disparities result directly from the imbalance between the demand for and supply of factor inputs. Regional variations in the unemployment rate lead to a higher *national* level of unemployment than would occur if these disparities were less severe. The rapid build-up of inflationary pressures at the end of the 1980s, for example, was primarily due to labour shortages in many southern labour market areas. If the demand and supply of labour were more evenly balanced between labour markets, the economy would be able to operate at a higher *aggregate* level of demand without this leading to inflationary pressures. Spatial disparities in unemployment therefore lead to a permanent loss of output. In addition, regional differences in growth have helped to worsen the problem of congestion in the South East, particularly in and around Greater London.

Since regional economic disparities result in substantial economic costs to the economy, this is a sufficient reason for asking what role the government should play in seeking to reduce these disparities. In particular, we need to ask whether current regional policy is adequate and if not what changes are needed to make it more effective.

References

Armstrong, H and Taylor, J (1985), *Regional Economics and Policy*, Philip Allan, Oxford.
— (1988), *Regional Policy and the North–South Divide* (Employment Institute, London).
— (1990), *Regional Economic Policy*, Heinemann, Oxford.
Ashcroft, BK (1988), 'External Takeovers in Scottish Manufacturing: the Effect on Local Linkages and Corporate Functions', *Scottish Journal of Political Economy*, 35.
— Love, JH and Scouller, J (1987), *The Economic Effect of the Inward Acquisition of Scottish Manufacturing Companies 1965-80*, Industry Department of Scotland, ESU Research Paper 11.
— and Love, JG (1988), *The Regional Interest in UK Mergers Policy*, Regional Studies, 22.
— and Love, JH (1989), 'Estimating the Effects of External Takeovers on the Performance of Regional Companies: The Case of Scotland 1965 to 1980', *Environment and Planning A*, 21.
Balchin, PN (1989), *Regional Policy in Britain: The North–South Divide*, Paul Chapman, London.
Birch, D (1979), *The Job Generation Process*, MIT Program on Neighbourhood and Regional Change, Cambridge, Massachussetts.
Blackaby, DH and Manning, DN (1990), *Earnings, Unemployment and the Regional Employment Structure*, Regional Studies, 24.

Bover, O, Muellbauer J and Murphy A (1989), 'Housing, Wages and UK Labour Markets', *Oxford Bulletin of Economics and Statistics*, 51, March, pp97-162.

Confederation of British Industry (November 1989), *Trade Routes to the Future*.

Dixon, RJ and Thirlwall, AP (1976), *Regional Growth and Unemployment in the United Kingdom* Macmillan, London.

Harris, RD (1988), 'Market Structure and External Control in the Regional Economies of Great Britain', *Scottish Journal of Political Economy*, 35.

— (1988b), *Technological Change and Regional Development in the UK: Evidence from the SPRU Database on Innovations*, Regional Studies, 22.

— (1989), *The Growth and Structure of the UK Regional Economy 1963-85* Gower.

Hughes, GA and McCormick, B (1987), 'Housing Markets, Unemployment and Labour Market Flexibility in the UK', *European Economic Review*, 31.

Lewis, J and Townsend, A (1989), *The North-South Divide* Paul Chapman, London.

Love, JH (1990), *External Takeover and Regional Economic Development: A Survey and Critique*, Regional Studies, 24.

Martin, R (1989), *The Growth and Geographical Anatomy of Venture Capitalism in the United Kingdom*, Regional Studies, 23.

Mason, C (May 1987), 'Venture Capital in the United Kingdom: A Geographical Perspective', *National Westminster Bank Review*.

— and Harrison, RT (1989), 'Small Firms Policy and the "North-South" Divide in the United Kingdom: The Case of the Business Expansion Scheme', *Transactions of British Geographers*, 14.

Minford, P (1985), *Unemployment: Cause and Cure* Basil Blackwell, Oxford.

Moyes, A and Westhead, P (1990), *Environments for New Firm Formation in Great Britain*, Regional Studies, 24.

Pissarides, C and Wadsworth, J (1989), 'Unemployment and the Inter-Regional Mobility of Labour', *Economic Journal*, September.

Smith, D (1989), *North and South: Britain's Economic, Social and Political Divide* Penguin, London.

Storey, DJ and Johnson, S (1988), *Are Small Firms the Answer to Unemployment?*, Employment Institute, London.

Walsh, J and Brown, W (March 1990), 'Regional Earnings and Pay Flexibility', NEDO Conference on *Reducing Regional Inequalities*. See Chapter 6 of this book.

3
*Regional Policy and Market Forces: A Model and an Assessment**

Patrick Minford and Peter Stoney

In this chapter we describe regional developments over the last few decades in terms of a new regional model, based on the Heckscher-Ohlin inter-regional trade model, allowing for mobility of certain factors of production. The model is described formally in Appendix 1 to this chapter; the treatment in the paper is diagrammatic and verbal. We begin with an informal description of the model's mode of analysis (section I). Section II discusses regional developments in the light of this analysis. Section III then reviews current and future developments in policy. Section IV finally applies this to future regional prospects, where the qualitative assessment is described by some numerical projections (the model has not to date been estimated econometrically, except for the sub-region of Merseyside to North Wales – see Appendix 3).

I Introduction: the theoretical framework of analysis

To understand regional behaviour it is crucial to separate out those productive elements whose prices are determined internationally or nationally from those whose prices are fixed regionally.

A firm deciding where to locate in the regions will not be influenced by the nationally or internationally fixed elements in its costs; these are common costs wherever it goes. It will be influenced by the elements which vary regionally – its 'regionally avoidable costs'; these can be quite a small proportion of its total costs, so that variations in them which are

---

* We are grateful for useful comments from the participants in the NEDO Conference on Reducing Regional Inequalities, especially Alex Bowen, Walter Eltis, Ken Mayhew and Peter Simmons.

Reducing Regional Inequalities

small in relation to total value added acquire a crucial significance in terms of regional location.

Capital and raw material costs are fixed in world markets. The costs of skilled labour, which, typically home-owning, is free to migrate anywhere in the country where rewards are more alluring, are determined nationally; there will be one going wage, adjusted for any differences in the cost or quality of life, for such workers across the country.

Prices are fixed regionally for those elements which cannot move from the region. Those are land with planning permission for non-agricultural use and unskilled labour (whose mobility is restricted by tenant subsidies and controls). Agricultural land and commons are not relevant for our purpose; their price is determined by the size of the agricultural sector relative to their supply, and that size is controlled, as is the land supply, by official intervention such as the European Common Agricultural Policy. This price can be considered a reserve price for land with planning permission; and will only affect the price of that land when the reserve is breached.

The prices of land with planning permission, or henceforth land for short, and of unskilled labour, henceforth labour for short, are fixed by their marginal value in industrial use. Let us assume that across the world economy are traded goods industries where economies of scale have been exploited and competition prevails (perhaps between giant multinationals, perhaps between a host of smaller firms); there is a going best-practice technology in each. It will follow that there is a wage rate and a land rental rate that will be just competitive for these industries if they establish locally.

Suppose we divide these industries for convenience into two groups – manufacturing and services. Manufacturing is more (unskilled) labour-intensive than services. Then a decline in manufacturing prices relative to services prices will reduce wages and raise land rentals. So will a shift in technology away from unskilled labour, in favour of mechanisation. Both of these will produce unemployment as wages fall relative to unemployment benefits. They will also raise land prices relative to the reserve cost of agricultural land, and increase the pressure for development.

Now consider how transport costs or other locational cost disadvantages can alter land and labour prices in the north and the south, which we will distinguish conventionally by the line drawn from the Wash to the Severn. If the north's transport costs to the UK's main markets go up relative to the south's, then the competitive wage and land price in the north falls in proportion to the loss in value added this produces (net of capital, raw materials and skilled labour). For most industries this

Regional Policy and Market Forces: Model and Assessment

proportion can be very large. For example, suppose a computer company can cut its costs by 3 per cent of its gross product price by moving to the south; if only 10 per cent of its value added goes on land and unskilled labour, then it can afford to pay 30 per cent more for those elements. For the northern economy, drops on such a scale would again produce manual unemployment and counteract the upward pressure on land prices from the decline of manufacturing.

We have just sketched out the basics of our explanation for the rise in manual unemployment, especially in the north (hit by the double blow of falling relative manufacturing prices and rising relative transport costs); also for the large rise in relative land prices in the south. We can also explain urban dereliction as the result of inner urban land prices being driven down below the value of non-industrial use.

To these basics we should add the roles of local authorities and unions. Local government raises taxes whose incidence falls variously on land rentals (business rates) and on the cost of skilled labour (personal rates). The former can drive the net price of land below its non-industrial use value, driving business out. The latter acts like a tax on the value added available for land and unskilled labour (like transport costs above).

Unions operate in what we shall call the protected sector, where competition from abroad is kept at bay and so wages can be marked up by monopoly unions. This sector includes government and non-traded goods and services such as restaurants, electricity production and telecommunications; also any parts of industry with quota protection against imports, such as the car industry and some electronics. In this sector the only limit to the wages that can be set for an industry is the loss of jobs as the industry contracts. Union mark-ups reduce jobs and, we assume, encourage people to spend extra time searching for a union job rather than take a low-paid non-union job; this raises unemployment.

It is a fact that in the north local authorities have pursued more aggressive rating policies and unions have been stronger than in the south. This has aggravated northern unemployment and dereliction.

A final factor is the size of the protected sector itself. After 1979 government policy was to reduce protection of industry and to cut the size of the public sector. While this enabled taxes to be cut and the economy as a whole to benefit, the contraction in the protected sector was particularly marked in the north (an obvious example being the coal industry, whose employment has more than halved, mostly in the north); this meant that people previously employed at high protected wages now found themselves faced with low-paid unskilled jobs, and many opted instead for benefits.

Reducing Regional Inequalities

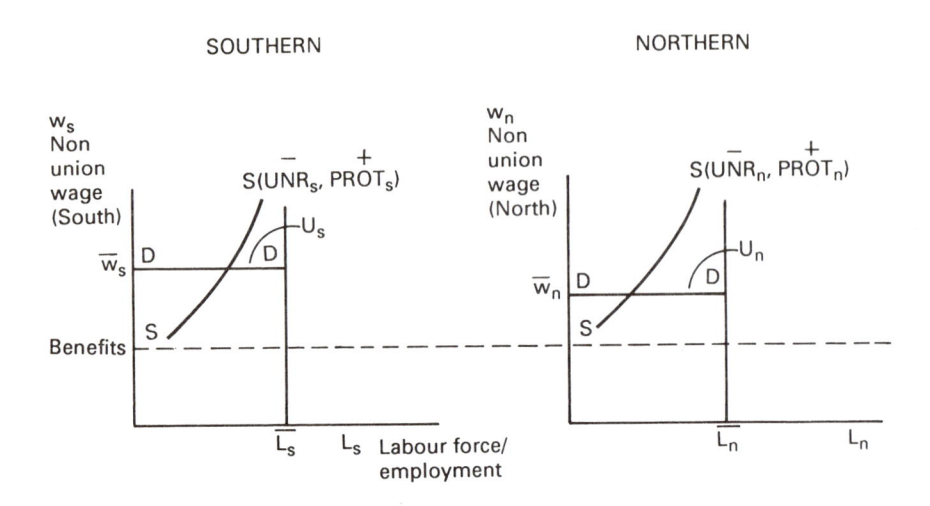

Note: $\bar{L}$ = available labour force; $\bar{w}$ = traded sector wage offer; U = unemployment; UNR = union mark up; PROT = size of protected sector – signs over these last two indicate effect on SS rise in each

Figure 3.1 *Typical regions' manual labour markets*

The whole method of analysis may be summed up by simple diagrams showing supply and demand for unskilled labour and land in north and south.

Take unskilled labour first – see Figure 3.1. In the north, the competitive wage drops, while in the south it rises; this mainly drives up real wages in the south, with little effect on unemployment as wages are already high relative to benefits. But in the north, the drop pushes wages close to benefits with the main effect coming therefore on unemployment. We get unequal and nationally higher unemployment because of the benefit system and the distorted rented housing market's interaction with the shift in regional conditions.

Land too – see Figure 3.2 – goes out of industrial use as its price falls, relative to its price in reserve uses (agricultural, urban and rural commons). Thus, there is also a sort of unemployment of land, creating the dereliction so familiar in the north. But, of course, the policy implications of such lack of use are different; from an economic viewpoint it is appropriate provided there are no distortions, especially in the planning process and in environmental property rights. The policy issue then boils down to assessing these distortions.

Regional Policy and Market Forces: Model and Assessment

We shall make much use of these two figures to illustrate the discussion which follows. They show between them the total use of specifically regional resources in a region according to its positioning in the spectrum of economic factors we have loosely labelled as 'north–south'; they also focus on the two key policy areas, unemployment and the overall pressure on planning land.

The rest of a region's profile depends on its structure – the share of the protected sector, and within the traded sector the split between manufacturing and services.

One might expect the share of the protected sector to remain broadly constant at constant relative prices; for example if there is a shift to the south in the traded goods sector, bringing more land and labour into use, services such as local authority provision or telephones will expand in line, being essentially directly linked to these market-determined activities. These are 'gravity' or 'centre-periphery' effects.

However, relative prices will stimulate relocation. As the costs of land and labour rise in the south because of the pressures driving the traded sector southwards, the protected sector which is not driven by the same pressures will react by relocating northwards those parts that are not directly linked to southern production; for example headquarters and

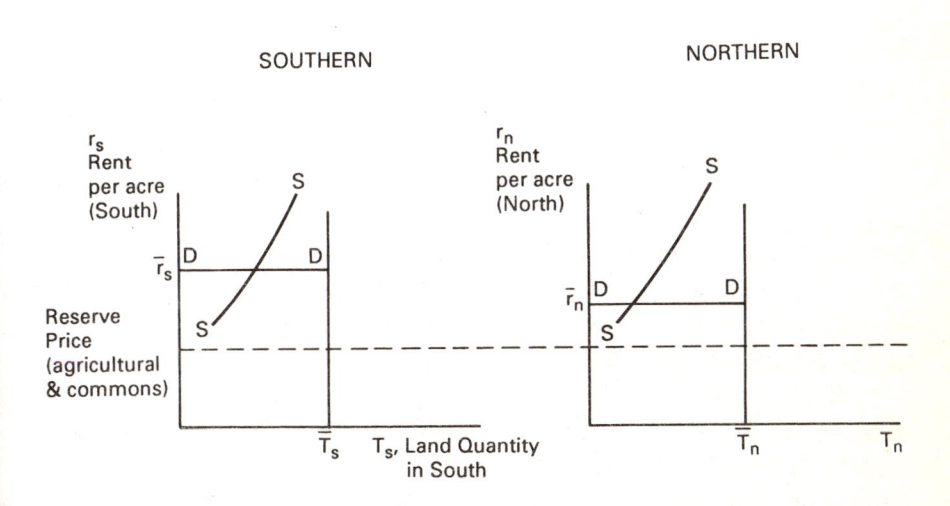

Note: $\bar{T}$ = available land without planning permission: $\bar{r}$ = traded rent offer

Figure 3.2 *Regional land market (industrial and housing land)*

Reducing Regional Inequalities

other overhead activities will be pushed to where land and labour are cheaper. Hence, the share of the protected sector in the north will tend to lean against the regional pull outwards of the traded sector.

Table 3.1 *Manufacturing and services factor intensity (%)*

	Share of Semi/Unskilled Labour	
	In Employment[1]	In Value added[2]
Manufacturing	21.2	11.5
Other industrial	18.5	7.4
Services	10.9	5.9
	Share of Rent in Value Added[3]	
Industrial and commercial companies	1.44	
Financial companies	2.95	
Public sector	4.69	

Notes:
1. Source: 1979 Labour Force Survey
2. As in 1 and 3, and assuming semi/unskilled wage = 80% of average wage
3. Source: 1988 National Accounts for 1987

Within the traded sector, the share of manufacturing will be set by the supply of unskilled labour relative to that of land, since manufacturing is about twice as manual labour-intensive as services, and considerably less land-intensive – Table 3.1. Even with high northern unemployment reducing the supply of manual labour (as manual wages dropped relative to benefits), the north has remained on this basis the natural location for manufacturing; as measures to bring back these unemployed into supply (for example, by worktesting and retraining, if necessary backed by benefit withdrawal for non-participation) take effect, then this comparative advantage will be enhanced. Anyone who needs visual conviction should visit Trafford Wharf in Manchester and Canary Wharf in London's Dockland on successive days.

Given total land and labour use, and its structure in these ways, the regional GDP, capital stock, skilled labour use, raw material and transport flows, can all be deduced from input-output analysis if desired, with the input-output matrix determined by the technological and relative price factors we have already discussed. For particular areas of policy, such calculations will be important – for example, transport infrastructure. But in this chapter we shall treat these matters qualitatively rather than quantitatively for the most part, in view of its primary policy focus on unemployment and planning.

The rest of this chapter applies this analytical framework in more detail to past regional developments since 1979 (section II), to recent and potential future policy changes (section III), and finally to prospective trends into the next century (section IV).

The approach taken is exclusively economic in nature. No account is taken of sociological factors, for example, such as possible prejudices or value judgements about the region which industrialists might have or social attitudes among groups of regional residents. Our implicit view is that regional development or lack of it is an economic problem and is likely to be understood with the tools of economic theory and empirical investigation.

This approach will not exercise a universal appeal. Some will argue that social or geographical forces have driven the 'periphery' further from the 'centre'. We are not impressed with such arguments. In the first place, we know of no systematic empirical evidence that any such forces have grown so much more powerful in the last two decades as to account for the sharp increase in unemployment differentials between north and south, and between skilled and unskilled labour. Second, while we freely acknowledge the importance of such forces in many contexts, we are impressed in this one here by the power of market forces (relative prices) to shift *marginal* decision-makers; it is shifts at the margin on which the economist relies to produce sharp changes in interesting aggregates. Unemployed manual workers in the north, for example, represent only some 1 per cent of national value added; yet as we argue repeatedly in this chapter, they *are* the bulk of the regional problem. It should not be beyond the wit and scope of policy makers to liberate sufficient market dynamics to deal with such a marginal problem. We will deal in some detail with existing market obstructions and policies that would remove them.

II Regional developments 1979–89

Regional unemployment

We begin with unemployment. It is not widely realised that the rate of unemployment has hardly changed among non-manual home-owning

Reducing Regional Inequalities

Table 3.2 *National employment by occupation and tenure*

Household heads who were working or seeking work†, by tenure	Managerial and professional	Intermediate and junior non-manual	Skilled manual	Semi-skilled	Unskilled	Total
Owner-occupiers						
1977/78						
All household heads (thousands)	2,119	1,529	2,236	591	179	6,654
Out of work (thousands)	28	39	57	26	11	161
Proportion out of work (per cent)	1.3	2.6	2.5	4.4	6.1	2.4
1984*						
All household heads (thousands)	2,718	1,588	2,586	790	181	7,862
Out of work (thousands)	6.3	46	117	50	22	297
Proportion out of work (per cent)	2.3	2.9	4.5	6.3	11.9	3.8
Local authority tenants						
1977/78						
All household heads (thousands)	195	402	1,437	657	356	3,047
Out of work (thousands)	12	27	119	58	58	274
Proportion out of work (per cent)	6.2	6.7	8.3	8.8	16.3	9.0
1984*						
All household heads (thousands)	114	259	844	502	233	1,951
Out of work (thousands)	23	24	130	108	45	330
Proportion out of work (per cent)	20.0	9.4	15.4	21.6	19.3	16.9
All tenures						
1977/78						
All household heads (thousands)	2,675	2,355	4,146	1,472	700	11,348
Out of work (thousands)	53	92	204	102	89	540
Proportion out of work (per cent)	2.0	3.9	4.9	6.9	12.7	4.8
1984*						
All household heads (thousands)	3,158	2,182	3,791	1,469	473	11,073
Out of work (thousands)	100	129	293	192	79	791
Proportion out of work (per cent)	3.2	5.9	7.7	13.1	16.7	7.1

* Households heads whose last job was 3 or more years ago omitted from 1984 results. Looking for work excludes any people aged 65 and over and includes people who were not looking for work because they thought no jobs were available.
† Seeking work excludes any people aged 65 and over, and includes those not seeking work because they thought no jobs were available.

Source: Labour Force Survey, 1984 1977/8

workers whose wages are well above benefits and who can move as they wish, given market wages and house prices. Table 3.2, based on the 1977/8 and 1984 Labour Force Surveys, shows the picture nationally for unemployment across occupational groups and housing tenure. It is

Regional Policy and Market Forces: Model and Assessment

notable that unemployment among non-manual groups with their own homes barely rose. Since the vast majority (93 per cent) of non-manual workers own their own homes, much the same was true of non-manual workers generally.

Table 3.3 *Regional wages and unemployment*

Non-Manual Labour	1979		1987	
Region	Wage (SE=100)	Unemployment (%)	Wage (SE=100)	Unemployment (%)
South East	100.0	1.37	100.0	3.57
East Anglia	97.7	0.65	95.8	1.90
South West	90.5	2.25	90.4	2.78
Average for 'South'	98.3	1.85	98.0	3.25
West Midlands	99.6	1.47	95.8	3.97
North West	98.6	2.00	97.4	5.00
East Midlands	100.1	1.61	94.2	3.85
Yorkshire and Humberside	101.6	1.71	97.0	5.16
North	100.6	1.65	97.4	3.43
Wales	101.7	2.09	96.6	4.72
Scotland	96.8	2.01	94.1	6.00
Average for 'North'	99.6	1.82	96.4	4.76

Manual Labour	1979		1987	
Region	Wage (SE=100)	Unemployment (%)	Wage (SE=100)	Unemployment (%)
South East	100.0	6.3	100.0	12.9
East Anglia	95.3	6.2	93.5	13.1
South West	89.6	7.6	90.1	12.8
Average for 'South'	97.6	6.6	97.3	12.9
West Midlands	97.4	6.8	91.4	18.4
North West	97.7	10.5	93.9	20.0
East Midlands	107.1	6.5	100.3	15.0
Yorkshire and Humberside	100.4	8.0	100.4	16.3
North	100.8	12.4	95.0	20.8
Wales	99.4	9.9	92.1	19.9
Scotland	96.4	11.5	93.1	21.5
Average for 'North'	99.4	9.7	95.1	19.0

Sources:
Labour Force Surveys
New Earnings Surveys
Building Societies Association

Reducing Regional Inequalities

What was true nationally also held at the regional level, as is shown by Table 3.3. Non-manual unemployment not only did not rise much from 1979 to 1987 in any of the regions, it was also strikingly similar across all regions. This was so at a time when total unemployment rose sharply nationally and became highly unequal across regions.

The rest of the picture is filled in when one turns to manual unemployment. Nationally their unemployment rate rose sharply; this rise was particularly marked among those who did not own their own homes (38 per cent of all manuals and 62 per cent of unskilled manuals), where the rate jumped to 17.2 per cent from 9.1 per cent. Regionally, there were marked differences, with particularly sharp rises in the north.

When one turns to earnings one finds further evidence of the mobility of non-manual workers. Table 3.3 also shows that their earnings corrected for differential house costs differ little across regions; the figures are presumably roughly equalised by mobility.

Those for manual workers (not corrected for the modest differences in council house rents) show greater inequality. Given that we know that unions attempt to keep their members' wages regionally equal by national negotiation, this implies a bigger inter-regional discrepancy between non-union wages, consistent with considerable immobility among these workers. This is brought about by the subsidisation of rents among sitting tenants, private or council; should they move they lose this subsidy since they must rent privately in the open unsubsidised market where little is available at high cost.

Table 3.4 *Regional land price trends (1980 prices, £000 per hectare)*

	Housing Land		Agricultural Land	
	1979	1987	1979	1987
North	72.30	100.50	2.16	1.54
Yorkshire and Humberside	48.63	77.30	2.60	2.04
East Midlands	42.15	114.00	3.07	2.40
East Anglia	52.90	135.50	3.28	2.44
South East	124.10	352.30	3.15	3.03
South West	92.62	161.65	2.70	2.58
West Midlands	94.40	170.20	3.07	2.70
North West	61.22	122.85	2.93	2.32
Wales	36.60	50.20	1.95	1.68
Scotland	n/a	n/a	n/a	n/a

Regional Policy and Market Forces: Model and Assessment

Table 3.4 *(Cont)   Agricultural land prices (average price: £ per hectare)*

	1979	1980	1981	1982	1983	1984	1985	1986
Standard Region								
North	n/a	n/a	2,085	2,010	2,428	2,173	2,245	2,340
Yorkshire and Humberside	n/a	n/a	3,042	2,891	3,009	2,543	2,540	3,086
East Midlands	2,601	3,730	3,348	3,578	3,789	4,053	4,091	3,631
East Anglia	n/a	n/a	3,408	3,865	3,887	4,529	4,549	3,699
South East	2,667	3,360	3,733	3,506	3,600	3,940	4,180	4,591
South West	2,291	3,326	3,259	3,297	3,468	3,840	4,042	3,910
West Midlands	2,577	3,734	3,823	3,627	4,250	4,559	4,453	4,090
North West	n/a	n/a	3,393	3,706	3,521	4,111	3,888	3,519
Wales	1,655	2,418	2,212	2,032	2,261	2,727	2,467	2,543
England and Wales	2,316	3,039	3,162	3,098	3,321	3,496	3,586	3,507

Sources: Inland Revenue Statistics, Housing Statistics

So we observe unemployment concentrated on manual workers whose wages are close to benefit levels and on northern workers in particular whose wages are lower and who cannot move because there is not a free rented market. Non-manual workers whose supply of labour is unaffected by benefits and who can move in a free market for privately-owned homes experienced little unemployment nationally or regionally.

This underlines the important fact that the north–south unemployment problem is not a general one, but a problem of manual workers only. Indeed, recent surveys have found that skilled labour shortages are serious even in Merseyside, the county with the second highest unemployment rate in the country.

Regional land prices and use

Since land is perforce immobile with a reserve price provided by alternative agricultural and commons uses, its behaviour is much like that of manual labour. In the south its price has risen sharply as rising demand met inelastic supply because of the limits placed by planning control. In the north its price has fallen so that there has been a net withdrawal from active industrial use. Table 3.4 shows the well-known price trends, while Table 3.5 gives some figures on regional use.

Migration

Our model suggests that only modest migration from north to south will be observed. Among the unskilled the regulated rented house market

Reducing Regional Inequalities

Table 3.5 *Regional land use*

	Housing Stock ('000 dwellings)		Industrial (million m²)		Commercial and distribution (million m²)		Agricultural ('000 hectares)	
	1979	1987	1979	1987	1979	1987	1979	1987
North	1,183	1,246	15.0	14.3	13.9	16.0	1,060	1,047
Yorkshire and Humberside	1,857	1,956	31.6	29.8	24.5	28.6	1,108	1,105
East Midlands	1,440	1,554	22.1	23.4	17.5	21.7	1,247	1,240
East Anglia	718	808	8.1	8.3	11.2	13.7	1,015	1,016
South East	6,479	6,686	55.1	52.7	89.4	105.3	1,736	1,726
South West	1,660	1,850	14.8	15.3	21.8	25.4	1,857	1,844
West Midlands	1,896	2,019	389	36.9	23.1	29.1	987	979
North West	2,441	2,533	47.7	42.9	35.3	40.0	459	456
Wales	1,057	1,128	9.2	n/a	10.5	n/a	2,077	1,515
Scotland	1,973	2,062	n/a	n/a	n/a	n/a	5,546	5,375

Sources: Commercial and Industrial Floorspace Statistics, Regional Trends

prevents it. Among the skilled mobility exists and continues until real wages are equalised, but the limited availability of industrial and housing planning land in the south will constrict the capacity of industrial activity to accommodate them.

The migration figures bear this out. Though particular parts of the

Table 3.6a *Inter-regional migration, 1980-6 ('000)*

Net Migration from regions, 1980-6:	1980	1980	1981	1982	1983	1984	1985	1980-6
North	-9	-8	-4	-7	-8	-9	-7	-52
Yorkshire and Humberside	-6	-5	-5	-7	-6	-12	-11	-52
East Midlands	+8	+4	+3	+5	+5	+6	17	+48
East Anglia	+14	+10	+14	+14	+13	+19	10	+94
South East	+1	+9	+4	+3	+6	-17	-3	+3
Greater London	-46	-32	-43	-34	-33	-56	-49	-284
Rest of the South East	+47	+41	+38	+37	+39	+39	+46	+287
South West	+28	+19	+26	+35	+34	+41	+47	+287
West Midlands	-10	-12	-13	-15	-13	-9	-8	-80
North West	-21	-20	-21	-21	-23	-16	-27	-149
Wales	+2	+2	+2	0	+3	+5	+5	+19
Scotland	-8	-2	-6	-4	-10	-8	-14	-52
Northern Ireland	n/a	n/a	-3	-4	-3	-3	-6	-19

Source: Regional Trends

Table 3.6b *Migration, population and labour force trends, 1979-87*

	Population (% Change)	Migration 1980-6 (% of population)	Labour Force (% change)
North	-1.6	-1.7	-
Yorkshire and Humberside	-0.4	-1.1	+3.1
East Midlands	+3.7	+1.2	+9.2
East Anglia	+8.1	+4.7	+22.2
South East	+2.2	-	+9.4
South West	+5.8	+6.3	+11.5
West Midlands	-10.3	-1.5	+2.8
North West	-2.0	-2.3	-4.2
Wales	+1.1	+0.7	-5.7
Scotland	-1.7	+1.0	0.4

Source: Regional Trends

north (eg Liverpool) and indeed the south (eg Greater London) have lost population, they have done so to contiguous areas for the most part; net inter-regional migration has been limited, considering the regional imbalances – see Tables 3.6a and 3.6b.

Interestingly, the figures reveal some evidence of cross-patterns in retired and non-retired migration, with labour force growth moving differently from migration, sometimes with opposite sign as in Wales, Scotland, Yorkshire and Humberside, and West Midlands. People retiring appear to be net immigrants to Wales and Scotland, and net emigrants from the South East and East Anglia, in particular. This cross-migration has allowed greater flexibility in the sizes of employed labour forces across the regions. Judging from employment trends, this flexibility has been concentrated on non-manual labour.

Industrial structure

Table 3.7 reveals that the north is more dependent on manufacturing than the south within the traded sector, but that in common with the south its share of manufacturing has declined over the last decade.

Our figures for the protected sector (see Table 3.8) are rough; we have included, out of the regional product categories available, agriculture, energy and water supply, public administration and defence, education and health, and construction. The protected sector share is of course smaller in the north than in the south, mainly because of the location of government activity. The decline of coal and the relative decline of

Reducing Regional Inequalities

Table 3.7 *Manufacturing intensiveness of regions (ratio of manufacturing to unprotected services)*

	1979	1987
Greater London	0.3342	0.2213
Rest of South East	0.6641	0.4540
South West	0.6474	0.5147
East Anglia	0.7690	0.6426
North	1.0334	0.7446
Yorkshire and Humberside	0.9708	0.6191
East Midlands	1.1089	0.8568
West Midlands	1.3019	0.9069
North West	0.9854	0.7261
Wales	0.8367	0.7472
Scotland	0.7453	0.5382
Northern Ireland	0.6778	0.5399
UK	0.7368	0.5308

Source: CSO, Regional Accounts

education, health and public administration have reduced the share of the protected sector in the South East, Wales and East Midlands over the last decade; but relocation of public sector offices has raised the North West share.

Output, profits and capital

With the abolition of exchange controls in 1979 the return on capital in Britain began to rise towards equality with international rates. This process was complete by the mid 1980s. Regional profit rates, in so far as can be judged from the share of profit in value added, have behaved similarly across the regions, as one would expect with total internal capital mobility (see Table 3.9).

The sharp fall in manufacturing in the early 1980s is reflected in a disproportionate fall in northern GDP (see Figure 3.3) with its larger share of manufacturing. However, after this shake-out and the corresponding drop in manual employment, northern GDP has risen about as fast and in some regions faster than the south.

The picture confirms the story of the unemployment figures. The problem of the north is that of unused manual labour resulting from a contraction in manufacturing. Growth and profitability are equal to rates in the south and real incomes of non-manual workers are the equal of

Table 3.8 *Regional industrial structure*

Manufacturing Sector – per cent of GDP

	1971	1975	1979	1983	1987
North	32.94	33.23	32.45	28.69	27.24
Yorkshire and Humberside	34.38	32.01	31.88	24.92	25.00
East Midlands	37.68	33.79	33.29	29.35	30.75
East Anglia	26.47	24.48	27.07	24.31	25.84
Greater London	21.67	17.81	17.73	14.91	13.41
Rest of South East	27.37	24.37	26.14	22.49	21.42
South West	26.98	24.70	24.39	21.84	21.74
West Midlands	44.49	40.47	39.41	32.89	33.01
North West	36.18	34.48	35.22	30.02	29.00
Wales	27.92	27.96	26.39	23.36	26.58
Scotland	29.23	27.27	27.42	23.67	22.17
Northern Ireland	29.30	26.11	22.41	18.67	18.38
UK	30.62	28.03	28.11	23.97	23.40

Protected[1] Sector – per cent of GDP

	1971	1975	1979	1983	1987
North	37.51	38.38	36.15	38.96	36.18
Yorkshire and Humberside	34.16	37.28	35.28	39.51	34.63
East Midlands	34.43	38.20	36.69	39.42	33.36
East Anglia	40.88	42.82	37.73	38.54	33.95
Greater London	28.31	31.35	29.22	29.74	26.00
Rest of South East	35.63	38.53	33.29	34.92	31.41
South West	39.22	40.50	37.94	39.80	36.02
West Midlands	27.53	30.84	30.31	34.41	30.59
North West	29.13	31.11	29.04	32.94	31.06
Wales	41.91	40.78	42.06	44.09	37.84
Scotland	37.04	39.42	36.23	38.69	36.64
Northern Ireland	39.00	41.94	44.53	48.75	47.57
UK	33.59	36.18	33.74	36.20	32.51

Reducing Regional Inequalities

Table 3.8 *(Cont)*

Unprotected[1] Services [2] – per cent[3]

	1971	1975	1979	1983	1987
North	29.55	28.39	31.40	32.34	36.58
Yorkshire and Humberside	31.46	30.71	32.84	35.57	40.38
East Midlands	27.90	28.01	30.02	31.24	35.89
East Anglia	32.65	32.70	35.20	37.15	40.21
Greater London	50.02	50.84	53.05	55.35	60.59
Rest of South East	36.99	37.10	40.58	42.59	47.18
South West	33.80	34.81	37.67	38.36	42.24
West Midlands	27.99	28.68	30.27	32.70	36.40
North West	34.70	34.41	35.74	37.04	39.94
Wales	30.17	31.25	31.54	32.55	35.57
Scotland	33.72	33.10	36.35	37.63	41.19
Northern Ireland	31.70	31.94	33.06	32.58	34.04
UK	35.79	35.79	38.15	39.82	44.08

[1] Protected sector consists of: agriculture, energy and water supply, construction, ownership of dwellings, forestry and fishing, public administration and defence, education and health services.
[2] Unprotected services sector: distribution, hotels, catering, repairs, transport and communication, financial and business services, etc, other services.
[3] GDP is before adjustment for financial services (ie this adjustment is added back to GDP).

Source: CSO, Regional Accounts

their southern counterparts. It is as if a very specific part of the north's supply side had been lopped off, but the rest was functioning normally.

Underlying causes

The decline in the manufacturing terms of trade

The sharp fall of manufacturing in the early 1980s followed a steady decline in its share of GDP from 1974. The origins of the difficulties go back further still, as Figure 3.4 showing the ratio of manufactured to service sector prices – its 'terms of trade' (based on national accounts value-added and output volumes) – reveals. There was a long and steep decline in these terms of trade from 1955 to 1980, totalling about 40 per cent. Since 1980 the trend has levelled out and is lately even rising a little.

Though manufacturing growth kept pace with GDP until 1974, the

Regional Policy and Market Forces: Model and Assessment

Table 3.9 *Regional profits as a share of GDP (%), 1971–87*

	1971	1975	1979	1983	1987
North	19.77	18.66	19.96	16.64	19.46
Yorkshire and Humberside	19.18	17.93	19.43	15.60	19.98
East Midlands	18.33	17.03	17.08	14.47	19.19
East Anglia	17.58	16.38	17.62	14.68	18.73
Greater London	17.19	13.14	17.00	12.09	17.26
Rest of South East	14.10	14.51	16.10	12.02	15.89
South West	15.85	14.58	16.46	12.04	15.70
West Midlands	18.42	15.92	17.46	14.10	20.11
North West	20.35	17.95	20.92	16.07	20.80
Wales	17.54	14.55	19.24	15.76	19.48
Scotland	20.51	16.58	18.81	15.39	18.67
Northern Ireland	18.28	14.91	13.03	9.96	14.50
UK	17.72	15.73	17.80	13.82	18.12

growth of services (excluding public administration and defence) was faster, so that in line with these relative price trends manufacturing fell relative to services. But the fall was delayed, especially in the 1970s, by policies of intervention designed to strengthen and protect manufacturing. With the general withdrawal of those policies from 1979, this delayed adjustment was concentrated in a short period.

This relative price trend implied that manual workers, in whom manufacturing is intensive, would suffer a fall in relative wages outside the protected sector. This fall again was concentrated in the same short period. While it was a national phenomenon, its impact on the north was more devastating, because the north has a relatively large population of manual workers compared with the south and is consequently more concentrated in manufacturing.

This factor alone, therefore, would have reduced northern GDP disproportionately and while raising unemployment rates among manual workers by much the same amount in all regions would have raised overall unemployment rates more in the north.

On the land side, it would have raised land prices in all regions (since services are relatively intensive in land), just as it lowered manual wages in all. Its proportionate effect on the stock of available land for industrial use might therefore be expected to be much the same across regions. Figure 3.5 illustrates this.

Reducing Regional Inequalities

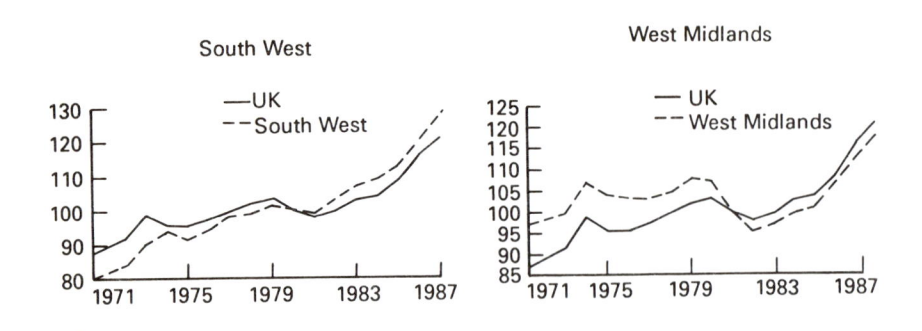

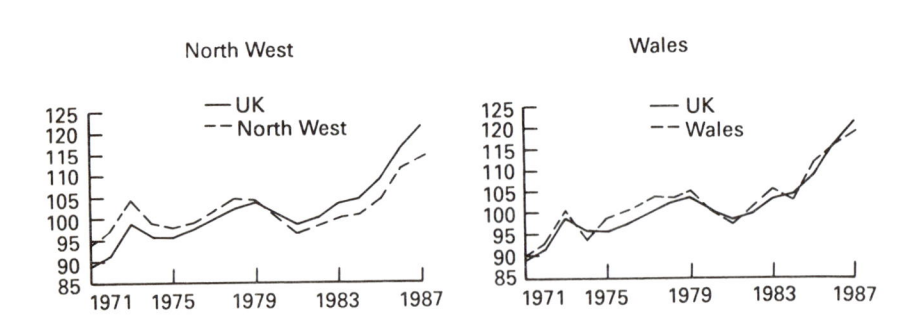

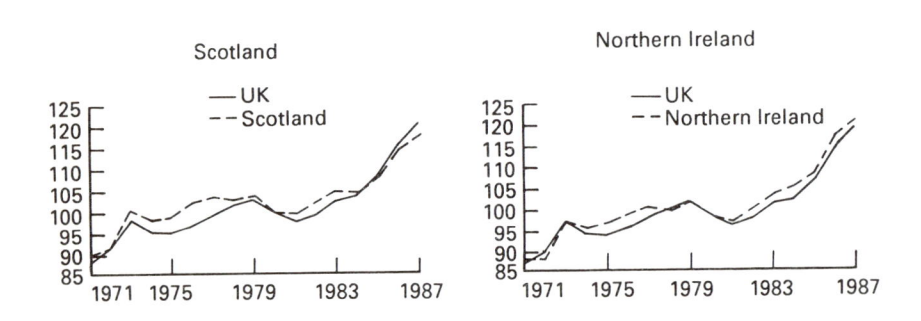

Source: CSO, Regional Accounts

Figure 3.3 *Index of regional GDP (1980=100)*

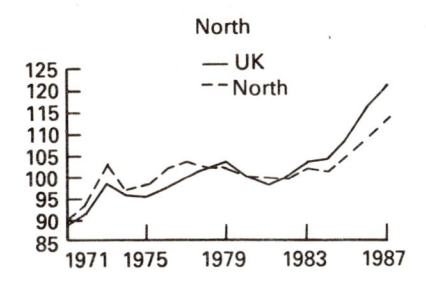

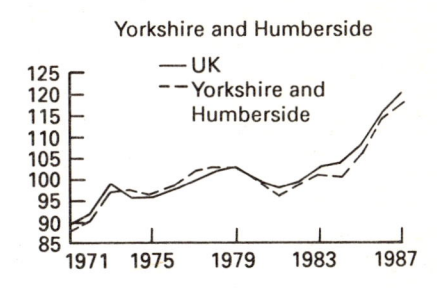

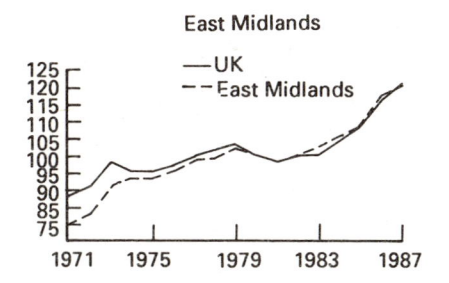

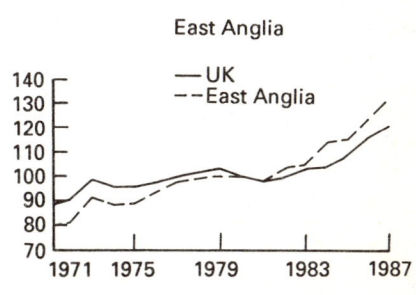

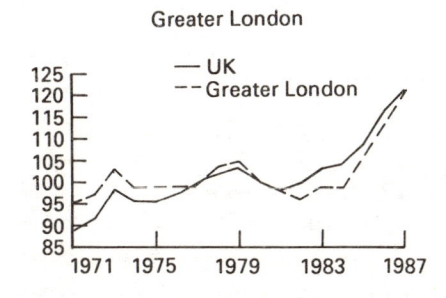

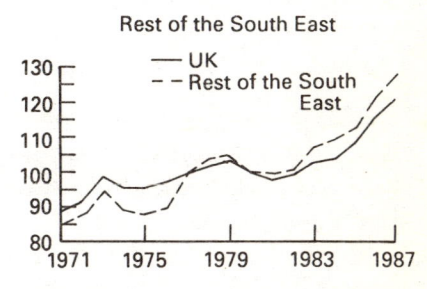

Regional Policy and Market Forces: Model and Assessment

Reducing Regional Inequalities

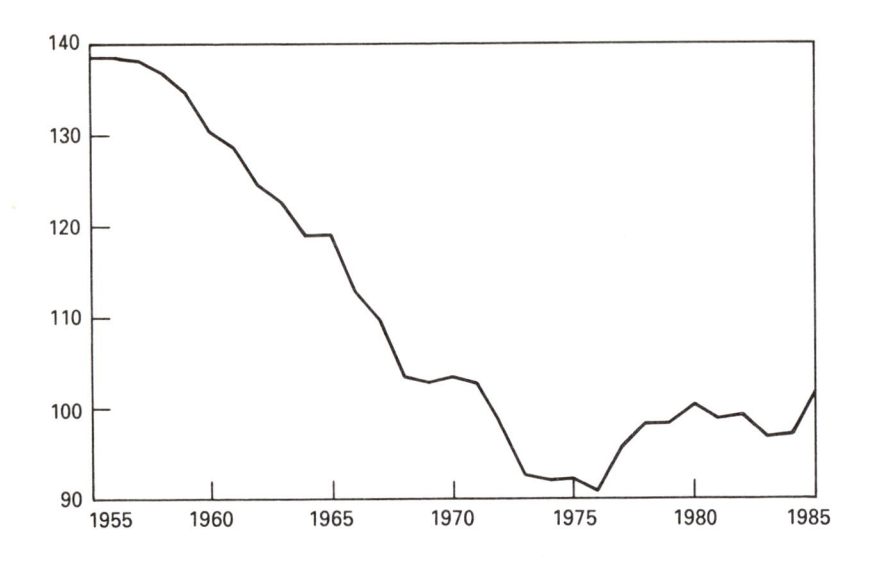

Figure 3.4 *Terms of trade (1980=100) – manufactures/services, traded sector*

Source: CSO, IMF

Technological shift away from manual labour
This too is a well-known national trend resulting from mechanisation and the information technology revolution. The share of unskilled labour in value added has declined no less than 20 per cent between 1977 and 1984 alone (see Table 3.10).

The effect of this trend would again be to lower the marginal product and wage of manual workers in all regions equally, with similar effects on manual unemployment rates. However, again the effect on overall unemployment rates and output would be greater in the north, as seen in the top half of Figure 3.5.

The increasing transport cost disadvantage of the north
Much has been made in discussions of the north–south divide of the attraction exercised by Europe; the orientation of our trade towards Europe and away from the US and the Commonwealth enhances southern locations close to the Continent and downgrades northern locations close to the deep sea ports like Liverpool serving the north Atlantic, African and Asian trade.

This is not quite accurate. Abstracting from the immense complexity

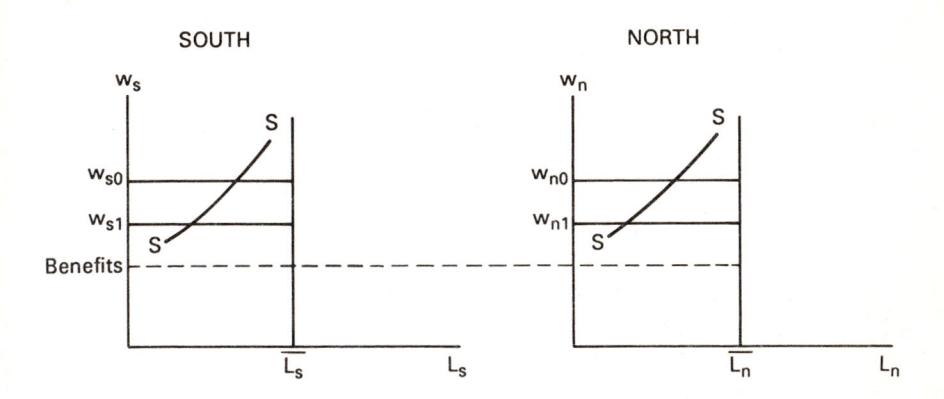

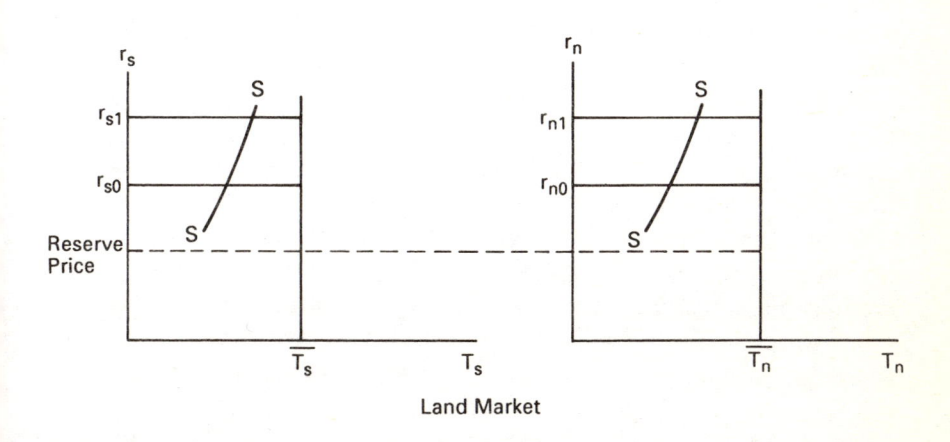

Regional Policy and Market Forces: Model and Assessment

Figure 3.5 *The effect of falling manufacturing/services terms of trade: manual labour wages, w fall from w_0 to w_1 in north and south, land prices/rents rise from r_0 to r_1 in north and south*

Reducing Regional Inequalities

Table 3.10 *Semi/unskilled labour intensity – whole economy*

	Semi/unskilled labour relative to total employment (%)[1]	Income from employment as a share of GDP (income based)[2] %	Semi/unskilled wages relative to average wages[3]	Share of semi/unskilled labour in costs (%)
1977	18.3	67.2	81.9	10.1
1984	16.3	64.6	77.4	8.1

[1] Source: Labour Force Surveys
[2] Source: National Accounts, 1988 Edition
[3] Assumes earnings of semi/unskilled on average 90% of manual earnings in both years.

Source: New Earnings Surveys

of different cargoes and transport modes, one can distinguish between deep sea and short sea cargoes, and within each in increasing order of labour-intensiveness among containers, bulks (such as grain and oil, piped out at port rather than crane-handled), and semi-bulks (such as sacks or packs of commodities requiring handling).

Twenty years ago deep sea ships were small and containers were barely in use; internal road transport was expensive relative to both sea and rail. The optimal deep sea voyage consisted of frequent stops at ports as close as possible to final inland destinations; any distance inland was covered by rail. All traffic, deep sea and short sea, required substantial use of labour in ports; the dock labour scheme used to cover the vast bulk of trade and labour-intensive methods gave the dockers' union immense power. In these circumstances, transport costs to/from any foreign destination/source were dominated by high port costs and were similar from most points in the British Isles.

Today, the scene has changed dramatically, mainly because of technological change and only partly because of shifts in trade. Deep sea ship size has increased and to achieve the 3.4p per container-mile it costs by sea requires a ship to have a full load that can only be provided if cargo to a large number of destinations is combined; hence ships calling in the UK typically carry large amounts of continental cargo. (For more details of transport costs and our sources see Appendix 2.) This tendency is reinforced by the relative decline in UK deep sea trade at the expense of continental short sea. The result is that deep sea ships mostly run from foreign source to the Channel, where they will call in at one or two UK ports and several continental ports. Part loads to or from the north are

Regional Policy and Market Forces: Model and Assessment

sent by road (80p per container-mile) or rail (30p) to avoid the lengthy northward diversion to Liverpool or Hull – uneconomic for a lightly loaded ship.

These ports have also been more expensive than their southern competitors because of both high pilotage charges and the now defunct dock labour scheme (which was undermined in the south by non-scheme ports like Felixstowe and Portsmouth). This has been a further factor driving cargoes south, particularly semi-bulks, with relatively high labour costs in handling.

For the time being deep sea shipping conferences subsidise the differential costs of shipping containers from the north, by charging as if the container were being sent through the nearest northern port. However, this 'grid' practice could be in restraint of trade (to prevent northern ports creating competition on a new container route) and may not outlast the vigilance of the Director of Fair Trading. Furthermore, on bulks and semi-bulks there is no such arrangement so that northern costs are higher.

On short sea routes (mainly for the continent) the main change is the lift-on lift-off (lolo) container and roll-on roll-off (roro) trailer traffic; larger ships are not generally economic on these routes. The other

Table 3.11 *Transport cost differentials, non-oil trade, Manchester-London*

	£ per 22-Ton Load	% Assuming Value of £25,000[2]	Approximate Share of Total UK Trade
Deep Sea			
Containers	–	–	7
Bulks and Semi-bulks	250[1]	1.00	27
Short Sea			
Containers	265	1.06	8
Bulks and Semi-bulks	250	1.00	58
Weighted Average	234	0.935	100

[1] Assumes Manchester load goes through Dover to Antwerp/Rotterdam because of lower port costs. Source: Appendix 2
[2] Assumes full load value of £30,000, with 85% average load factor

Source: Trade sources

Reducing Regional Inequalities

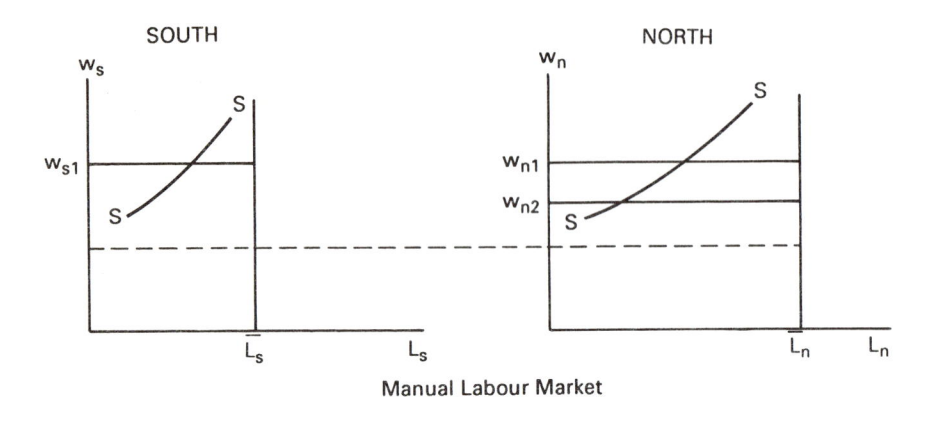

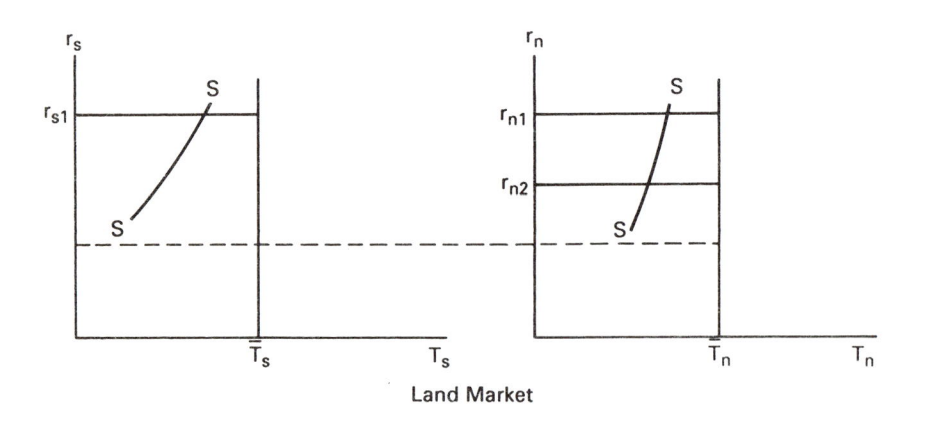

Figure 3.6 *The effect of the northern transport cost disadvantage – lowering land and labour rewards equiproportionately in the north from r_{n_1}, w_{n_1} to r_{n_2}, w_{n_2}*

change is the competition in the south from non-scheme lolo and roro ports like Dover and again Felixstowe.

On bulks and semi-bulks, the high cost of northern scheme ports makes routing via road or rail through a southern port marginally economic. Even on containers with less labour handling the same appears to be true. The result is that a northern producer has to absorb overland charges for an extra 300-odd miles compared with his southern competitor.

Cost comparisons are not easily come by in this area. However, we have obtained some indicative calculations by enquiries from the trade (see Table 3.11). On average they suggest that northern locations could be paying 1 per cent of gross value more per one way trip than southern. In the traded sector value added averages some 50 per cent of gross value and typically both inputs and outputs will require transporting; thus total transport costs as a fraction of value added could be around 4 per cent higher in the north than in the south. Since about 10–20 per cent of value added in manufacturing is spent on land rental and unskilled labour, the regionally avoidable costs, and probably less for services, transport costs are imposing a 20–40 per cent (4/.2–4/.1) tax on northern land and unskilled labour. This is the extent by which both wages and land rentals are driven down in the north.

Much, if not all, of this premium would go if there were real competition in northern port and shipping conference activities. Abolition of the dock labour scheme should now eliminate the incentive to go overland to southern ports except for deep sea containers; for these, cessation of the grid would allow Liverpool and Hull to compete for container services. A competitive through rail service via the Channel Tunnel (for part loads) could combine with this to allow a revival of the northern container routes on similar terms to their southern competitors.

As things are currently, the effect of northern transport disadvantages is illustrated in Figure 3.6. A wedge is driven between northern and southern land and unskilled labour prices of up to 40 per cent.

The protected sector and union power
Large areas of our economy are protected from foreign competition either by natural barriers or by government regulation. Internal competition in both the goods and labour markets can nevertheless keep costs and prices down so that the barriers are inoperative. However, in practice, this is unusual because either goods market monopoly is exploited or unions organise on the input side; either way costs and prices can be raised by such exploitation to levels only limited by the

Reducing Regional Inequalities

decline in final demand – and the gains can be distributed to workers, management and shareholders in these industries according to some bargaining procedure.

It is no accident that managements and shareholders of large unionised firms are often opposed to reductions in union power, because this may act as an effective barrier to entry by market contestants. Since the abolition of resale price maintenance and the advent of active competition policy, union power has been the main remaining vehicle for firms to retain monopoly power in these protected markets, the other one being government regulation (eg in public procurement).

Table 3.12 *Regional unionisation rates*

Year/Region	N	YH	EM	EA	SE	SW	WM	NW	W	SC	NI
1963	0.475	0.493	0.503	0.430	0.452	0.453	0.476	0.474	0.501	0.468	0.476
1964	0.479	0.498	0.509	0.435	0.456	0.457	0.482	0.480	0.505	0.473	0.483
1965	0.482	0.501	0.511	0.438	0.459	0.460	0.485	0.483	0.507	0.475	0.485
1966	0.479	0.498	0.508	0.434	0.455	0.456	0.482	0.480	0.504	0.472	0.481
1967	0.476	0.496	0.506	0.433	0.455	0.454	0.485	0.480	0.501	0.469	0.479
1968	0.479	0.499	0.509	0.436	0.458	0.457	0.491	0.483	0.504	0.471	0.480
1969	0.490	0.510	0.519	0.449	0.472	0.471	0.505	0.496	0.515	0.483	0.493
1970	0.519	0.540	0.549	0.476	0.500	0.500	0.540	0.526	0.546	0.513	0.523
1971	0.525	0.547	0.556	0.482	0.504	0.505	0.549	0.532	0.552	0.518	0.528
1972	0.534	0.559	0.569	0.490	0.512	0.513	0.557	0.542	0.561	0.527	0.539
1973	0.543	0.570	0.580	0.496	0.521	0.520	0.568	0.554	0.570	0.536	0.549
1974	0.554	0.582	0.592	0.505	0.531	0.530	0.580	0.566	0.582	0.547	0.562
1975	0.564	0.595	0.604	0515	0.544	0.539	0.594	0.580	0.590	0.558	0.570
1976	0.577	0.609	0.617	0.528	0.557	0.552	0.611	0.594	0.603	0.571	0.581
1977	0.590	0.621	0.628	0.541	0.571	0.565	0.625	0.607	0.616	0.584	0.592
1978	0.600	0.631	0.638	0.551	0.582	0.575	0.637	0.619	0.626	0.595	0.602
1979	0.610	0.638	0.645	0.560	0.592	0.585	0.649	0.627	0.636	0.603	0.609

Source: Minford *et al* (1987)

We have poor information both on the extent of government-induced protection and on union power by regions. However, up to 1979 we were able to compile an index of regional union power based on industrial union membership figures from Professor Bain of Warwick University; these were weighted by industry shares in each region to give an estimate of regional union penetration. The index showed that there was systematically higher penetration in the north than in the south (see Table 3.12). This appears consistent with other data on strike incidence (higher also in the north) and patterns of behaviour across regions in national strikes, where northern workers seem to come out earlier and

Regional Policy and Market Forces: Model and Assessment

stay out longer than their southern counterparts. The dock labour scheme discussed above is a clear example of this; in the north competition from fellow dockers in non-scheme ports is feeble compared with the south, yet there is no particular lack of natural port facilities with which to repeat the Felixstowe phenomenon.

In previous work (Minford *et al*, 1988) we made an attempt to estimate the effect of this higher northern union power on unemployment. For 1979 we suggested a differential effect raising the overall northern unemployment rate by around 1 per cent. There seems little reason to believe this has changed over the 1980s.

We reviewed above the way in which the share of the protected sector has moved in the regions. It seems that government reaction to incentives to redeploy its activities to cheaper northern locations has been fairly weak, while its policies of deregulation have had a substantial direct effect in contracting the size of the northern protected sector.

Local authority behaviour

Until the recent Reform Act local authorities levied business and personal rates. The effect of personal rates, which because of rebates fell mainly on skilled workers (and the retired), was to alter the relative cost to firms of these workers across regions. Since skilled wages (net receipts to the worker in real terms) will be equalised across the country, this acts essentially like transport costs above, given that there is likely to be little scope for substitution to or from skilled labour to reduce the impact on costs. For example, suppose higher rates add 2.5 per cent to northern living costs of skilled workers; avoidable regional costs are 10–20 per cent of value added, while skilled labour is about 60 per cent of value

Table 3.13 *Domestic rates as a tax (rates paid pa as a percentage of house prices)*

	SE	EA	EM	N	NW	SW	W	WM	YH
1978/9									
1979/80									
1980/1	0.81	0.78	0.80	0.95	0.92	0.72	0.66	0.92	0.80
1981/2		0.89	0.89	1.09	1.07	0.78	0.76	1.11	0.90
1982/3		0.96	1.16	1.33	1.29	0.93	0.93	1.36	1.21
1983/4		0.94	1.10	1.30	1.27	0.91	0.82	1.25	1.12
1984/5		0.93	1.11	1.25	1.27	0.89	0.86	1.20	1.12
1985/6		0.88	1.14	1.40	1.39	0.88	0.88	1.33	1.17
1986/7		0.96	1.26	1.54	1.46	0.91	0.92	1.40	1.30
1987/8	0.78	0.92	1.23	1.40	1.46	0.84	0.92	1.31	1.27

Sources: CIPFA, BSA Bulletin

Reducing Regional Inequalities

added, in the traded sector. Then land and unskilled labour costs in the north have to fall to offset this by 7.5-15 per cent (=2.5 x 0.6/0.2-0.1).

Under pressure from declining revenues and a vocal electorate largely unskilled with little to lose from higher rates, northern authorities have raised personal rates more than in the south (see Table 3.13). Northern rates are typically 0.7 per cent higher as a fraction of house prices, which in turn are about three times average wages, so living costs are indeed about 2 per cent higher for northern skilled workers, as in the example above. This has contributed therefore a further (7.5-15 per cent) tax on northern land and unskilled labour prices, driving them down by this much, just as illustrated for transport costs in Figure 3.6.

As for business rates, their incidence is on net land rentals. Gross-of-rate rentals are set by what the traded sector will pay, as above. But higher business rates reduce the return to the land owner of industrial use and so diminish the supply of industrial land, as in the bottom half of Figure 3.6. This creates an artificially distorted unemployment of land or dereliction.

Where the supply of industrial land has contracted substantially and cannot fully satisfy even the needs of the protected sector, then this sector will pay above the traded sector price to satisfy its needs. In these circumstances higher business rates will raise the price of protected sector land and cause a contraction in this sector, with a consequent rise in unemployment.

Table 3.14 *Non-domestic rates as a percentage of approximate market rental value (%)*

	1980	1981	1982	1983	1984	1985	1986	1987
SE	51.06	60.68	69.29	65.51	61.61	57.62	50.05	45.79
EA	49.83	56.30	59.70	57.34	55.98	52.36	55.49	49.19
EM	47.09	51.59	64.64	60.44	59.91	60.73	65.94	63.62
N	57.89	65.31	77.77	75.25	71.56	79.54	85.07	76.93
NW	49.63	57.12	66.91	65.06	64.31	69.45	71.33	71.16
SW	50.90	55.28	63.96	61.88	59.76	58.92	59.45	54.49
W	61.43	66.95	71.23	62.55	64.47	65.63	67.64	66.75
WM	47.57	56.05	66.99	61.13	57.89	63.60	65.68	61.12
YH	49.91	54.66	70.56	64.62	63.85	65.68	71.58	69.24

Sources: CIPFA Housing and Construction Statistics

Again, northern local authorities under financial pressure have found it attractive to levy high business rates when the reduced supply of land and so of industrial activity is largely compensated by the central government through rate support grant and the higher unemployment bill is paid for direct by central government (see Table 3.14). In Minford *et al* (1988) we created an index of regional business rates; its average difference between north and south had an estimated effect on the unemployment differential of around 1 per cent.

III The current and future policy background to future trends

Introduction

We can think about development in each region as driven by the size of its land and labour supply at the competitive prices ruling. The limits placed by these on southern development are already clearly indicated by rising skilled wages and land prices; even unskilled labour in many parts of the south is probably now fully employed. The problem in the south is that of setting the planning system to release the optimal amount of land for development; below we argue that this is best done in theory by moving to an auctioning method for planning consents – in practice it may be possible to move closer to such a system by other politically more acceptable means.

In the north, the problem is one of manual unemployment. With the uniform business rate (UBR) and rate revaluation there will be little lack any more of viable land development. Indeed, already the north is crawling with developers offering housing, shopping and leisure schemes for derelict land sites. What is missing is the mechanism to convert all this activity into the eradication of unemployment.

Wages are low relative to benefits in these areas, driven down by the mechanisms described above. This reduces labour supply among the unskilled. Many lead a twilight existence on benefits, but move around to jobs held furtively elsewhere. The Monday and Friday trains tell a clear enough story.

One possibility is that more will move as the housing market eases up; some may buy their council houses, sell up and move out, others may decide to move to newly available rented accommodation in the south as council rents go up and private rents come down (or become available).

Another possibility is that more will retrain, acquiring better productivity and so higher potential wages. This is an area where the Employment Training Scheme (ET) can be effective if made compulsory like the Youth Training scheme. Another mechanism of retraining is the inward investor like Nissan in Tyne and Wear; urban development corporations (UDCs) can be more active in wooing such entrepreneurial investors with suitable packages, and introductions to eligible union (or non-union) partners.

A final possibility is a crackdown on job refusal by the long-term unemployed, such as was intended by Restart, but has never been carried out in the north with any seriousness – these measures appear to be prosecuted most vigorously in the south on the grounds that the jobs are obviously there and the public will thus more easily understand. Yet it is in the north that there is the greatest need for such measures, for it is here that wages are closest to benefit levels and therefore the temptations to abuse the system are at their most intense. The general increase in prosperity in the north may, as in the south, make this progressively easier politically.

Current trends and the effects of enacted policy measures

A number of measures with substantial regional implications have reached the statute book in the last few years. Before evaluating current trends we discuss the direction of effects of each measure, with quantitative comment where possible.

Local finance reform

The poll tax or community charge will basically remove the large discrepancies (due to widely differing rates) of local tax on skilled labour; this will (see p136 above) eliminate a premium on southern location of some 10 per cent with a corresponding fall in the manual wage differential.

The uniform business rate (UBR) combined with rateable value revaluation will lower business rates in the north – possibly by as much as 50 per cent in some cases – and correspondingly raise them in the south. This will increase the supply of industrial land for development in the north and reduce it in the south; it will also lower northern land prices in the protected sector, stimulating that sector and lowering unemployment. The UBR acts like a congestion tax/dereliction premium, pushing activity northwards much as the protected sector reacts to the north–south land price differential by relocating northwards. It is a self-liquidating tax/premium in so far as once the many distortions that have opened up the southern land premium have been removed, the UBR's effect in this respect will disappear.

The UBR also has a role, which is not self-liquidating, of removing from local authorities the power to tax business, a power which as public choice theory suggests is likely to be abused. It still leaves those authorities which wish to give an incentive to business the power to do so by other means if their voters approve; therefore it acts as a ceiling on net business taxation. Later we shall discuss the proposal to allow councils to auction planning consents (or measures moving in an equivalent direction); if this were to go ahead, councils on behalf of their voters will have complete freedom to control the supply of industrial land in line with local preferences – an undistorted and free system of planning. Hence the UBR does not interfere with the council's freedom to compete for business and development.

Minford *et al* (1988) suggested business rate differentials pushed up northern unemployment by 1 per cent. The UBR should remove this effect.

Housing reform

The reforms effectively deregulate new private tenancies, while leaving sitting tenants alone (except that controlled tenancies cannot be passed to heirs any more). The intention in the council sector seems to be to improve management efficiency by opting out to housing associations and steadily to increase rents to market rates over a three-to-five year period perhaps. At the same time the drive to sell council houses is continuing.

As we saw earlier, around half of manual unemployed are council tenants, while nearly 40 per cent of manual workers are. These percentages are higher still in the north. To enable normal mobility incentives to operate, council rents will have to reach market levels. Hence these reforms will take some five years to have their full effect on manual migration and so unemployment. Nevertheless, the opening up of the rented sector in the south could have some immediate effects on those who currently migrate weekly for work.

The long-term effects on manual wages and unemployment are illustrated in Figure 3.7. Both would be equalised across regions. Because there is some available stock of under-utilised rentable accommodation, the reforms should increase available housing units without putting pressure on planning land restrictions. Therefore, some successful net migration of manual labour should occur, unlike in the case of non-manuals. Where national manual unemployment would then settle would depend on the level of benefits relative to the manual wage payable by the traded sector, and also on the toughness of worktesting under the benefit regime.

Reducing Regional Inequalities

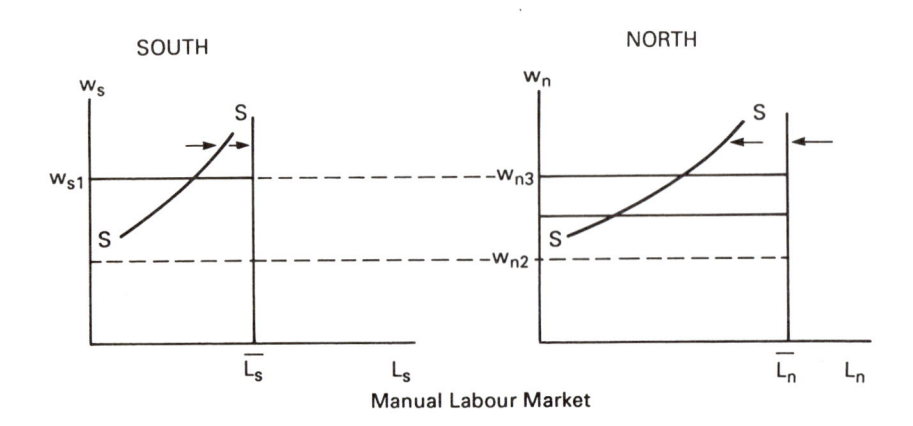

Arrows indicate that free migration will somewhat raise labour supplies in South, lower than in the North.

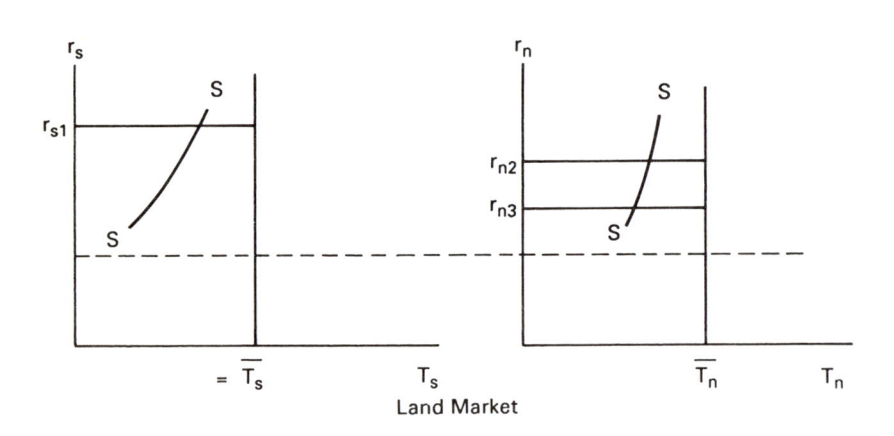

Figure 3.7 *Effect of manual labour migration: real wages (adjusted for living costs) are equalised in north and south, driving up wages, lowering land prices in north: to w_{n_3}, r_{n_3}*

140

Regional Policy and Market Forces: Model and Assessment

It is of interest to examine the effects on land prices. Once manual labour can migrate, the only immobile regional factor, or regionally avoidable cost, is land. This will now bear the entire brunt of any differential locational advantages of the south; thus, if these continue as above, then the housing reforms would widen land price differentials as shown in Figure 3.7. This lends some urgency to the further reforms, to be discussed below, designed to eliminate the distortions which artificially favour the south.

Restart and the training schemes

A crucial part in perpetuating manual unemployment is played by the benefit system. The supply curves of manual labour in our stylised diagrams all flatten off as non-union wages fall, representing the fact that people will not take jobs that pay too little, relative to the dole.

Hence, it is necessary to have a return to the Beveridge principle that the dole is only payable to those who cannot help themselves, either by taking an available job or going on training courses offered to them. This is now Department of Employment policy; though theoretically employment training is still not compulsory for the jobless, one assumes that failure to take up a place will be taken into account in the award of benefit.

Assuming this goes ahead, the supply curve of manual labour will steepen, reducing manual unemployment rates. Furthermore, training will increase the average productivity of manual workers and so raise the average wage offered in the traded sector, again reducing unemployment.

Table 3.15 *National unemployment trends from 1982*[1]

	Workforce '000	% change	Workforce in employment '000	% change	Employees in employment '000	% change	Unemployed '000	% change
1982	26.61		23.89		21.39		2.55	9.6
1983	26.63		23.61	-1.2	21.05	-1.6	2.79	10.5
1984	27.31	+2.6	24.22	+2.6	21.22	+0.8	2.92	10.7
1985	27.82	+1.9	24.61	+16	21.49	+1.3	3.03	10.9
1986	28.06	+0.9	24.75	+0.6	21.57	+0.4	3.11	11.1
1987	28.29	+0.8	25.30	+2.2	21.81	+1.1	2.82	10.0
1988	28.16	-0.5	25.74	+1.8	22.09	+1.3	2.36	8.4

[1] Mid year figures

Source: *Employment Gazette*

Reducing Regional Inequalities

The effect of this programme on the national and regional unemployment rates is likely to be dramatic. Already it is clear that the rapid fall in unemployment since August 1986 is due to the Restart complex; it is no coincidence that no sooner did that programme go national in July 1986, than unemployment began its steady fall from what had become a plateau for four years.

The Labour Force Survey confirms that the fall in benefit claimants, the official measure of unemployment, is matched by the fall in unemployment as measured by the Labour Force Survey. This suggests that the labour market is working more efficiently to create jobs. The growth of employment averaged 2 per cent pa in 1987-8 (see Table 3.15).

Union power and the protected sector

The four trade union laws have according to the most recent evidence been having an effect on the union mark-up over non-union wages – Figure 3.8 shows the evidence from the New Earnings Survey as estimated by Layard *et al* (1978, updated). Metcalf (1988) surveys other evidence confirming that unions have in the past had a negative effect on productivity and profitability, which is now diminishing. Our own regional evidence suggested that in 1979 about a 1 per cent unemployment differential could be attributed to differential union power; our

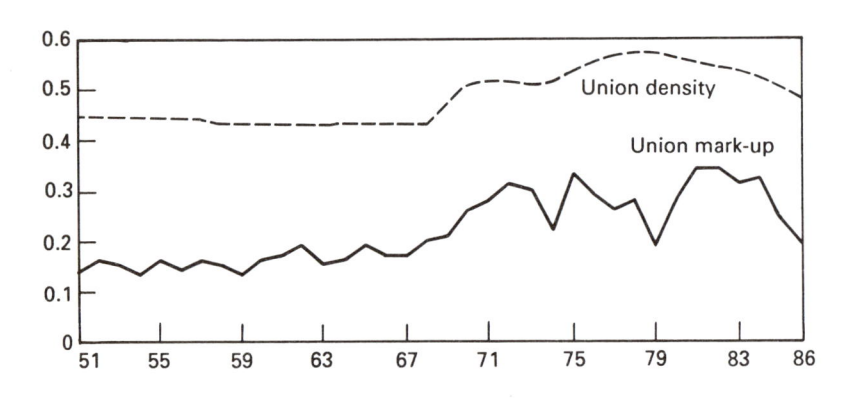

Source: Liverpool Macroeconomic Research Group Centre for Labour Economics, LSE

Figure 3.8 *The union mark-up*

Regional Policy and Market Forces: Model and Assessment

earlier national estimates in Minford *et al* (1983) indicated around one million unemployed (around 4 per cent) were so through union power.

While measures so far may not fully restore labour market competition in the protected sector, it seems clear that they are moving in that direction at a slow but perceptible pace.

Turning to protection of the goods market, deregulation and privatisation have substantially increased goods market competition within the protected sector. As health and education, and other parts of the public services, are explosed to market forces, this will continue.

Between these two forces, of deregulation and reduction of union power, competitive conditions within the protected sector should become the rule rather than the exception. This should lower unemployment in all regions, but particularly the north where lack of competition has been the greatest.

Regional aid, UDCs and intra-regional policy

Regional aid in its traditional forms of generalised regional grants has been essentially abandoned – with good reason. The problem with it is that it was neither universal in a typical northern region or even area, nor permanent in the sense that a recipient in a marginal activity could rely on its repetition when the capital stock needed renewing. To reduce northern manual unemployment, aid must raise the manual wage payable in the traded sector – to all for all time so that marginal workers are induced to supply more labour. This it is not designed to do. Hence, it has a temporary effect, if that (assuming that for a time it locates marginal workers); long term it has none at all.

Present policy is to use modest amounts of public money to facilitate market forces, given that planning procedures of possibly hostile local government and other distortions may stand in the way. UDCs are part of this, as are discretionary grants and consultancy aid to new businesses by DTI.

Success in this policy requires that those administering the funds and the planning understand the nature of the market distortions. The danger is that they will impose an inappropriate interventionary pattern of their own on existing distortions and worsen the situation. Currently, the north is awash with proposals for 'development' – of shopping, office, leisure and housing – slung together by consortia aware that government money may ease the initial costs and enable a profit to be made short term, leaving the areas with the risks that long term the facilities will be redundant.

The foregoing analysis has shown that the comparative advantage of

Reducing Regional Inequalities

the north lies in manufacturing, given its large stock of manual labour. While this is perfectly compatible with ancillary investment in the protected sector, government agencies should be careful not to subsidise investment which cannot be validated by reference to a gap created by market distortions. Instead it should provide necessary infrastructure, planning assistance, etc for genuine market-led opportunities. These agencies may also have a chance, by using their leverage, to break up existing distortions (eg insisting on open shops or non-union agreements).

It is too early to say whether the new grant arrangements are working satisfactorily in this sense or not. It may be necessary to give UDCs guidelines on projects which should and should not be assisted. Since the problems of the north are those of manual unemployment and urban dereliction, public money should only be awarded if either manual jobs will be created or if the urban environment will be improved. Other projects should go ahead purely with private money since their costs and benefits are reflected in market prices.

UDCs should also give priority to support of training of manual workers and to strengthening non-union and single-union no-strike set-ups. UDCs have considerable leverage over industrial organisations in their area of influence and that leverage should be used to reduce manual market distortions.

The analysis applied here to the region can also be applied at the sub-regional level; manual labour is as immobile within regions outside travel-to-work areas as it is between regions, for the same reasons. UDCs (and before them New Towns) are intra-regional policy instruments; and the guidelines set out above are designed to ensure that they do not distort comparative advantage and reflect as far as possible in their planning decisions the preferences of local people – as if there were some viable quasi-auctioning system for consents. The main problem with UDCs in this respect is that they are cut off from the local political process and risk becoming bureaucracies answerable only to a central government which may be nervous of being seen to interfere too directly in the local scene. Designed to be a tool to speed up local development and bypass obstructive local authorities, they may end up frustrating that very development unless closely supervised by that central government.

The risk for the UDC is thus that it may be neither a representative locally-based body nor an effective direct arm of central government. It could then acquire an independent life of its own, motivated by the aims of its bureaucracy; its activities could then be inimical to the interests of the area's residents as well as the aims of central government.

Future policy choices

As we will argue in section IV, the prospects for northern growth and falling manual unemployment are encouraging. Measures taken to date to liberalise the housing and labour markets are having the desired effect in combination with the national resurgence of activity as a result of the general supply side programme. However, there is still room for further measures to reinforce these trends. In this sub-section we pull together our comments on future policy directions to which we have already referred in discussion scattered through the chapter. In spite of the repetition involved such a section should be useful as a focus for policy discussion.

The manual labour market

The present programme must continue vigorously to revive the Beveridge worktest procedures, fraud inspection and training, which should be compulsory for those on the dole with no other job prospect. This programme is succeeding in returning large numbers of unemployed to the official labour market. It is in fact the main instrument for curing unemployment, given that the incentives to remain on the dole, perhaps with some illegal employment, are for manual workers frequently substantial. The example of Switzerland shows that it is possible to have quite generous benefits at the same time as low unemployment, when there is strict screening of eligibility. The new principles must be to (never again being allowed to lapse) 'generously assist those demonstrably unable to help themselves, and be tough in checking on that inability'.

The rented sector needs to be freed as rapidly as possible by pushing council rents to market levels. The working of the new Rent Reform Act needs to be monitored to see that the rent tribunals do not reintroduce restrictions by case law, as was done in the case of licences before; also that registration does not restrict the supply of new landlords willing to take advantage of the new arrangements. Meanwhile, the council house sale programme remains an extremely effective immediate way of creating mobility among an immobile tenant class.

Union power remains a problem, especially in the north. In competitive sectors unions have no leverage. Therefore, one major way to reduce union power pockets in the north is by introducing competition within the protected sector, whether by contracting-out or privatisation or deregulation. Government departments should negotiate locally, insisting on value for money, given local conditions.

The size of the protected sector in the north will increase through market forces as land prices in the south drive these businesses north. But it is useful in all this if government itself reacts to the same market pressures in a normal way. Government departments should evaluate moving decisions in a commercial manner; if it pays them to move north they should do so, but they should not move out of a special desire to help the regional problem, because their activities being very low in manual labour intensity are not subject to distortions of any significance. If they move artificially they will only reduce national productivity.

Transport
Our work shows clearly that the north suffers from a locational disadvantage for transport reasons. However, those reasons are made not by nature or technology mainly, but rather by monopoly practices.

The dock labour scheme has raised the cost of short sea cargo for the north, because northern ports have been predominantly scheme ports, with little competition from non-scheme ports, owing apparently to the strength of the Transport and General Workers Union in the north. Their high labour costs may well have been matched by poor management practices, as the scheme reduced competition which in turn protected the latter.

In deep sea shipping the north has suffered from a combination of uncompetitiveness, again from the dock labour scheme, and the existence of shipping cartel practices which continues. These latter currently subsidise northern traffic, awarding the same rates as in the south, which eliminates any disadvantage in deep sea transport of northern non-port industries in the short run. However, it prevents the development of northern deep sea ports, themselves a potential growth point; and in the long term by preventing this development it frustrates possible rival and cheaper deep sea routes that could be used by northern business.

To illustrate the effect of this combination of scheme and shipping cartel, it is instructive to imagine what a Felixstowe operation in Liverpool free to bid for shipping lines plying the world trade routes to Europe might achieve. With efficient rail transport for those small parts of their cargoes that are to be transshipped to the Continent, a number of lines could be attracted to set up a service bringing mainly northern and Midland cargo to Liverpool, saving time on steaming to the Channel, possibly Channel congestion, and of course the road/rail mileage to and from the Channel port northwards. The port would revive and the users would receive a service cheaper than their current one.

The landbridge from the north through the Channel Tunnel to the

Regional Policy and Market Forces: Model and Assessment

Continent and back is an idea that has been studied by various bodies currently. At present it is handicapped by the two monopoly practices of scheme and shipping lines just discussed. The lines will not divert from their collectively-convenient three or four-port run down the Channel; they would rather divert the northern and Midlands traffic to those ports than steam up to Liverpool, where in any case costs are higher. This implies that as far as deep sea traffic is concerned the Channel Tunnel will do little under present institutions. Rotterdam will not take extra cargo and send it north by rail, simply because the ship can more cheaply drop it in Ipswich or Portsmouth. Liverpool will not send cargo by rail to the Continent because the shipping lines find it cheaper to send it direct through Rotterdam.

The key to change is again competition. In this case besides the scheme and the shipping lines, rail competition will be necessary to improve the capacity to serve the landbridge innovation. The point is that once a port attracts a ship with its major cargo for north/Midlands use, the landbridge becomes attractive for marginal offloads, to avoid expensive extra steaming south.

The achievement of competition in these areas is within the power of the government. The scheme has now been abolished. Our analysis suggests that abolition will significantly contribute to northern revival in the manual/manufacturing areas that are its comparative advantage.

The shipping conferences' practices in the UK are within the jurisdiction of the DTI and could be investigated by the Director of Fair Trading, for possible referral to the Monopolies and Mergers Commission (MMC). British shipping firms breaching future MMC guidelines could be disciplined and British customers of foreign shipping firms in breach likewise.

Rail is under consideration for privatisation. Various sources of competition can be envisaged, including allowing track on different routes to be owned by different operators and allowing different services to operate on the same track. There do not appear to be economies of scale in rail, but rather economies of scope (more trucks and carriages per engine), which argues for small companies and diversity of competition.

Finally, in the transport area there is air, where the development of Manchester airport has been held up by British Airways's desire to push traffic via its Heathrow hub, together with the British Airports Authority's (BAA) interest in the same thing (Manchester being owned not by BAA, but by Manchester City Council). While this is more of an irritant than a major factor in northern development, competition here too could help usefully to lower executive costs in the north.

Planning

It is sometimes not realised that planning consents are a sort of market in property rights in the environment. These rights cannot be individually allocated without prohibitive cost (imagine a developer negotiating individually with 100,000 residents affected by a large new industrial project); the local authority is a fairly cheap intermediary who releases these rights in return for expected benefits to its voters.

The general voter, not in the area, may also have rights related to a development; it may be for a strategic resource or a national park, for example. But mostly the benefits and costs to the general public are reflected in the market prices faced by the developer. It is locals' rights that are not reflected, except in the planning process.

Viewed in this way, it becomes apparent that there is nothing wrong in principle with planning controls. Recalcitrant residents of Crawley, for example, who object to a new factory close to their houses and lobby local politicians are indulging in part of the political-market process necessary to balance gains to the developer (and his general private clients) against losses of environmental rights. If the planning process is working efficiently, the development will proceed to the point at which the marginal gain to the developer just equals the marginal loss to residents; this will happen if the developer has some mechanism to compensate the losers, for example by offering some 'planning gain' to the local community in addition to any gains in jobs and business spin-off that will occur naturally with the project. (The Coase theorem (Coase, 1960) on property rights states that there will be efficient resource allocation, provided the rights are owned by someone or somebody, leaving others to trade with that owner; of course governments in practice like the ownership and hence enjoyment of its fruits to go to those who legitimately have the right. But from an allocative point of view this does not matter. The present vesting of rights in local residents through their authority seems roughly to satisfy legitimacy, as well as being adequate for allocative efficiency.)

In a pure free-market system this could be achieved by an explicit auction of planning consents, with the proceeds being earmarked for the community coffers. This is politically unlikely to be feasible, as it could be too easily misunderstood and seen as bribery – whether it would actually make bribery of officials easier, if large amounts of money were changing hands, is debatable, as the present system is itself vulnerable and open auctions would be more easily monitored than the present committee processes.

Accepting that such an explicit system is unlikely, we can still look for ways of making the present system work better, by making planning gain a more flexible negotiating instrument. Rules could be drawn up for a wider class of acceptable planning gain transactions, with even particular types of cash transfer – eg to subsidise a particular local service – permissible when carefully monitored.

The benefits from moving cautiously along this road would be that it would open up new land for development without as much political expenditure of capital by central government against local resident lobbies. In effect the market would perform its usual anonymous solvent role, luring the residents into agreement in cases where the rewards were large and could be shared by the developer. With the market operating more flexibly, this market test of development would obviate awkward central decisions about how much land to release in the south by dictat. Thus, it would have a double advantage to the centre – saving both management time and political capital.

In what follows we shall assume that some such quasi-market system is in place and will attempt to judge what its pressures will produce in the way of land supplies in the north and south. To anticipate somewhat, it is fairly apparent that extra land in the south will be increasingly expensive, whereas in the north large supplies are available at low price. Hence, the major future expansions of the economy are likely to be located in the north. We now turn to a more detailed account of these future tendencies.

IV Current and future regional trends

The model used here and set out formally in Appendix 1 has not been estimated econometrically (though we have estimated a sub-regional model for Merseyside to North Wales, some notes on which are attached in Appendix 3). The assessment that follows is therefore essentially qualitative, based on the preceding analysis and our judgements of actual and likely policy developments and of trends in other underlying causal processes. The one set of quantitative model-based projections we do have and use are the forecasts to 2000 of the Liverpool World and UK Models; however, we do also add some quantitative projections or guesses of regional trends to illustrate and bring into focus our qualitative discussion of regional prospects. The forecasts were made in early 1989; and the short-term details would alter somewhat if redone today. However, the trends would not alter and so, though not updated, they still represent our basic views.

Reducing Regional Inequalities

Table 3.16 *Central and pessimistic forecast UK and world economic forecast (% pa)*

	\multicolumn{5}{c}{Central}	\multicolumn{4}{c}{Pessimistic}
	1988	1989	1990	1990-5	1995-2000	1989	1990	1990-1995	1995-2000
Group of Seven:									
GDP Growth	4.0	3.3	3.1	3.1	3.2	3.3	0..4	2.1	2.1
Inflation	2.9	3.4	2.6	2.4	2.3	3.4	4.9	2.7	2.7
World Trade Volume Growth	5.5	4.5	4.2	4.3	4.4	4.5	0.4	3.0	3.0
UK									
GDP Growth	5.4	3.8	3.7	4.2	4.1	0.9	2.5	2.5	2.5
Inflation	5.0	4.0	2.8	2.2	0.6	6.8	5.2	4.0	3.3
Unemployment (end of period)									
– million	2.1	1.6	1.2	0.6	0.6	1.8	1.6	1.0	0.6
– rate (%)	7.8	5.7	4.2	2.0	2.0	6.4	5.6	3.5	2.0

Source: Liverpool Macroeconomic Research Group

Our central forecasts for the world and UK economies are shown in Table 3.16. These show growth continuing at around 3 per cent for the OECD as a whole and rather more than this for the UK, where supply side reforms are expected to proceed, maintaining the higher productivity growth seen since 1979. UK unemployment drops continuously and by the mid-1990s the rate has reached the percentages we last saw in the early 1960s.

We also consider a pessimistic alternative, in which there is a prolonged world slowdown in the early 1990s as a result of trade tensions. But the discussion of regional prospects assumes the central case; modifications for the pessimistic case can be deduced easily enough.

With land already in short supply in the southern regions and much the same being true of manual labour, the strong growth we see in the UK economy can only be satisfied if it spreads to the north where availability of both is plentiful. Land supplies in the north will be activated by rising land prices; in any case so great is the availability of land that supply is likely to be highly responsive to quite small rises in price.

As for labour, underlying our optimism on unemployment is our view that manual workers will be compulsorily retrained under the Restart/ET programme and that their productivity will thereby rise, increasing the gap between wages and benefits. With benefits being held constant in real terms and eligibility tests being tightened, as well as tougher policing and penalties on fraudulent claiming, the effective gap will widen further still. Finally, we expect the growth in productivity to average 3–4 per cent per annum, steadily increasing the gap year by year as open sector manual wages rise in line with this productivity trend.

This picture is not fundamentally modified by our expectation that manual labour will gradually become more mobile with the greater flexibility in the rented sector and increasing home ownership. As this happens, manual wages will tend to be equalised in real terms across the country, just as non-manual wages are now; they will hence in the very long term cease to be regionally avoidable costs. Ultimately, land prices will take on all the north–south locational (transport) differential; only the differential will be wider because it is now on a narrower portion of value added.

So, in short, as manual labour joins the migratory business, the pressures on firms to relocate in the north will remain as intense because the resulting increased land price differential will offset the elimination of the manual real wage differential.

Our expectation is that some but not massive migration by manual workers will take place because of southern land shortages. Instead, firms will migrate northwards to take advantage of cheaper land and also (because of lower housing costs) cheaper wages gross of housing costs.

Rising productivity across the whole economy will increase the price open sector firms will pay for land in all parts of the country, but the land supply response will be limited in the south, and very substantial in the north. Hence, in this longer perspective where manual labour is mobile, land will still continue to direct growth northwards.

It may be asked whether the supply of land in the south could not be more responsive to price. This depends on the flexibility with which the planning consent price or quasi-price mechanism is allowed to work. The way to achieve maximum supply response is to permit a high degree of flexibility as discussed above. Then local residents will be highly conscious of the financial advantages of development as well as its environmental disadvantages; some trade-off will be achieved which must be more favourable to development than, when as now, little financial advantage is perceived by the average resident.

Even assuming the greatest flexibility however, it is our view that the southern response must be considerably smaller than the northern. The sort of large-scale expansion required by our forecasts simply does not seem to be capable of accommodation by the southern planning system, given the sort of environmental lobbying and revealed preferences seen so much in recent years by southern residents.

Reducing Regional Inequalities

There is another element that in our view will speed up the northward migration of business. This is the spread of competition in transport, especially the ports. The dock labour scheme being abolished and a recent Department of Transport (1986) paper pointed to increasing cost consciousness by shippers which could put pressure on northern port managements to innovate. Certainly, our researches seem to show that there is no reason (except lack of competition) why northern ports should not compete with southern on deep and short sea routes. Such competition could erode to disappearance the transport differential which has driven the wedge in our story between north and south open sector wages and land prices. The implication is that business would be willing to pay more for northern locations, increasing labour and land supplies yet further in these areas.

Turning to the structure of regional economies, the two main questions are the split between manufacturing and services and the size of the protected sector. Relative to the south, the north will continue to be dependent on manufacturing because of its large stock of manual labour. Though services have grown faster than manufacturing in the north as well as in the south, this has reflected the general decline in the manufacturing terms of trade producing a general decline in the wages the open sector could pay for manual workers; hence a general contraction of manual employment across the country. In some parts of the north this produces disproportionately rapid growth in services because it is from a low base.

The point remains that relatively the north is manufacturing-intensive; that is its comparative advantage. As the manual unemployed are coaxed, pressured and retrained back into employment that comparative advantage will be enhanced. The archetypal vision of the northern renaissance is Nissan in Tyne and Wear, Sony in South Wales, and IBM in Greenock.

This underlines the dangers in UDCs using public money to excite property developers into building yet more shops, offices, leisure facilities and houses, if these are to be at the expense of bedrock industrial development which is the north's specialism. UDCs' activity must go with the grain of comparative advantage, otherwise they will simply hold up the developments which would have gone ahead without them.

The size of the protected sector in north and south will depend quite substantially on the reaction of government agencies to relative land prices. Most other parts of the protected sector are fairly closely tied into the size of other sectors, obeying the rule we enunciated earlier that the protected sector will grow in proportion to the rest.

Table 3.17a Regional profiles: North

	1979	1983	1987	1990	Central 1995	2000	1990	Pessimistic 1995	2000
Manufacturing % of GDP	32.5	28.7	27.2	27.0	27.0	27.0		(as for central)	
Services	31.4	32.3	36.6	36.0	35.0	34.0			
Protected Sector	36.2	39.0	36.2	37.0	38.0	39.0			
Real GDP (1980=100 Prices)	102.00	101.80	113.80	128.0	159.0	197.0	126.0	143.0	164
Manual Employment	754,840		622,126						
Manual Unemployment (%)	12.4		20.8	12.0	5.0	2.0	15.0	8.0	2.0
Non-Manual Employment	544,988		643,557						
Non-Manual Unemployment (%)	2.1		4.5	3.0	1.5	2.0	3.5	2.0	2.0
Land Supply									
Housing (stock of dwellings '000)	1,183	1,219	1,246						
Industrial (million sq metres)	15.0	14.3	14.3						
Commercial and Distribution (mill sq m)	13.9	15.2	16.0						
Agricultural (thousand hectares)	1,060	1,049	1,047						
Earnings (housing adjusted)									
Manual (SE=100)	100.8	85.5	95.0						
Non-Manual (SE=100)	100.6	96.6	97.4						
Land Price in Real Terms[1]									
Housing	72.3	56.0	100.5						
Agricultural	2.16	1.92	1.54						
Population (millions)	3.13	3.10	3.08						
Civilian Labour Force (millions)	1.44	1.37	1.44						

[1] 1980 prices, £000 per hectare

Reducing Regional Inequalities

Table 3.17b *Regional profiles: Yorkshire and Humberside*

	1979	1983	1987	1990	Central 1995	2000	1990	Pessimistic 1995	2000
Manufacturing % of GDP	31.9	24.9	25.0	25.0	25.0	26.0		(as for central)	
Services	32.8	35.6	40.4	40.0	39.0	38.0			
Protected Sector	35.3	39.5	34.6	35.0	36.0	36.0			
Real GDP (1980=100 Prices)	102.8	101.5	118.4	134.0	167.0	213.0	132.5	151.2	177.0
Manual Employment	1,278,516		1,051,000						
Manual Unemployment (%)	8.0		16.3	9.0	4.0	2.0	12.0	8.0	2.0
Non-Manual Employment	854,720		1,027,750						
Non-Manual Unemployment (%)	2.2		5.2	3.0	1.0	2.0	4.0	2.0	2.0
Land Supply									
Housing (stock of dwellings '000)	1,857	1,914	1,956						
Industrial (million sq metres)	31.6	29.8	28.4						
Commercial and Distribution (mill sq m)	24.5	27.2	28.6						
Agricultural (thousand hectares)	1,108	1,097	1,105						
Earnings (housing adjusted)									
Manual (SE=100)	100.4	100.4	100.4						
Non-Manual (SE=100)	101.6	97.6	97.0						
Land Price in Real Terms[1]									
Housing	48.63	65.60	77.30						
Agricultural	2.60	2.38	2.04						
Population (millions)	4.92	4.91	4.90						
Civilian Labour Force (millions)	2.25	2.22	2.32						

[1] 1980 prices, £000 per hectare

Table 3.17c *Regional profiles: East Midlands*

	1979	1983	1987	1990	Central 1995	2000	1990	Pessimistic 1995	2000
Manufacturing % of GDP	33.3	29.4	30.8	32.0	33.0	34.0		(as for central)	
Services	30.0	31.2	35.9	35.0	34.0	33.0			
Protected Sector	36.7	39.4	33.4	33.0	33.0	33.0			
Real GDP (1980=100 Prices)	102.3	102.6	120.4	137.0	172.0	216.0	135.0	155.0	180.0
Manual Employment	963,227		900,139						
Manual Unemployment (%)	6.5		15.0	8.0	4.0	2.0	10.0	7.0	2.0
Non-Manual Employment	724,517		843,046						
Non-Manual Unemployment (%)	2.1		4.4	3.0	1.0	2.0	3.5	2.0	2.0
Land Supply									
Housing (stock of dwellings '000)	1,440	1,503	1,554						
Industrial (million sq metres)	22.1	23.3	23.4						
Commercial and Distribution (mill sq m)	17.5	20.5	21.7						
Agricultural (thousand hectares)	1,247	1,239	1,240						
Earnings (housing adjusted)									
Manual (SE=100)	107.1	101.4	100.6						
Non-Manual (SE=100)	100.1	94.6	95.8						
Land Price in Real Terms[1]									
Housing	42.15	67.60	114.00						
Agricultural	3.07	2.99	2.40						
Population (millions)	3.80	3.86	3.94						
Civilian Labour Force (millions)	1.74	1.77	1.90						

[1] 1980 prices, £000 per hectare

Table 3.17d Regional profiles: East Anglia

	1979	1983	1987	1990	Central 1995	2000	1990	Pessimistic 1995	2000
Manufacturing % of GDP	27.1	24.3	25.8	26.0	26.0	26.0		(as for central)	
Services	35.2	37.2	40.2	41.0	42.0	42.0			
Protected Sector	37.7	38.5	34.0	33.0	32.0	32.0			
Real GDP (1980=100 Prices)	100.2	105.2	132.5	148.0	182.0	223.0	145.0	164.0	186.0
Manual Employment	459,279		418,423						
Manual Unemployment (%)	6.2		13.1	6.0	2.0	2.0	8.0	4.0	2.0
Non-Manual Employment	375,793		487,933						
Non-Manual Unemployment (%)	1.4		3.6	2.5	1.0	2.0	3.0	2.0	2.0
Land Supply									
Housing (stock of dwellings '000)	718	967	808						
Industrial (million sq metres)	8.1	8.3	8.3						
Commercial and Distribution (mill sq m)	11.2	13.1	13.7						
Agricultural (thousand hectares)	1,015	1,013	1,016						
Earnings (housing adjusted)									
Manual (SE=100)									
Non-Manual (SE=100)									
Land Price in Real Terms[1]									
Housing	52.9	55.1	135.5						
Agricultural	3.28	3.07	2.44						
Population (millions)	1.86	1.93	2.01						
Civilian Labour Force (millions)	0.81	0.86	0.99						

[1] 1980 prices, £000 per hectare

Table 3.17e Regional profiles: South East

	1979	1983	1987	1990	Central 1995	2000	1990	Pessimistic 1995	2000
Manufacturing % of GDP	22.2	19.2	17.9	18.0	18.0	17.0			
Services	46.3	48.2	53.1	54.0	55.0	56.0			
Protected Sector	31.4	32.6	29.0	28.0	27.0	27.0			
Real GDP (1980=100 Prices)	103.2	102.5	124.5	137.0	165.0	196.0	135.0	149.0	163.0
Manual Employment	3,479,353		3,112,414						
Manual Unemployment (%)	6.3		12.9	6.0	2.0	2.0	8.0	4.0	2.0
Non-Manual Employment	4,217,946		4,952,517						
Non-Manual Unemployment (%)	1.7		3.9	2.0	1.0	2.0	3.0	2.0	2.0
Land Supply									
Housing (stock of dwellings '000)	6,479	6,686	6,924						
Industrial (million sq metres)	55.1	54.2	52.7						
Commercial and Distribution (mill sq m)	89.4	100.7	105.3						
Agricultural (thousand hectares)	1,736	1,731	1,726						
Earnings (housing adjusted)									
Manual (SE=100)	100	100	100						
Non-Manual (SE=100)	100	100	100						
Land Price in Real Terms[1]									
Agricultural	124.1	159.0	352.3						
Housing	3.15	2.84	3.03						
Population (millions)	16.95	17.11	17.32						
Civilian Labour Force (millions)	8.36	8.50	9.15						

[1] 1980 prices, £000 per hectare

Reducing Regional Inequalities

Table 3.17f *Regional profiles: South West*

	1979	1983	1987	1990	Central 1995	2000	1990	Pessimistic 1995	2000
Manufacturing % of GDP	24.4	21.8	21.7	21.0	20.0	19.0		(as for central)	
Services	37.7	38.4	42.2	44.0	46.0	49.0			
Protected Sector	37.9	39.8	36.0	35.0	34.0	32.0			
Real GDP (1980=100)	101.50	106.90	128.10	143.0	176.0	216.0	140.0	159.0	180.0
Manual Employment	940,498		921,888						
Manual Unemployment (%)	7.6		12.8	6.0	2.0	2.0	8.0	4.0	2.0
Non-Manual Employment	902,779		1,125,766						
Non-Manual Unemployment (%)	2.4		4.1	2.0	1.0	2.0	3.0	2.0	2.0
Land Supply									
Housing (stock of dwellings '000)	1,660	1,763	1,850						
Industrial (million sq metres)	14.8	15.4	15.3						
Commercial and Distribution (mill sq m)	21.8	24.3	25.4						
Agricultural (thousand hectares)	1,857	1,848	1,844						
Earnings (housing adjusted)									
Manual (SE=100)	89.60	89.60	90.10						
Non-Manual (SE=100)	90.50	90.20	90.40						
Land Price in Real Terms[1]									
Agricultural	92.62	95.60	161.65						
Housing	2.703	2.740	2.581						
Population (millions)	4.34	4.42	4.59						
Civilian Labour Force (millions)	1.83	1.90	2.04						

[1] 1980 prices, £000 per hectare

Table 3.17g *Regional profiles: West Midlands*

	1979	1983	1987	1990	Central 1995	2000	1990	Pessimistic 1995	2000
Manufacturing % of GDP	39.4	32.9	33.0	33.0	34.0	35.0		(as for central)	
Services	30.3	32.7	36.4	36.0	35.0	34.0			
Protected Sector	30.3	34.4	30.6	31.0	31.0	31.0			
Real GDP (1980=100)	107.1	99.8	118.4	134.0	169.0	212.0	132.0	152.0	177.0
Manual Employment	1,414,936		1,128,133						
Manual Unemployment (%)	6.8		18.4	13.00	4.0	2.0	14.0	8.0	2.0
Non-Manual Employment	926,307		1,100,835						
Non-Manual Unemployment (%)	2.1		4.6	3.0	1.0	2.0	4.0	2.0	2.0
Land Supply									
Housing (stock of dwellings '000)	1,896	1,961	2,019						
Industrial (million sq metres)	38.9	38.3	36.9						
Commercial and Distribution (mill sq m)	23.1	27.2	29.1						
Agricultural (thousand hectares)	987	980	979						
Earnings (housing adjusted)									
Manual (SE=100)	97.4	91.0	91.4						
Non-Manual (SE=100)	99.6	94.3	95.8						
Land Price in Real Terms[1]									
Housing	94.4	105.3	170.2						
Agricultural	3.07	3.36	2.70						
Population (millions)	5.8	5.18	5.20						
Civilian Labour Force (millions)	2.49	2.47	2.56						

[1] 1980 prices, £000 per hectare

Table 3.17h *Regional profiles: North West*

	1979	1983	1987	1990	Central 1995	2000	1990	Pessimistic 1995	2000
Manufacturing % of GDP	35.2	30.0	29.0	29.0	29.0	30.0		(as for central)	
Services	35.7	37.0	39.9	39.0	38.0	37.0			
Protected Sector	29.0	32.9	31.0	32.0	33.0	33.0			
Real GDP (1985 Prices)	104.0	99.6	114.0	128.0	159.0	197.0	126.0	143.0	164.0
Manual Employment	1,483,842		1,271,077						
Manual Unemployment (%)	10.5		20.0	12.0	5.0	2.0	14.0	8.0	2.0
Non-Manual Employment	1,344,250		1,399,241						
Non-Manual Unemployment (%)	2.3		4.2	2.0	1.0	2.0	3.0	2.0	2.0
Land Supply									
Housing (stock of dwellings)	2,441	2,484	2,533						
Industrial (million sq metres)	47.7	45.4	42.9						
Commercial and Distribution (mill sq m)	35.3	38.2	40.0						
Agricultural (thousand hectares)	459	457	456						
Earnings (housing adjusted)									
Manual (SE=100)	97.7	94.1	93.9						
Non-Manual (SE=100)	98.6	95.5	97.4						
Land Price in Real Terms[1]									
Housing	61.22	92.84	122.85						
Agricultural	2.93	2.78	2.32						
Population (millions)	6.50	6.41	6.37						
Civilian Labour Force (millions)	3.07	2.95	2.94						

[1] 1980 prices, £000 per hectare

Table 3.17i *Regional profiles: Wales*

	1979	1983	1987	1990	Central 1995	2000	1990	Pessimistic 1995	2000
Manufacturing % of GDP	26.4	23.4	26.6					(as for central)	
Services	31.5	32.6	35.6						
Protected Sector	42.1	44.1	37.8						
Real GDP (1980=100)	104.1	105.3	118.3	133.0	165.0	204.0	131.0	149.0	170.0
Manual Employment	620,443		534,393	00.00	00.00	00.00			
Manual Unemployment (%)	9.9		19.9	14.0	5.0	2.0	16.0	9.0	2.0
Non-Manual Employment	510,337		541,585						
Non-Manual Unemployment (%)	2.6		5.0	4.0	1.0	2.0	5.0	2.0	2.0
Land Supply									
Housing (stock of dwellings)	1,057	1,099	1,128						
Industrial (million sq metres)	9.2	n.a	n.a						
Commercial and Distribution (mill sq m)	10.5	n.a	n.a						
Agricultural (thousand hectares)	2,077	1,504	1,515						
Earnings (housing adjusted)									
Manual (SE=100)	99.40	93.40	92.10						
Non-Manual (SE=100)	101.70	96.60	96.60						
Land Price in Real Terms[1]									
Housing	36.6	56.0	50.2						
Agricultural	1.95	1.79	1.68						
Population (millions)	2.81	2.81	2.84						
Civilian Labour Force (millions)	1.23	1.18	1.16						

[1] 1980 prices, £000 per hectare

Reducing Regional Inequalities

Table 3.17j *Regional profiles: Scotland*

	1979	1983	1987	1990	Central 1995	2000	1990	Pessimistic 1995	2000
Manufacturing % of GDP	27.4	23.7	22.2	22.0	22.0	22.0		(as for current)	
Services	36.4	37.6	41.2	42.0	43.0	44.0			
Protected Sector	36.2	38.7	36.6	36.0	35.0	34.0			
Real GDP (1980=100)	104.1	105.3	118.3	133.0	165.0	204.0	131.0	149.0	170.0
Manual Employment	1,206,631		1,030,023						
Manual Unemployment (%)	11.5		21.5	14.0	5.0	2.0	16.0	9.0	2.0
Non-Manual Employment	1,045,740		1,037,045						
Non-Manual Unemployment (%)	2.8		6.0	4.0	1.0	2.0	5.0	2.0	2.0
Land Supply									
Housing (stock of dwellings '000)	1,973	2,005	2,062						
Industrial (million sq metres)	na	na	na						
Commercial and Distribution (mill sq m)	na	na	na						
Agricultural (thousand hectares)	5,546	5,472	5,375						
Earnings (housing adjusted)									
Manual (SE=100)	96.40	94.30	89.80						
Non-Manual (SE=100)	96.80	95.80	94.10						
Population (millions)	5.20	5.15	5.11						
Civilian Labour Force (millions)	2.43	2.40	2.42						

Table 3.17k *Regional profiles: Northern Ireland*

	1979	1983	1987	1990	Central 1995	2000	1990	Pessimistic 1995	2000
Manufacturing % of GDP	22.4	18.7	18.4	18.5	19.0	19.0		(as for central)	
Services	33.1	32.6	34.0	33.5	33.0	33.0			
Protected Sector	44.5	48.8	47.8	48.0	48.0	48.0			
Real GDP (1980=100)	103.8	105.2	122.6	138.0	173.0	215.0	136.0	156.0	179.0
Manual Employment	293,175		249,023						
Manual Unemployment (%)	9.8		28.4	24.0	9.0	4.0	26.0	15.0	8.0
Non-Manual Employment	257,707		295,629						
Non-Manual Unemployment (%)	3.2		3.9	3.0	1.0	2.0	3.5	2.0	2.0

Notes:
1. Agricultural prices, 1987=Sept 1986. Stock of Houses, 1987: Dec 1986, Industrial and Commercial floor space 1987: 1985.
2. LFS Manual unemployment figures adjusted by adding in 'not applicable' unemployed who appear from analysis of 1979 data tapes to be predominantly long-term unemployed and/or without educational qualifications.

Source: CSO, Regional Accounts

Reducing Regional Inequalities

This is not true of many specialised parts of government which can perform their functions anywhere, especially with modern communications. Dispersion here is already widely practised (Inland Revenue, HM Customs & Excise for example), but as government hives off more functions, the economic advantages of relocating may be more widely perceived. The appropriate policy, as discussed earlier, is to decide in a purely commercial way, weighing up disadvantages of distance from clients for example against the cost savings in staff (lower housing costs mean that non-manual too are cheaper) and rents. We expect some relocation from this source.

A further point which bears on the overall size of the northern regional economies is the location of those business elements in the open as well as protected sectors that can be detached from the operating end of the business. The obvious examples are company headquarters and R&D units. In large companies with operating activities widely dispersed there is a large choice of locations; these will be weighed up much as by government departments above (an example recently of such a move was the relocation of Shell Chemicals headquarters to Chester). For convenience, we include these effects within the protected sector. They are likely to boost the size of the northern regional economies to some degree.

In what follows we set out for each region our best guesses based on these arguments of how its key economic indicators will evolve under our central forecast. In a separate box we also supply the equivalent figures on the pessimistic variant; these show that the basic trends are unchanged but merely work themselves out at a slower pace (see Tables 3.17 (a)-(k)).

Bibliography

Coase, RH (1960), 'The problem of Social Cost', *Journal of Law and Economics*, 3, p1.

Department of Transport (1986), 'Liner Shipping and Freight Rates - A United Kingdom and Northern Europe Comparison,' London.

Layard, R, Metcalf, D and Nickell, S (1978), 'The Effect of Collective Bargaining on Relative and Absolute Wages', *British Journal of Industrial Relations*, November, 16, pp287-302.

Metcalf, D (1988), 'Trade Unions and Economic Performance: The British Evidence', Centre for Labour Economics, Working Paper No 320, LSE.

Minford, P, Davies, D, Peel, M and Sprague, A (1983), *Unemployment - Cause and Cure*, 1st ed, Martin Robertson, Basil Blackwell, Oxford.

Minford, P, Ashton, P and Peel, M (1987), 'The Housing Morass', IEA, London.

— (1988), 'The Effects of Housing Distortions on Unemployment', *Oxford Economic Papers*, 40, pp322-45.

Appendix 1
The Regional Model – A Technical Description

This appendix is intended to explain the model in technical terms; the main text contains a full intuitive explanation of its workings, which is quite adequate for discussion of its policy implications.

The regional model used in this chapter, which is an elaboration of the model in Minford *et al* (1988) can be thought of in two stages. First, there is the determination of the open sector's rewards for the regionally immobile factors of production, land and manual labour; also the structure of the open sector in each region will depend on the interaction of these rewards and the regional supplies of these factors. This part of the model is illustrated in Figure 3.App 1.1. Second, there is the size of

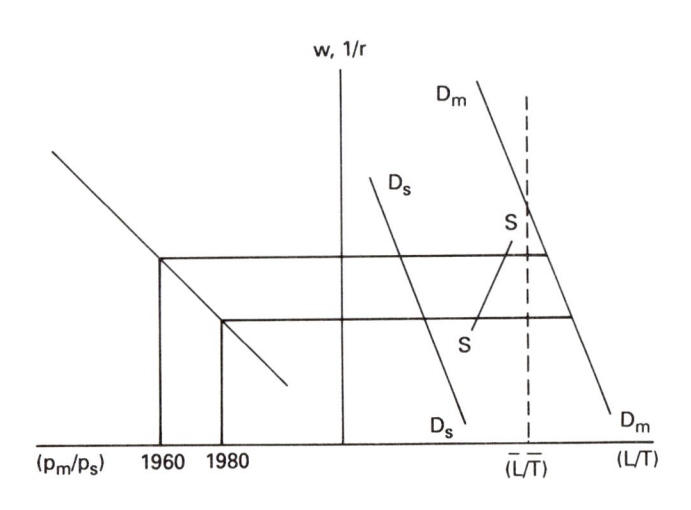

$p_s(p_m)$ = price of regionally avoidable value added in (manufacturing) services; w = real non-union wage for manual work; r = rental on industrial/housing land; L = manual employment ($\bar{L}$ = labour force); T = Land use ($\bar{T}$ = available land with planning permission)

Figure 3.App 1.1 *The structure of the traded goods sector in a region*

Regional Policy and Market Forces: Model and Assessment

the protected sector and of overall regional output and employment, including the open sector. This part is illustrated in Figure 3.App 1.2. The whole model is also set out in algebraic form in Table 3.App 1.1.

Figure 3.App 1.1 is an adaptation of the Harrod-Johnson diagram for the Heckscher-Ohlin trade model. As the 'regionally avoidable' value added in manufactures falls relative to services, so does the manual wage fall absolutely while the price of land goes up; this is because manufactures demand (right-hand quadrant) a higher ratio of labour to land at all factor prices. The relative supply of labour to land is shown by the SS curve. As wages fall and land prices rise the supply of labour drops while that of land rises, causing a contraction in manufacturing in all regions and a rise in services.

This diagram explains the structure of the open sector. But, in addition, there are forces affecting the total size of the open sector. Transport costs cause the absolute amount of regionally avoidable value added to fall in the northern region; hence both manual wages and land prices fall by an equal proportion in that region from this cause, which is to be added to the previous effects from moving relative prices.

This is captured on Figure 3.App 1.2 in the left-hand quadrant by the

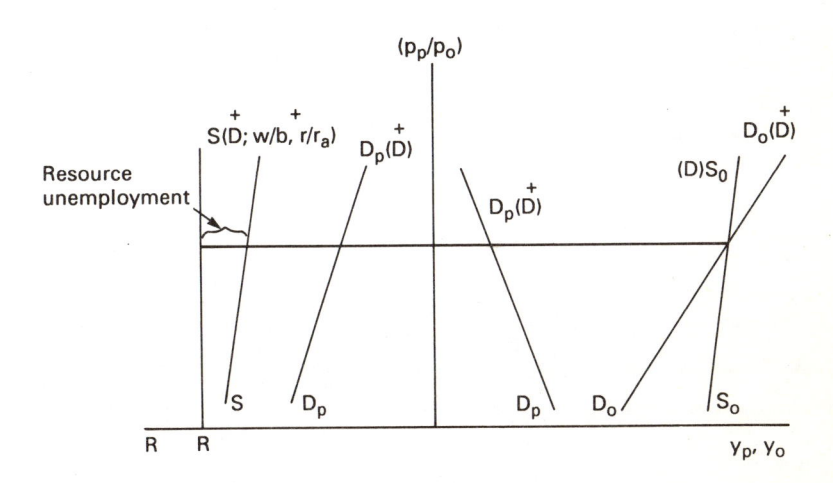

$p_p(p_o)$ = price of protected (open) goods; y = output; R = supply of regionally available resources (manual labour and land); $\bar{R}$ = available supply

Figure 3.App 1.2 *Equilibrium size (and structure between protected and open sectors) of typical region*

167

Reducing Regional Inequalities

SS curve which shows the overall supply of regionally immobile resources in the region. This curve shifts inwards as wages fall relative to benefits and as land rentals fall relative to reserve rentals of agricultural use. It also shifts outwards as total regional demand rises, because this raises demand for protected sector output which translates directly into higher protected output and jobs.

The curve is a declining function of the ratio of protected to open prices because this ratio reflects the monopoly power of local unions and land suppliers *vis-à-vis* their protected sector customers; the higher this ratio the less will those customers demand protected output for a given level of overall demand. As protected output declines, some of the resources laid off that were willing to work for the rewards in that sector will be unwilling to work at the lower rewards in the open sector.

The DD curves in the figure show demand for open and protected sector output in the right-hand quadrant; these both shift outwards with higher aggregate demand. Next, there is the derived demand for labour in the protected sector, DD in the left-hand quadrant, also shifting with aggregate demand.

Finally, the supply of open sector output, SS in the right-hand quadrant, is derived as the output produced with the labour supply not used in the protected sector; this is upward sloping because as relative protected prices rise the total supply of labour falls but less than the fall in the demand for labour in the protected sector. This SS curve shifts inward as aggregate demand, and so also protected labour demand, rises – for the same reason.

The diagram shows how aggregate demand and output is set in

Table 3.App 1.1 *The model algebra*

Regional factor demands:

(1), (2)
$$\begin{bmatrix} \log w^i \\ \log r^i \end{bmatrix} = \begin{bmatrix} \theta_{LM} & \theta_{TM} \\ \theta_{LS} & \theta_{TS} \end{bmatrix}^{-1} \begin{bmatrix} \log \bar{p}_M - (1-\theta_{LM}-\theta_{TM}) \log \bar{c}_M - \log \bar{t}_M^i \\ \log \bar{p}_S - (1-\theta_{LS}-\theta_{TS}) \log \bar{c}_S - \log \bar{t}_S^i \end{bmatrix}$$

The regional protected price:

(3) $\log p_P^i = \theta_{TP}(\log r^i + \bar{u}_T^i) + \theta_{LP}(\log w^i + \bar{u}_L^i) + (1-\theta_{TP}-\theta_{LP}) \log C_P + \bar{t}_P^i$

Regional factor supplies:

(4) $\log L^i = \log \bar{L}^i + l_1(\log w^i(1-\bar{t}_i) + \log \bar{b}) + l_2 \log L_P^i$

(5) $\log L_P^i = \log D_P^i - \sigma_P(\log w^i + u_L^i - \log p_P^i)$

Regional Policy and Market Forces: Model and Assessment

Table 3.App 1.1 *(cont)*

(6) $\log L_O^i = (\log L^i - k_L^i \log L_P^i)/(1-k_L^i)$

(7) $\log T^i = \log \bar{T}^i + n(\log r^i(1-\bar{t}_r^i) - \log \bar{r}_Q) + n_2 \log T_P^i$

(8) $\log T_P^i = \log D_P^i - \sigma_P(\log r^i + \bar{u}_t^i - \log p_P^i)$

(9) $\log T_O^i = (\log T^i - k_T^i \log T_P^i)/(1-k_T^i)$

Regional output and demand:

(10) $\log D_P^i = \gamma \log D^i - \eta(\log p_P^i - v_M^i \log \bar{p}_M - (1-v_M^i)\log \bar{P}_S)$

(11) $v_M^i = (\bar{p}_M y_M^i)/(\bar{p}_M y_M^i + \bar{p}_S y_S^i)$

(12), (13)
$$\begin{bmatrix} \log y_M^i \\ \log y_S^i \end{bmatrix} = \begin{bmatrix} \lambda_{LM} & \lambda_{TM} \\ \lambda_{LS} & \lambda_{TS} \end{bmatrix}^{-1} \begin{bmatrix} \log L_O^i + \lambda_{LM}\sigma_M(\log w^i - \log p_M) \\ \log T_O^i + \lambda_{LS}\sigma_S(\log r^i - \log p_M) \\ + \lambda_{LS}\sigma_S(\log w^i - \log p_S) \\ + \lambda_{TS}\sigma_S(\log r^i - \log p_S) \end{bmatrix}$$

(14) $\log y^i = (1-v_P^i)(v_M^i \log y_M^i + (1-v_M^i)\log y_S^i) + v_P^i \log D_P^i$

(15) $\log y^i = (NFI^i)/y^i + \log D^i$

(16) $v_P^i = (p_P^i y_P^i)/(\bar{p}_M y_M^i + \bar{p}_S y_S^i + \bar{p}_P y_P^i)$

Glossary

w = wage of manual labour*
L = quantity of manual labour (also subscript)
r = rate of interest
T = quantity of land (also subscript)
θ_{ij} = share of factor i in unit costs of product j
M, S, O, P = Manufacturing, Services, Open, Protected subscripts
C = Nationally common costs
t = technological and transport effects on costs
u = monopoly elements in land and labour costs in particular sector
λ_{ij} = share of factor i's supply (to open sector) absorbed by sector j
D, D_P = Total Demand, Protected Sector Demand
y = Output
b = Unemployment benefits
p = price*
NFI = net foreign interest payments (and other net transfers/long term capital outflows)

* all prices are deflated by a general consumer price index

Reducing Regional Inequalities

equilibrium – giving the 'natural rates' of output, employment and unemployment of resources. D, aggregate demand, has to settle where demand for open sector output is equal to its supply – the condition of external current account balance in the case illustrated where there are no net interest, profits and dividends, other net transfers or net long-term capital movements. This is assumed to be the normal case for a region – however, the equilibrium is easily adjusted for any such net flows, which are treated here as exogenous (though again it would be possible to endogenise them). The diagram consequently shows the equilibrium state. Protected output equals protected demand, by definition of the sector to include only goods and services which are prevented from being traded by natural or regulative barriers.

The model algebra is set out in the thirteen equations of Table 3.App 1.1. Equations (1), (2), (11), and (12) set out the Heckscher-Ohlin conditions on open sector production and costs, treating the factor cost and use shares as parameters. Constant returns to scale are assumed in this, the grounds being that international competition would have driven all trading firms to select the optimum scale to survive in the world market.

Equations (4)–(9) spell out the labour supply in total, with the split-off to the protected sector being given by its demand (as the more attractive place to work) and the open sector getting the rest.

Equations (3) and (10) are the price and demand functions for the protected sector. The former assumes constant returns to scale for simplicity but nothing hangs on this, though there is plenty of evidence for the UK showing insensitivity of margins to output and so supporting this assumption. Equation (10) is a conventional demand function.

Last, equation (13) states the current balance condition, since the model is being used primarily (and certainly in this chapter) to explain long-term trends or equilibrium states. The balance condition is put as that aggregate demand plus net transfers equals aggregate output; since protected demand equals protected supply this is equivalent to equality of open sector output and demand.

Appendix 2
*United Kingdom Export/Import Transportation Costs**

Introduction

Traffic movements are usually classified according to three different kinds of destination by distance travelled: near sea (eg Continental Europe); short sea (eg the Baltic); and deep sea (eg the Far East, North and South America). There are many costs involved in the movement of any given load of goods within any one of these categories of trip, chief among them being (not necessarily in order of importance) inland haulage costs, port charges (eg handling and pilotage costs as well as light dues), and sea freight rates. The methods used to levy these charges sometimes vary according to whether the traffic handled is either deep sea or near and short sea, according to whether it is carried by either containers or in bulk and/or semi-bulk forms, and according to whether the goods are either dry or liquid. From these variations in charges, there emerges a complex pattern of transportation costs across the whole spectrum of commodities carried throughout the world and it is necessary, therefore, to analyse costs separately by deep sea and by near/short sea in order to isolate the fundamental characteristics of charging mechanisms.

Deep sea traffic

Containers

Dry cargoes are containerised in either 20 or 40 feet units and the trend of container usage is upwards. Containers of either size usually carry up to about 20 tons of goods. Transportation costs at sea and inland are set mainly by the shipping lines by their grouping together into shipping conferences, which cartelise inland haulage costs according to a grid

Reducing Regional Inequalities

system. (There exist some non-conference lines – eg around ten in number operating out of Felixstowe – but these are dominated by the conference lines which fix inland haulage costs within the framework of the grid system). The grid is very complex in structure, but its fundamental effect is to charge an inland haulage cost to either an origin or a destination for a given cargo, irrespective of either entry or departure point out of the UK. For example, a shipper in Manchester might be charged the same haulage rate to Harwich as he would be to Liverpool. Because deep sea traffic is of the long-distance variety, the sea freight rate will have a much smaller relative impact on the shipper's total transport costs than either the inland grid rate and/or the port charges so that, as far as a shipper in the north is concerned, he would aim to minimise these last two charges. In practice, however, his choice of route and, therefore, his costs are controlled by the shipping conferences, because it is they who decide which UK port will be their base. Grid prices are not available for public consumption in any great detail so that Table 3.App 2.1 shows nothing other than a possible favourable effect for inland haulage costs for a northern shipper using a southern port. Not many shipping lines use northern ports as a base so that the overall effect of the grid is to favour southern ports.

Bulks and semi-bulks (dry and non-oil liquids)

On deep sea routes, there is no grid system. Consequently, a shipper of goods can make an effective choice of both inland haulage routes and of port of entry/exit in order to minimise total costs (again, on long-distance routes, sea freight rates can be safely ignored because their variability in terms of port of entry/exit will have a very small effect on total costs compared to both inland haulage and port charges). In theory, a northern shipper could find it cheaper to take goods to a southern port because higher port charges in the north could more than cancel out the smaller inland haulage costs confined to a northern route. In practice, evidence on this question is hard to ascertain, but we have been told by the trade that it is not unknown for a northern shipper to make savings of around £3 per ton by transporting goods to Dover, thence to either Antwerp or Rotterdam for transfer to a larger vessel bound for a long-distance destination, rather than by utilising a northern port in the UK for direct shipment to the ultimate intended destination.

For bulks and semi-bulks, there is the additional consideration that much is carried on smaller ships, which tend to use smaller ports, which in turn tends to favour northern ports other than Liverpool and on

Regional Policy and Market Forces: Model and Assessment

Humberside, and southern ports other than London, Southampton, Dover and Felixstowe.

Table 3/App 2.1 indicates the haulage cost differentials for a northern shipper using a southern port and a northern one, and they show that, for a 20-ton load, port (and other, eg sea rate) charges would need to be about £300 less for a northern shipper to save money by using a southern

Table 3.App 2.1 *Typical northern freight differential*

Inland Haulage by Road

	Deep Sea		Near/Short Sea Typical Tariffs	
	Containers (20'&40')(Lolo)	Bulks/Semi Bulks (Non-Oil)	Containers (20')(Lolo)	Bulks/Semi Bulks (Roro) (Unaccompanied Trailers)
North Grid System – same as for South	£1–£1.50 per vehicle mile (average load = 20 tons) Antwerp/ Rotterdam FOB charges up to £15/ton< British deep sea port. Port nearest to industrial location usually used	Birmingham– Dusseldorf £460/unit (up to 20 tons)	Manchester– Antwerp via South £310/20 ton average load Hull–Zeebrugge £312/20 ton average load
South Grid System – same as for North	same as for North	London– Dusseldorf £215/unit (up to 20 tons)	London– Antwerp via South £210/20 ton average load

Notes: Lolo = Lift On/Lift Off; Roro = Roll On/Roll Off
Acknowledgement: shipping and freight forwarding companies, in particular *Middlegate Shipping and Forwarding*, Knowsley, UK

Reducing Regional Inequalities

port. In practice, for bulks and semi-bulks, their massive volumes mean that the port nearest to the industrial location of production is usually used. Thus, for example, steel output from South Wales is often shipped through either Avonmouth or a southern Welsh port, chemicals produced in the North East are often placed via either Tees port or Immingham, etc. Indications, in our research to date, are that the differential on northern port costs is of the same order as the road transport differential if a load goes to a southern port.

Oil
Oil sea freight rates are fixed according to a world scale. Oil is usually landed at ports which are adjacent to refineries, which are located at strategic regional points within the UK, so that their output can be channelled locally. The oil companies ensure through their swap arrangements that a minimal amount of inter-regional haulage is necessary in order to meet demand. Thus, for example, Stanlow refinery supplies most of the petroleum spirit needs of the North West.

Near and short sea traffic

Containers
There is no grid system. Ferry companies compete against each other, which means that a shipper has a range of alternative available routes at a variety of prices. From our enquiries of the trade, Table 3.App 2.1 gives examples of these for both northern and southern-based shippers. Without a comprehensive survey of all prices, it is impossible to know how representative our examples are, but trade advice suggests that they at least approximate to the average. They show a price differential which favours shippers located in the south.

Bulks and semi-bulks (including liquid non-oils)
The same principles apply as for deep sea traffic, except that sea freight rates are more important because of the shorter distances travelled. Prices quoted in Table 3/App2.1 are typical examples from primary trade sources and, as in the case of containers, they are given as door-to-door prices. There is again a clear-cut advantage for a shipper to be located in the south, and there is quite a large incentive for a northern shipper not to use a northern port for Continental business, because the extra sailing time from Humberside and from Tyne-Tees pushes up the total transport cost.

Oil
The same charging principles apply as for deep sea traffic.

Table 3.App 2.2 *Rail freight rates*

Collection	Dusseldorf 20/40 (L)	Dusseldorf 40 (H)	Delivery Paris 20/40 (L)	Paris 40 (H)	Rotterdam 20/40 (L)	Rotterdam 40 (H)
London	£250	£250	£330	£340	£315	£325
Glasgow	£420	£420	£500	£515	£485	£500
Birmingham	£300	£300	£390	£405	£375	£390

Notes:
(L) Content weight to 18,000 KG
(H)   "   "   18-22,000 KG
Rates are based on using Railfreight owned boxes, are door to door, but exclude customs formalities

Rail freight

The above discussion about inland haulage costs relates to road transport. On near and short sea routes, there exists a similar cost advantage in favour of southern shippers, some typical examples being given in Table 3.App 2.2.

Note:
We gratefully acknowledge assistance from several shipping companies and freight forwarding agents, particularly Middlegate Shipping and Forwarding, Knowsley.

Appendix 3
An Econometric Model of the Merseyside to North Wales Sub-region

£ million, 1985 prices

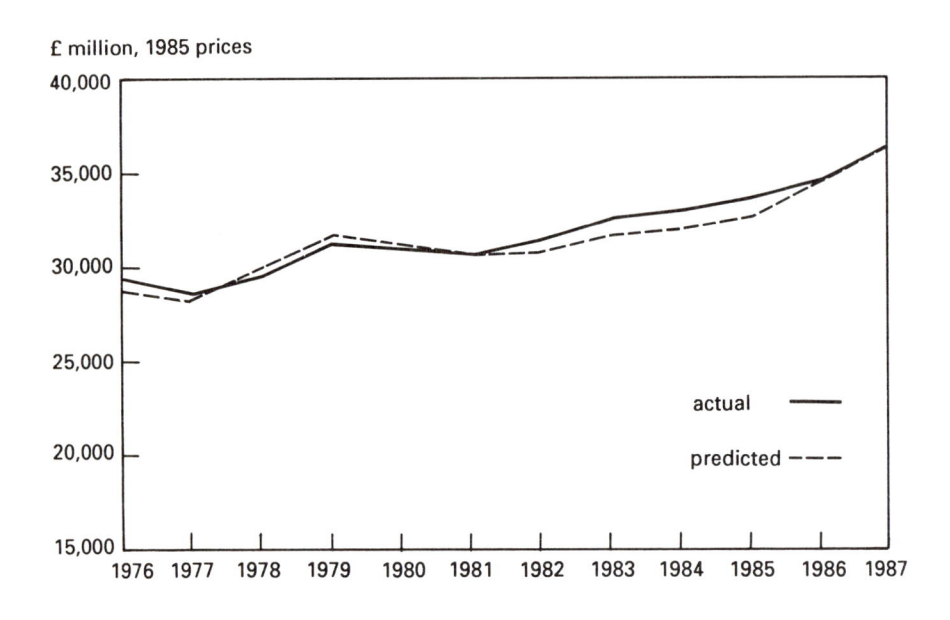

Figure 3.App 3.1a *Actual and predicted consumption expenditure plus manufacturing investment*

Figure 3.App 3.1 *Static predictions over the past: full model*

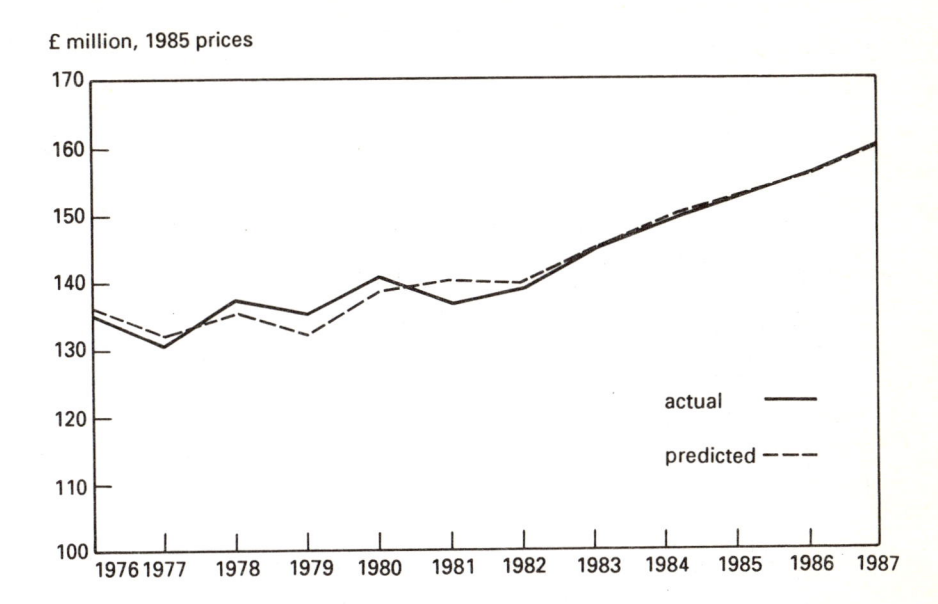

Regional Policy and Market Forces: Model and Assessment

Figure 3.App 3.1b *Actual and predicted average gross weekly wage*

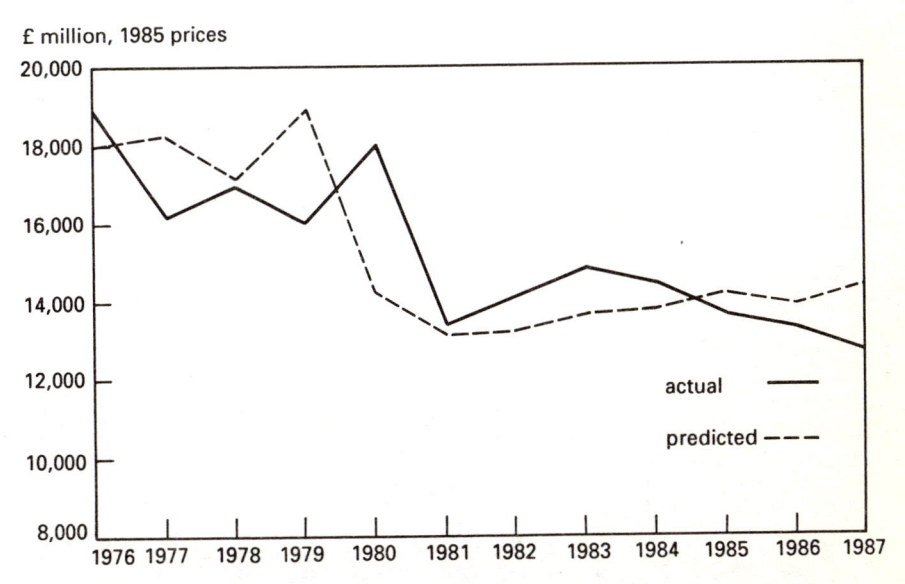

Figure 3.App 3.1c *Actual and predicted manufacturing production (£million – 1985 prices)*

Figure 3.App 3.1 *continued*

Reducing Regional Inequalities

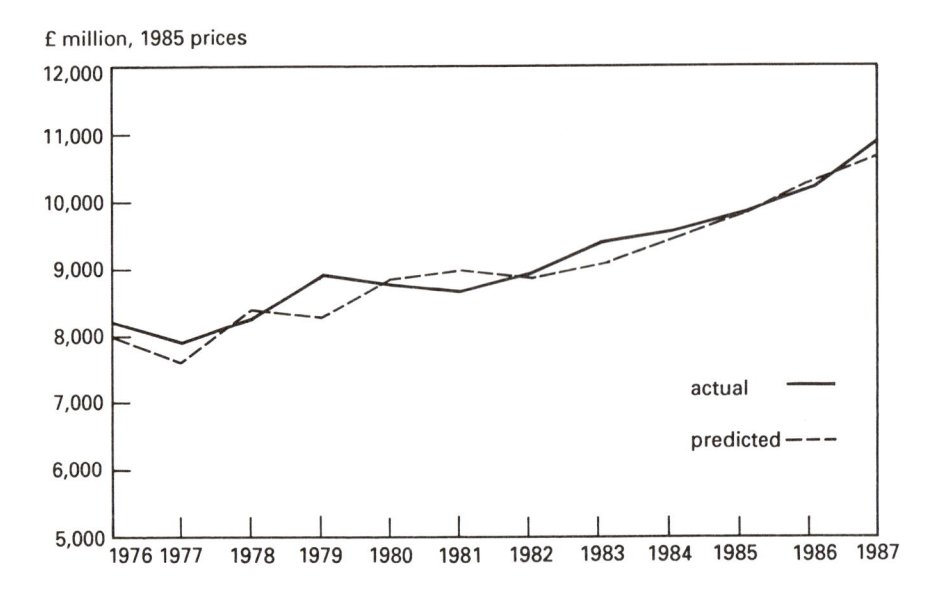

Figure 3.App 3.1d *Actual and predicted non-manufacturing GDP (£million – 1985 prices)*

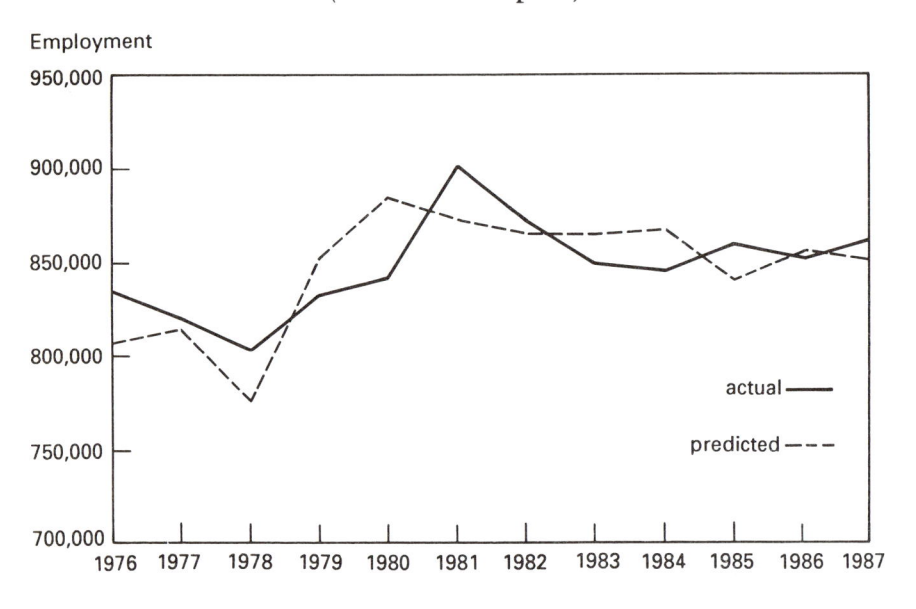

Figure 3.App 3.1e *Actual and predicted non-manufacturing employment*

Figure 3.App 3.1 *continued*

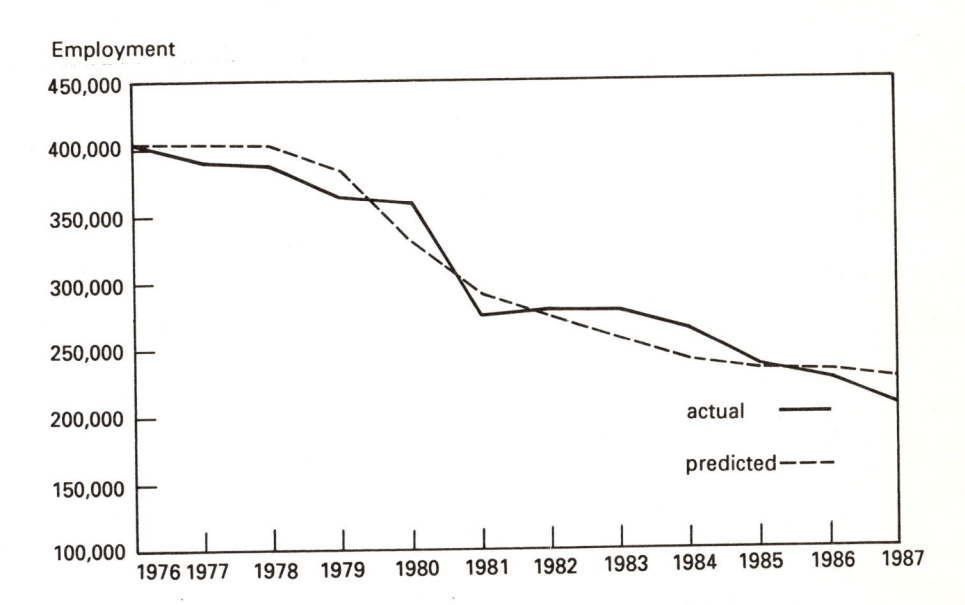

Regional Policy and Market Forces: Model and Assessment

Figure 3.App 3.1f *Actual and predicted manufacturing employment*

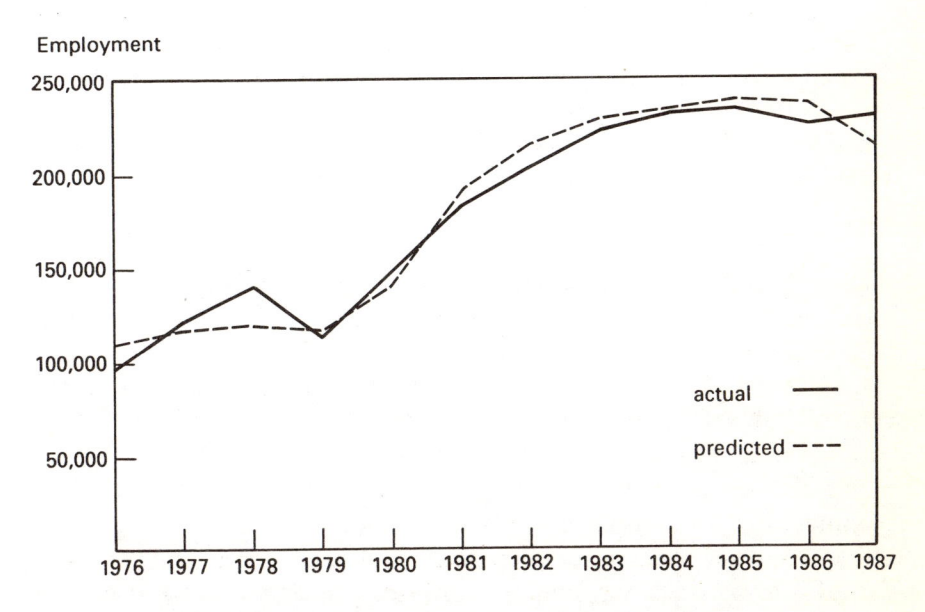

Figure 3.App 3.1g *Actual and predicted unemployment*

Figure 3.App 3.1 *continued*

Reducing Regional Inequalities

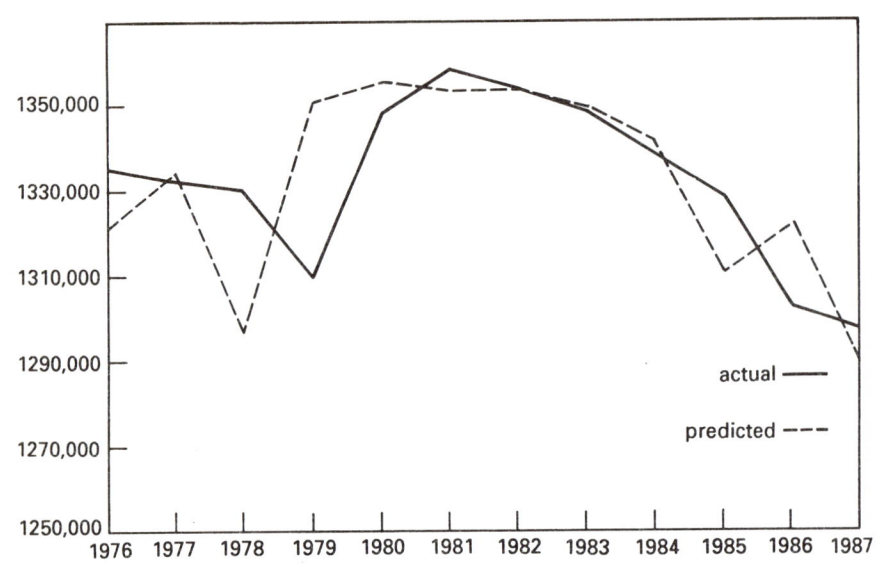

Figure 3.App 3.1h *Actual and predicted working population*

Figure 3.App 3.1 *continued*

This appendix presents a new model of the local Merseyside, Cheshire and North Wales sub-region, stretching from Merseyside through West Cheshire (including Warrington–Runcorn) and Deeside to all of North Wales. The framework of analysis is taken from the main body of this chapter.

In our model of the area we divide the regional economy into manufacturing, which we equate with the traded sector, and the rest, which we treat as the non-traded sector. Though some services in this 'rest' may be traded they are likely to be a relatively small element; tourism, which is a large industry, is strictly speaking internationally traded, but the international competition is rather weak, since the product is so differentiated.

We also assume that planning control does not significantly limit manufacturing expansion in the region. Planning authorities are generally anxious to encourage development because of jobs and revenue, and manufacturing does not require enormous amounts of land. It follows that the key constraint on development of manufacturing is the local manual labour supply (non-manual labour is mobile and imported as needed into the region from elsewhere at the going national price). It is

Regional Policy and Market Forces: Model and Assessment

manual labour on which the model focuses therefore.

The model has been estimated over the period 1976–87 and the equations fit reasonably, though on such a small sample there is inevitably a high margin of error around the estimated coefficients. When the model as a whole is asked to predict the past, it gives encouragingly accurate results, improving substantially on a simple no-change prediction. This suggests it should be useful in forecasting the future.

The model estimates

The estimated equations of the model are shown in Table 3.App 3.1. Because we only have limited data the individual equations are kept as simple as possible to maximise degrees of freedom. The model's starting point is equation 5 for demand in the area, determined by UK demand with an elasticity of 1.3 and a slight downward trend. But in the long run it converges on the equilibrium value of area demand, closing the gap by

Table 3.App 3.1 *The sub-region model (estimated over 1976–87)*

Short-run model

1. $\log(\text{manu}) = -12.5570 + \log(\text{lman}) + 1.9801\log(\text{wman}/\text{pm})$
  D.W.=2.16 $r^2 = 0.70$

2. $\text{Dlog}(\text{lnman}) = \exp[\ 3.3.921 + 0.291494\text{Dlog}(\text{ynman})^{-1} - 0.43194*$
  $\log(\text{w.rpi}/\text{ps}) - 0.27536\log\{(\text{lnman}) - (\text{ynman})\}_{-1}\ ]$
  D.W. = 2.31 $r^2 = 0.57$

3. $\log(\text{ynman}) = -5.1529 + 1.3768\log(\text{dd})$
  D.W.=1.49 $r^2 = 0.88$

4. $\log(w) = 0.075767 + \log(\text{wman}/\text{rpi}) + 0.0061174\text{trend}$
  D.W.=1.52 $r^2 = 0.96$

5. $\text{Dlog}(\text{dd}) = -0.0085012 + 1.2928\text{Dlog}(\text{UKdd}) - 0.13966\{\log(\text{dd}) - \log(\text{dd}^*)\}_{-1}$
  D.W.=1.61 $r^2 = 0.89$

6. $\log(u) = -61.976 - 1.7920\ \{\log(\text{wman}/\text{pm})-\log(\text{ben})\} -$
  $2.1372\log(\text{lnman}) + 8.0413\log(\text{wpop}) + 0.12851\text{trend}$
  D.W.=1.61 $r^2=0.93$

7. $\text{Dwpop} = 54698.0000 - 1043300.0000\ (\text{ur} - \text{UKur})$
  D.W.=2.89 $r^2 = 0.26$

8. $\text{lman} = \text{wpop} - \text{lnman} - u$

Reducing Regional Inequalities

Equilibrium model

1. log(manu*) = − 12.5570 + log(lman*) + 1.9801log(wman/pm)
2. log(lnman*) = (1/0.27536) (3.3921 + log(ynman*) − 0.43194log(w.rpi/ps))
3. log(ynman*) = − 5.1529 + 1.3768log(dd*)
4. log(dd*) = 1.0898 + 0.7043log(ynman*) + 0.2957log(manu*)
5. log(u) = − 61.976 − 1.7920{log(wman/pm)−log(ben)} − 2.1372log(lnman*) + 8.0413log(wpop) + 0.12851trend
6. lman* = wpop − lnman* − u*

Glossary

manu	=	manuf production in region (£mill-1985 prices)
lman	=	manuf employment in region
lnman	=	non-manuf employment in region
ynman	=	non-manuf gdp in region (£mill-1985 prices)
w	=	avg gross weekly real wage in region (all workers)
wman	=	average gross weekly nominal wage in region (manufacturing workers)
dd	=	consumption expenditure + manufacturing investment (£mill-1985 prices)
UKdd	=	consumption + investment + government expenditure (for the whole of the UK)
u	=	unemployment in region
ben	=	UB(1+LO) for the UK
ps	=	price of services (for the UK) 1985=100
pm	=	price of manufactures (for the UK) 1985=100
rpi	=	retail price index 1985=100 (for the UK)
ur	=	unemployment rate in region (for all workers)
UKur	=	unemployment rate (for the whole of the UK)
wpop	=	working population in region

14 per cent each year. This equilibrium value is set by the long-run model. This consists of all the model's other equations, and in place of this one an equation constraining area demand to be equal with area income: therefore in the long-run the area economy settles down at a level where it is neither over nor under-spending.

Area demand then, in equation 3, determines non-manufacturing (non-traded or services) output, with an elasticity of 1.4. (In this latest data set there is no price elasticity with respect to relative traded versus non-traded prices.)

Equation 2 relates the demand for labour in services to the output of services and to service labour costs relative to prices. The elasticity to these costs is set at 0.4 in the short run and 1.8 in the long run, showing a strong effect of higher labour costs on labour demand. The long-run elasticity to services output is constrained to unity, on the assumption of constant returns to scale. In the short run the elasticity is 0.3.

Wages in the area overall are set in equation 4. They are determined by the internationally set real manufacturing wage and a time trend – proxying factors, such as union power, determining the 'mark-up' of service wages over the manufacturing rate. We regard wages in services as being determined nationally, with high mobility of these workers between the area and the rest of the country.

Equation 6 relates the level of unemployment overall in the area to the level of real manufacturing wages compared to benefits (an elasticity of 1.8) and to the demand for labour in the service sector (with an elasticity of 2.1). The latter effect reflects the pull of the non-traded service sector, in which workers are assumed to be willing to work in preference to the internationally constrained manufacturing sector. Demand for labour in the service sector is always able to be satisfied by bidding up wages above the rate in manufacturing.

The amount of labour willing to work at manufacturing wages is affected by their level relative to benefits; if this gets too low, people make little effort to find a job in the manufacturing sector. However, should demand in the service sector pick up, then some more people may be induced to work.

The equation is completed by the area working population, whose decline tends to reduce unemployment, with an elasticity of 8.0 (implying that a fall in the labour force of about 13,000 reduces unemployment by roughly as much); and by a time trend, proxying other factors such as local union power.

This working population in turn – equation 7 – is affected by unemployment in the area relative to that in the UK as a whole. As this goes up, so emigration occurs. A 1 per cent rise in local unemployment (about 13,000) causes emigration of 10,400 a year, around 0.8 per cent of the working population. Putting this effect together with the effect of working population on unemployment in the last equation, we can see that a rise in unemployment is both stimulated and is contained by emigration.

Equation 8 is an identity which deduces from the difference between

Reducing Regional Inequalities

working population and unemployment how much labour supply is available, and deducts in turn that used in services. The remainder is the supply available for and employed in manufacturing.

This employment in manufacturing is then related to manufacturing output and labour costs in equation 1, reflecting the productive conditions there. The elasticity of the output/labour ratio to labour costs is 2.0, again a powerful wage effect. With real wages in manufacturing growing at 0.8 per cent pa over the sample period and relative prices of manufactures falling by 0.2 per cent pa, the elasticity of 2.0 implies that estimated productivity grew at around 2 per cent pa. Actual productivity growth 1976-87 was 2.7 per cent pa.

Forecasting the past

The equations' statistical properties are set out in Table 3.App 3.1 which also lists the full model. In the charts which follow we show how the model performs when it is asked to forecast the past. The forecasting of the past is for each period on the basis of last year's economic outturn and current information about the exogenous variables: that is, the model (all its equations operating together) is asked to 'predict' this year's outcome. Inspection of the charts shows that the model picks up the trends of the past decade reasonably well.

4
Regional Earnings and Pay Flexibility

Janet Walsh and William Brown

I Introduction

The notion that wage 'inflexibility' is a major factor explaining the persistence of regional disparities in unemployment has become an important strand of British government thinking. In particular, national, multi-employer collective bargaining and wages councils are regarded as mechanisms which prevent wages from fully responding to local demand-supply conditions. The objective of this chapter is to discuss changes in the institutional framework of wage determination and the imperatives behind the reorganisation of British pay bargaining arrangements. Our point of departure is the government's claim – made in the 1988 White Paper on Employment – that decentralised bargaining facilitates wage-setting behaviour which is more responsive to competitive conditions prevailing in regional labour markets.

The structure of the chapter is as follows. After a discussion of trends in regional earnings variations, the economic reasoning underlying the government's policy stance is outlined. We then examine survey evidence on pay bargaining in the private sector to ascertain whether local demand-supply factors have been influential in the restructuring of British bargaining arrangements. The final section examines the forces underpinning the increase in the magnitude of the South East pay differential. We focus in particular on the only explicit regional pay differential in the UK – the London allowance.

II Trends in regional earnings variations

An important feature of the behaviour of regional earnings of Britain was the steady convergence in wage rates in the 1960s and early 1970s (Moore and Rhodes, 1981; Lee, 1990). Figure 4.1 charts these trends by

Reducing Regional Inequalities

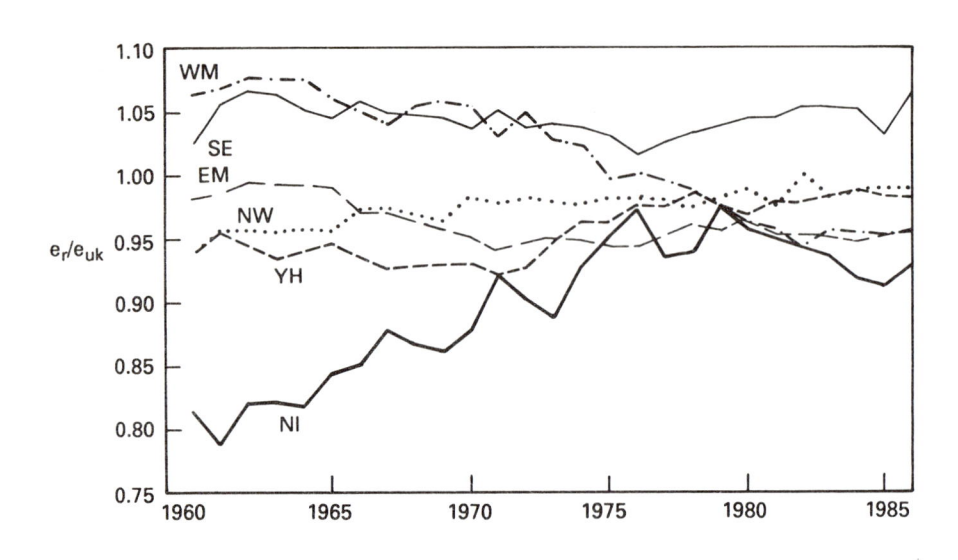

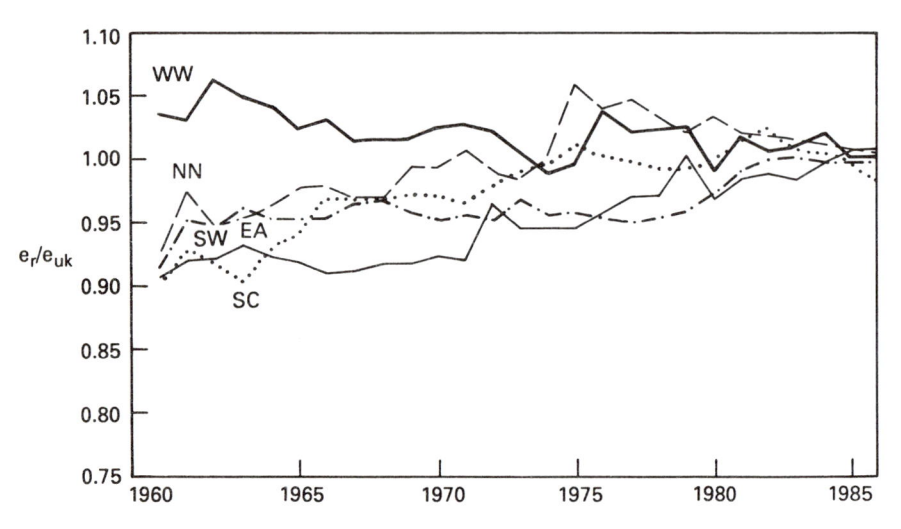

e_r/e_{uk} = Hourly earnings full-time males, manufacturing sector, region r (of 11 standard regions) relative to UK average

Source: Lee (1990)

Figure 4.1 *Regional convergence in earnings, 1960–86*

plotting the ratio of regional hourly earnings to hourly earnings in the UK as a whole for full-time male manual employees in the manufacturing sector over the period 1960–86. The rise in earnings in those regions lying below the UK average, and the relative fall in earnings in those lying above the UK average is illustrated by the convergence of the majority of the ratios to unity over the period 1960–75. These ratios stabilised in the late 1970s, but more recently there is some evidence that regional earnings have been diverging. In particular, there has been rapid growth in the earnings ratio in the South East (Lee, 1990:1).

Table 4.1 uses data from the New Earnings Survey to widen the scope to all adult full-time male employees. Adult male earnings for each region of Britain are expressed as a percentage of the unweighted average of all regions for the years 1970, 1975, 1979, 1985 and 1989. Over the period there has been a considerable reordering between regions. The West Midlands and Wales have suffered a downward slide in earnings relative

Table 4.1 *Average gross weekly earnings of adult full-time male employees by region expressed as a percentage of the unweighted average of all regional averages*

	1970	1975	1979	1985	1989
South East	113	110	110	116	123
East Anglia	93	95	97	99	100
South West	96	95	93	97	100
West Midlands	107	99	99	98	97
East Midlands	97	99	98	95	96
Yorkshire	96	99	100	98	96
North West	102	100	100	100	99
North	98	102	101	97	96
Wales	100	100	99	97	94
Scotland	96	102	102	103	99
Coefficient of variation:	5.5%	4.1%	4.0%	5.6%	7.9%
Coefficient of variation (excluding the South East):	3.7%	2.5%	2.5%	2.1%	2.1%

Source: New Earnings Survey

Reducing Regional Inequalities

Table 4.2 *South East regional earnings differentials, 1970-89*[1]

	1970	1975	1979	1985	1989
London area	116	115	114	121	129
Rest of South East	101	100	100	102	104
Britain	100	100	100	100	100

1. Earnings are expressed as a percentage of the *weighted* national average

Source: New Earnings Survey

to the average; East Anglia has shifted from a low earnings area to, in general, the middle rank of earnings and the South East has moved into a strong lead.

During the 1980s, the magnitude of the South East's earnings lead has increased. In particular, as Table 4.2 demonstrates, the magnitude of the

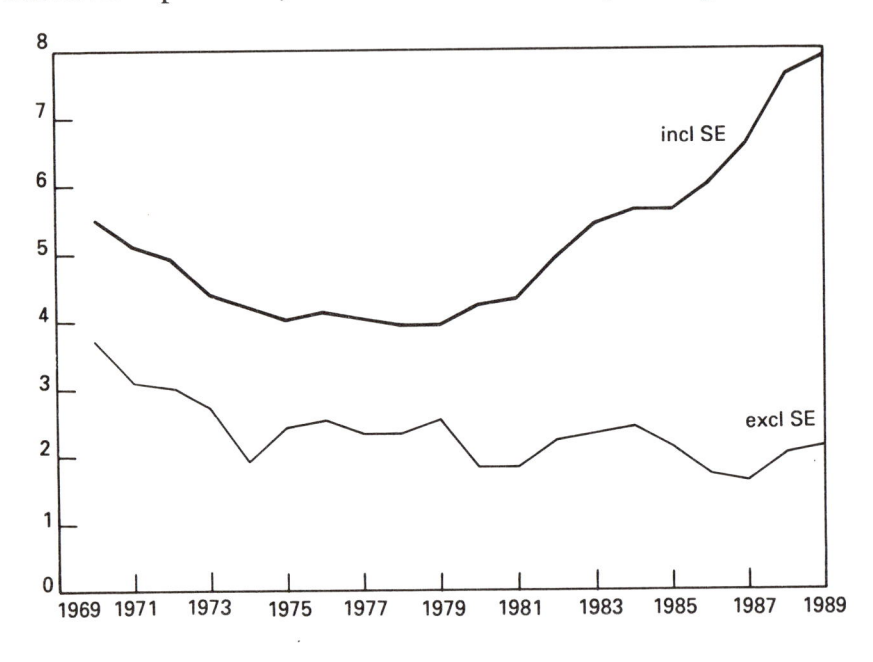

Source: New Earnings Surveys

Figure 4.2 *Coefficients of variation in earnings*

188

Regional Earnings and Pay Flexibility

Greater London differential has greatly accelerated, while the rest of the South East has made a far more modest advance over the weighted earnings average for Britain. Indeed, the South East factor more than explains the increase in inter-regional pay dispersion during the 1980s. Figure 4.2 illustrates this trend by plotting coefficients of variation on a yearly basis over the period 1970–89. If we exclude the South East, it is clear that the inter-regional pay dispersion narrowed in the 1970s and showed no tendency to widen in the 1980s.

Could these trends merely reflect changes in hours worked across regions or shifts in industrial structure irrespective of wage-setting behaviour? Lee (1990) has corrected the earnings series of 1960–86 for the effects of regional factors (see Figure 4.3). His results – which reinforce the earlier work of Moore and Rhodes (1981) – indicate that the larger part of the convergence in earnings has resulted from neither movements in hours worked across the regions nor from changes in industrial structure. According to Lee (1990:4):

> the primary element explaining the convergence is that generated through movements in the 'residual' element across regions, as

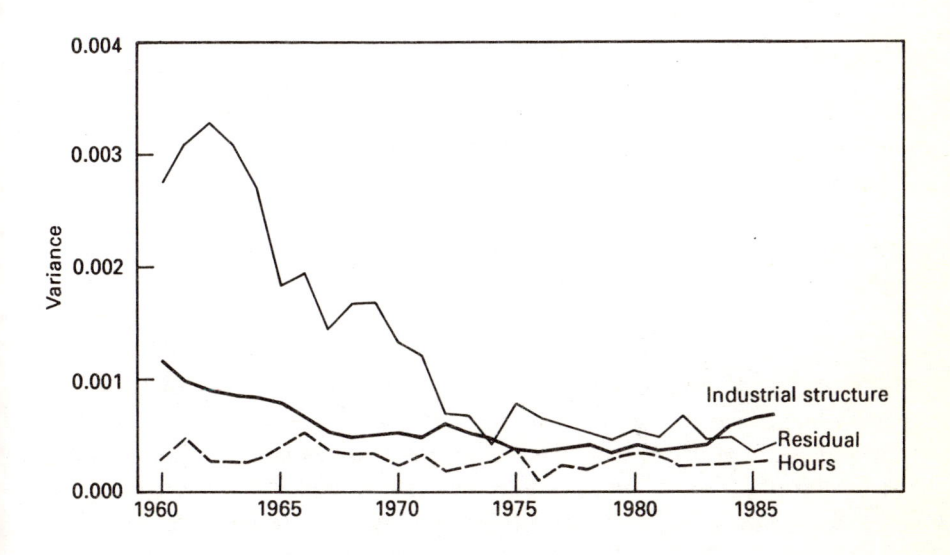

Source: Lee (1990)

Figure 4.3 *Decomposition of variability in relative earnings across regions*

actual earnings in each region have converged to the levels that might be expected, given their industrial mix and given the same extent of overtime working, using average UK rates.

The dominance of the 'residual' term in the reduction in variability across all the regions suggests that wage-setting behaviour has played an important role in the convergence of regional earnings.

III The policy context

The co-existence of a relatively stable inter-regional earnings structure and persistent regional unemployment differentials has been the focus of government concern (cf Department of Employment (1988), *Employment for the 1990s*). The 1988 White Paper on Employment points out that even in regions where the unemployment rate is particularly high, this is not reflected in the level of earnings. An important strand of government thinking is the notion that wage 'inflexibility' underlies regional unemployment disparities. Real wages are prevented from falling in areas of excess labour supply relative to other areas because of 'rigidities' in local labour markets.

A range of institutions allegedly obstruct the smooth functioning of regional labour markets, notably industry-wide collective bargaining arrangements and statutory wages councils. Multi-employer, industry-wide pay agreements, for example, establish wage rates which are insufficient to attract skilled labour in low unemployment areas, and allegedly restrict downward wage flexibility necessary to promote jobs in high unemployment regions. Such institutional 'rigidities' fracture the relationship between pay and local labour market conditions and severely constrict the inter-regional wage structure. Inter-regional labour flows are thereby hindered and geographical inequalities in unemployment tend to persist.

The government's policy response is to facilitate labour market deregulation. This is intended to 'free up markets', improve the functioning of the price mechanism and increase employment. In particular, the government has been prepared to legislate in order to remove any 'rigidities' associated with trade unionism. The most recent statement of labour market policy – the 1988 White Paper on Employment – outlines plans to curtail the closed shop and severely weaken wages councils. As far as wage-setting behaviour is concerned, the White Paper argues thus: 'many existing approaches to pay bargaining ... will need to change if we are to secure the flexibility essential to employment growth' (1988:23-4). The decline of multi-employer pay agreements and the trend towards company and plant level bargaining are cited as 'encouraging' but relatively recent developments. Decentralised pay structures, it is alleged, reflect a greater responsiveness to local labour market conditions.

To what extent are these arguments securely grounded in empirical evidence? Econometric investigations of the relationship between earnings and regional unemployment disparities tend to be inconclusive. The Bank of England (1988) found little systematic response of regional wages to variations in unemployment. In contrast, Blackaby and Manning's (1987) time-series study indicates that both the rate of unemployment and its change affect regional wage growth – after allowing for the effect of national cost-of-living changes – and that this relationship is broadly similar for all regions. Their results also lend support to the hypothesis that wage-inflationary pressures are institutionally transmitted across regional labour markets, irrespective of local labour market conditions.

The analysis by Lee (1990) found some evidence to support the hypothesis that wage determination is responsive, at least to some extent, to local labour market conditions. But his results also indicate that external influences, such as informal comparability mechanisms and national wage agreements, affect wage targets and that these operate independently of competitive conditions prevailing in local labour markets.

Despite these results to the contrary, the government persists in the view that national wage bargaining constitutes a key institutional mechanism which restricts the responsiveness of wages to local demand–supply conditions. Indeed, the notion that co-ordinated, multi-employer bargaining constricts the inter-regional wage structure currently dominates the debate on the future of pay bargaining within the public sector. In this context, the government has put intense pressure on public employers to dissolve national wage agreements and to 'regionalise' bargaining structures.

With these points in mind, we now turn to examine survey evidence on company pay structures in the private sector. In particular, does the fragmentation and decentralisation of pay bargaining arrangements indicate that wage-setting behaviour is now significantly more geared to competitive conditions prevailing in regional and local labour markets?

IV The decline of multi-employer agreements

The government argues that multi-employer, national wage agreements restrict the responsiveness of nominal wages in regional labour markets. Consequently, the decline in the proportion of the workforce covered by such agreements has been welcomed as an encouraging but recent development (Department of Employment, 1988:24). Contrary to the impression given in the 1988 White Paper, however, multi-employer agreements have been in decline since the 1950s, although this trend has clearly accelerated over the past ten years.

Multi-employer bargaining arrangements originated in the late nineteenth century when the normal form of collective agreement was regionally based, stipulating wage rates for all employees in an industry within a particular county, conurbation or town. Around the time of the First World War, and strongly encouraged by successive governments, they were increasingly merged together into national agreements. These were negotiated through bodies commonly termed national joint industrial councils (NJICs) which stipulated national wage rates and employment conditions for their respective industries. Each step towards centralising wage bargains was accompanied by a progressive reduction in differentials between the localities brought within one settlement (Robertson, 1961). By 1945, there was a fairly comprehensive set of national agreements covering over three-quarters of all employees, many of whom were not trade union members.

Since the 1950s the coverage and influence of industry-wide agreements (including wages councils, regional agreements, national joint industrial councils) has steadily declined. The Donovan Commission highlighted this phenomenon in 1968 by focusing on the problem of wage drift. This referred to the increasing divergence between basic pay rates stipulated in national agreements and actual earnings. In effect, the pay rates set by many multi-employer agreements – particularly those covering the engineering and chemicals industries – had become little more than 'minima' or 'safety nets' which were supplemented by workplace bargaining (Brown and Terry, 1978).

A number of industrial relations surveys conducted in the late 1970s and 1980s provide evidence of the decline of multi-employer agreements. The Warwick Survey, conducted in 1978, pointed to the dominance of single-employer over multi-employer pay agreements and reported a preponderance of bargaining at establishment as opposed to divisional and company level (Brown, 1981). The most comprehensive data for the 1980s are provided by the Workplace Industrial Relations Survey (WIRS2), carried out in 1984. The survey is a nationally representative sample of approximately 2,000 British establishments with 25 or more employees, in both the public and private sectors. The only important industries excluded from the survey were agriculture and coal mining.

The WIRS2 Survey (Millward and Stevens, 1986) demonstrates the continuing decline in the observance of multi-employer pay agreements. This applied to both manufacturing and services in the private sector and to manual and non-manual workforces, and is evident in replies about which bargaining level was most important in pay fixing (see Table 4.3). The full effect was masked by bias introduced by the rapid decline in establishment size in manufacturing in the early 1980s; the probability of factories following multi-employer agreements is greater in the smaller size ranges.

There was an even stronger decline over the period in the proportion of the sample reporting that multi-employer agreements had some influence on pay. Of those employers reporting that pay is the subject of bargaining, those stating multi-employer agreements to be the most important covered 30 per cent of manuals and 13 per cent of non-manuals in manufacturing, and respectively, 50 per cent and 40 per cent in private services. Given that bargaining was reported to be used for approximately 80 per cent of all employees in manufacturing and a little under 50 per cent in services, one can deduce that, even including

Table 4.3 *Most important levels of bargaining influencing pay increases in the private sector, 1984 (1980)*[1]

Private manufacturing = PM; Private services = PS

	Manual PM	Manual PS	White Collar PM	White Collar PS
National/regional	40 (41)	54 (57)	19 (18)	36 (43)
Company/divisional	20 (15)	33 (28)	36 (29)	52 (36)
Plant-establishment	38 (41)	11 (9)	42 (49)	11 (7)
Other answer	1 (1)	3 (3)	3 (2)	10 (*)
Not stated	1 (2)	- (2)	1 (*)	- (4)

Note:
1. These are based on the assessments of managers and testify to the growing importance of company/divisional bargaining over pay rates in the early 1980s

Source: Millward and Stevens (1986: 232, 238)

Reducing Regional Inequalities

Table 4.4 *The fragmentation of bargaining on individual items of pay and conditions: all sectors manufacturing, 1986*

Items	Level at which item is bargained (% of employees)		
	Single-employer only	Multi-employer only	Multi-level
Basic pay	87	4	9
Bonuses	98	0	2
Overtime	73	18	11
Shift pay	76	13	11
Sick pay	95	1	3
Hours of work	64	25	11
Holidays	67	20	13
Pensions	99	1	0
Any item	100	29	21

Source: CBI 1988:45

construction, multi-employer bargaining was the principal means of fixing pay for only about one private sector employee in five. There was also a continuing decline in the number of workplaces reporting multiple level pay bargaining.

A CBI survey, conducted in 1986, also showed a pronounced shift towards single employer bargaining arrangements in private manufacturing compared to a similar survey carried out in 1979. Single employer bargaining featured strongly in the determination of all issues, particularly basic wages, bonuses, sick pay and pensions (see Table 4.4). As far as pay was concerned, the data reveal that in 1986, 87 per cent of employees in plants with collective bargaining had basic rates negotiated at establishment or company level.

Moreover, only 56 per cent of all those companies which adhered solely to a multi-employer agreement in 1979 continued to do so by 1986. In line with the WIRS2 survey, there had been a rationalisation of wage bargaining arrangements, with a decline in the extent of two-tier or multi-level bargaining in manufacturing and a reduction in the impact of national agreements on plant-level negotiations (see Table 4.5).

The collapse of industry-wide agreements has accelerated in the 1980s, with about 16 multi-employer bargaining groups disintegrating since 1986 alone. As Table 4.6 demonstrates, these included the councils for

Regional Earnings and Pay Flexibility

Table 4.5 *Matched sample, 1979-86, changes in the structure of bargaining: rates of pay*

Level at which pay rates are determined	1979	1986	Percentage change 1979-86
Establishment	208	260	+25
Company	68	86	+26
National	48	46	-4
Multi-level	149	65	-56
Items not bargained	1	17	n/a
Total Matched sample	474	474	n/a

Source: CBI 1988:49

Table 4.6 *Multi-employer negotiating bodies dissolved since 1986*

Negotiating body	Numbers covered
Banking Joint National Councils	200,000
Bus and Coach Industry	35,000+
Cinema Exhibitors Association/BETA National Agreement	13,000
Cotton Textile Industry	12,000
Independent Television National Agreements	12,000
Joint Industrial Council for the Scottish Milk Trade (JIC)	6,500
Joint Industrial Council for the Scottish Packing Case Industry	1,000
Joint Negotiating Council of the Scottish Banking Industry	16,000
Multiple Food Trade Jt Cttee	110,000
NJIC for Cement Manufacturing	4,300
Craftsmen's Ctte for Cement Manufacturing	1,800

Reducing Regional Inequalities

NJIC for Corn Trade	10,000
NJIC for Paint, Varnish, Lacquer Industry	10,000
National Newspapers NPA Agreement	15,000
Provincial Newspapers NS/NUJ National Agreement	5,500
Wholesale Sunday Newspaper Distribution National Agreement	1,000
Engineering and Shipbuilding National Agreement	600,000
Total	1,053,100

Source: IRS Employment Trends, 23.5.89

multiple food retailing, commercial television, cotton textiles, cement, cinemas and banking. Nine agreements have been reduced in coverage or influence either because of dwindling membership or continual difficulties in reaching agreements; they continued to survive but in a revised form. Merchant shipping, roadstone quarrying, dairying, cable-making, hosiery and wholesale meat were among the industries affected. Several factors emerge as important in the recent dissolution of industry-wide agreements. A number of large employers have broken ranks after a prolonged period of negotiations and concluded separate settlements (eg cotton textiles, cement industry). In other industries (eg banking), multi-employer agreements have collapsed because employers wished to regulate pay and conditions free from external interference in the context of a tightening labour market.

V The decentralisation of bargaining: a paradox?

On the face of it, then, we have a paradox. Contrary to government thinking, the narrowing of inter-regional pay differentials (excluding the South East) has been associated with a long-term trend towards the decentralisation and fragmentation of pay bargaining arrangements. Moreover, despite extensive deregulation of labour markets in the 1980s, including the legislative dismantling of controls over wages at national and industry level, regional pay averages – outside the South East – continue to converge.[1] How can these contradictory phenomena be explained?

[1] Examples are the weakening of the wages councils, the rescinding of the fair wages resolution and the repeal of Schedule II of the Employment Protection Act 1975.

The spread of single-employer bargaining arrangements is commonly connected to the growth of large, multi-plant corporations. Using a transactions-cost framework, Aoki (1984) argues that as production processes, skills and knowledge become firm-specific, there will be a trend away from uniform setting of wages at industry or district level. Although this initially takes the form of supplementary workplace bargaining, with multi-employer agreements setting minimum rates and conditions: 'as the size of firms becomes larger, internalising more specific employment structures and more diverse activities, firms could benefit from breaking away from multi-employer bargaining, settling for higher wages, and running their own industrial relations' (1984:136).

Analyses of the determinants of bargaining structure are consistent with this argument. Multi-employer bargaining tends to predominate in industrial sectors with a large number of small establishments, and with relatively low capital requirements and ease of entry. It is also significantly associated with multi-unionism (Deaton and Beaumont, 1980; Booth, 1989). Small establishments derive important scale economies from such arrangements: they enable trade unions to be dealt with on a collective basis and facilitate the standardisation of pay and employment conditions. The destabilising impact of inter-firm competition based on wage cutting can therefore be neutralised (Craig *et al*, 1982). Thus, the recent dissolution of multi-employer agreements in several industries (weaving side of cotton textiles, multiple food retailing), was quickly followed by a number of companies re-establishing joint negotiating machinery because they lacked the resources for extensive domestic negotiations. A number of employers and unions have also sought to restore joint regulation in their industries following the legislative weakening of statutory wages councils in 1986 (eg licensed clubs, lace finishing and the flax and hemp industry).

In contrast, multi-plant firms, foreign-owned firms, and high concentration industries tend to be associated with single-employer bargaining. Using data from WIRS2, Booth (1989) found that companies with over 50,000 employees were more likely to bargain at corporate level. Corporate bargaining was also associated with companies which have job evaluation schemes and performance-related pay schemes, such as profit-sharing, share ownership or value-added schemes.

Contrary to government thinking, it appears to be changes in competitive strategy and the internal organisation of large, multi-plant companies that are the major reasons underlying the decentralisation of wage bargaining. It is not regional or local labour market factors. The shift away from centralised, functionally organised corporations (the U-form) towards multi-divisional structures (the M-form) is widely viewed as a critical development (Marginson *et al*, 1988). When firms diversify into different product markets, the adoption of an M-form structure enables management to act 'strategically' on a number of fronts. Corporate planning becomes the exclusive responsibility of top level management whilst operational decisions are decentralised to divisional level. This has important implications for pay determination. A study of the nine multi-divisional companies over the period 1983-7 (Purcell and Ahlstrand, 1989) concludes that overall business strategy, such as the devolution of corporate responsibilities into separate budget, profit and product centres, was the most significant determinant of changes in bargaining structure:

> Structural changes included the creation of operating units as separate limited companies.... Principally, the stated aim has been to tie industrial relations and bargaining outcomes to the business performance of divisions or the operating units. (1989:09)

The degree of decentralisation appears to be related to a range of factors, including the extent of product diversification and the intensity of competition. A company level survey conducted in 1985 indicates that in the case of highly diversified companies - with establishments or divisions producing a heterogeneous range of goods/services - pay determination is significantly more likely to occur at establishment level. National structures for pay determination - multi-employer or single-employer - tend to be associated with firms which produce relatively standardised products and services. According to Marginson *et al* (1988:150) companies which compete primarily within the UK, rather than internationally, appear to be the most interested in taking wages 'out of competition' on an intra-enterprise or intra-industry basis.

Despite complex variations in the degree of decentralisation, a common theme emerges from the survey evidence: employers are restructuring their internal pay-fixing arrangements along product market lines so that pay can be directly linked to the performance of specific business units, whether they be profit centres, divisions, bargaining units or the individual establishment. Lucas, Pilkingtons, National Freight, GEC, Philips and Cadburys Bournville are examples of the many firms that have decentralised pay bargaining on the basis of profit-related product centres, rather than local or regionally differentiated labour markets. According to one survey (IRS Employment Trends, 19.12.89):

> many large multi-site employers ... are reluctant to respond to

localised labour market problems by introducing a number of parallel pay structures, primarily for reasons of flexibility. Local structures are administratively complex to manage centrally, and hinder redeployment. The durability of a pay supplement in the form of a location allowance provides sufficient pay flexibility for many organisations. (p6)

In any case, establishment pay bargaining – which supposedly facilitates a greater responsiveness to local labour market conditions – is often an illusion. Plant management, for example, may only be given freedom to negotiate within certain strict financial limits or have to check with the divisional or headquarters management before arriving at a final decision (Brown, 1981). In the company level survey of multi-divisional corporations, two-thirds of establishments (where establishment was the most important level of pay bargaining) reported that higher levels of the organisation had a policy on pay settlements or issued pay guidelines (Marginson *et al*, 1988). This indicates that even the most decentralised company tends to co-ordinate pay settlements at higher corporate levels. The fragmentation of bargaining arrangements is therefore not necessarily synonymous with localised determination of pay.

Single-employer bargaining is an attempt to bring wage determination under the direct control of individual employers. When the Midland Bank broke away from the Federation of London Clearing Bank Employers in 1985, it explicitly acknowledged that single-employer pay arrangements would enable the company 'to introduce new and revised employment practices at our own pace and geared to our own requirements' (IDS Study No 434, May 1989). Employers have been motivated by the desire to gain greater control over unit labour costs within their enterprises and not by any wish to differentiate wages on a regional basis.

Developments in the textile industry provide a case in point. As the modest recovery of British textiles ran aground in the late 1980s and competitive pressures intensified, two of the largest employers – Coats Viyella and Courtaulds – took steps to decentralise their pay bargaining structures in various sectors of the industry (Walsh, 1989). Coats Viyella, for example, broke away from the Hosiery National Joint Industrial Council (NJIC) in January 1989 and established a bargaining structure based on 13 'profit combines' – essentially local subsidiary companies comprising both single and multi-site establishments. Subsequent pay deals (which in 1989 approximated NJIC rates) were now to be linked to productivity concessions. The company stated that the major impetus behind decentralisation was the need to establish 'a something for something' philosophy with the relevant trade unions. In practice, the change in bargaining level meant that the overall cost of the 1989 pay settlement was less than that under the national agreement (IRS Employment Trends, 19.12.89).

Significantly, Cahill and Ingram's (1987) analysis of the CBI Pay Databank indicates that over the years 1979–86, productivity concessions were twice as common in bargaining groups with over 100 workers than they were in those with under 100. Over the period 1980/1 to 1984/5, where unions were recognised, nearly one-third of wage settlements involved productivity-enhancing concessions. It is probable that decentralised bargaining has contributed to the spread of concessionary agreements by enabling firms to negotiate complex performance-related payment systems at local and establishment levels.

It is no coincidence that employers perceive pay settlements to be increasingly affected by corporate performance, as reflected in profitability and productivity indicators, and less responsive to conditions prevailing in external labour markets. Gregory, Lobban and Thomsons' (1987:144) principal conclusion regarding the CBI Pay Databank was that:

> the most important influence on pay settlements over the period [1979–84], both upwards and downwards, has been profits, their importance as a downward pressure diminishing and as an upward pressure rising as corporate profits first weathered the crisis of 1980–1, then showed a recovery.

The importance of 'external' factors, eg comparability nationally and with the industry, locality and national agreement, fell sharply in 1980, with only a minor recovery subsequently. Econometric analyses of these data sets (Gregory *et al*, 1987; Blanchflower *et al*, 1988) also demonstrate that wage-rates significantly depend on the employer's financial performance.

Summary

So far we have examined the trend towards decentralised pay bargaining from a micro or company-level perspective. We have argued that when freed from the ties of national agreements, firms do not, generally, rush to exploit the pay contours of local labour markets. Firms that decentralise their pay bargaining tend to do so according to their product divisions, irrespective of where the particular production facilities may be based geographically. This is a result of long-term organisational

Regional Earnings and Pay Flexibility

changes associated with the widespread devolution of business decision-making structures.

The analytical significance of these developments can only be comprehended if we reject the neo-classical assumption that all firms react passively to conditions in the external labour market and that wages are given to employers by anonymous market forces. The *active* role of employers in structuring pay and employment conditions stems from the open-ended nature of the employment relationship. Unlike other commodities, wages are advanced in return for the capacity to work rather than some pre-specified quantity of performed labour. Once workers are hired, therefore, employers need to devote considerable resources to securing the motivation and co-operation of the workforce so that commercial objectives, the minimisation of production costs, can be obtained. Payment systems and wages are one among several important devices, including screening and hiring techniques, internal job hierarchies, and systems of work organisation, that are used by employers to elicit the motivation and productivity of employees. Within this framework, single-employer bargaining constitutes a response by individual employers to heightened competitive pressures as they attempt to exert greater (independent) control over unit costs of production.

This does not imply that the determination of wages and employment conditions is completely independent of conditions in the external labour market or the economy. In fact we argue in the next section that these factors are currently exerting a destabilising influence on wage structures in the South East. It does suggest, though, that the textbook model of firms as passive wage-takers – which appears to inform government thinking – is considerably wide of the mark.

So far the role of trade unions has been omitted from the discussion. But do not unions seek to equalise regional pay differentials? Is not a major function of 'combine committees' of shop stewards to press comparability claims between different parts of geographically dispersed, multi-plant companies? To some extent this is true. But both the willingness and the ability of trade unions to pursue these objectives are profoundly affected by the bargaining structures within which they are obliged to operate. We have argued that these have altered in response to changes in the ways in which firms have structured themselves. It is probable that the recession and the industrial relations legislation of the 1980s have diminished the capacity of trade unions to pursue these objectives. But their behaviour has undergone greater alteration as a consequence of changes in employers' bargaining practices (Brown and Wadhwani, 1990). Furthermore, these changes have not been exclusive to unionised firms. A similar pattern of product-based divisionalisation is to be found in many large non-unionised firms. In the context of regional pay determination, therefore, we argue that the role of trade unions is reactive, rather than proactive, and that a satisfactory explanation can only come from an understanding of employer behaviour.

VI Skill shortages, housing and the South East pay differential

Payment systems and bargaining arrangements are not static institutions but are reshaped by management in response to internal (profitability, productivity) and external (labour shortages) pressures. Analyses of datasets covering the first half of the 1980s (eg WIRS2, CBI pay databank) indicate the importance of internal influences, notably, corporate performance and productivity indicators, in shaping employees' remuneration. In the late 1980s, however, external labour market factors, particularly skill shortages, have re-emerged as a key upward pressure on pay settlements. With rising employment rates, shortages of skilled labour have increased markedly (OECD, 1989:20) and although econometric evidence for manufacturing points to the relatively limited impact of skill shortages on overall wage growth, it has been argued that this factor explains most of the increase in wage inflation since 1987 (Sentance and Williams, 1989).

Survey evidence indicates that labour shortages are now an acute problem for many employers. According to the CBI, the proportion of employers reporting difficulties in recruiting specific types of skilled employees has risen from just under 55 per cent in 1986 to more than 70 per cent in 1989 (CBI, 1989:8). Shortages are particularly pronounced in engineering and the textiles, clothing, footwear and leather industries. As Table 4.7 indicates, the need to recruit/retain labour emerged as an important determinant of pay settlements in both private manufacturing and services in the year to July 1989.

Employers have restructured their salaries and benefits packages in response to these developments. A recent survey of 1,100 personnel managers by the Institute of Personnel Management found that two-thirds of companies had raised basic pay and over one-third had introduced performance-related pay schemes in an effort to address staff shortages over the past two years (European Industrial Relations Review, 1989). Analyses of the growth of occupationally related 'fringe' benefits also suggest that an important contributory factor is the need to attract and retain labour (Mann, 1989).

Regional Earnings and Pay Flexibility

Table 4.7 *Pay settlement pressures in private manufacturing and services*

Manufacturing 'Very important' upward pressures (%)			
Major factors:	1986/7	1987/8	1988/9
Cost of living	26	23	59
Recruit/retain	14	22	30
Profit levels	20	25	20
Labour productivity	15	19	17
Order levels	11	15	11

Services 'Very important' upward pressures (%)		
Major factors:	1987/8	1988/9
Recruit/retain versus risk of redundancy	54	49
Profits	26	18
Cost of living	23	52
Labour productivity	17	9

Source: CBI Employment Affairs Report, November/December 1989

Recruitment and retention difficulties are, however, most severe in the South East. Labour shortages – exacerbated by the relatively high cost of housing in the region – have put particular pressure on internal pay structures in both the private and public sector. In the 1990s these problems will be compounded by demographic factors, notably the projected fall in the number of young people entering the labour market. The interaction of skill shortages, high housing costs and demographic shifts underpin the growing divergence of South East pay from the overall regional average. The next section examines the manner in which employers have sought to reorganise their salaries and benefits packages in response to these problems. We focus in particular on the fate of the only explicit regional pay differential in the UK – the London allowance.

The London allowance

The London allowance was originally intended to compensate employees for the relatively high cost of living in the South East. The Pay Board

Reducing Regional Inequalities

Report on London Weighting in 1974 put forward the notion of 'cost compensation' based on comparability of real earnings between employees in London and elsewhere. Thereafter, the Department of Employment published annual indices on the basis of which the recommended allowance could be updated. In 1982 this index was abolished, as was an alternative index produced by Incomes Data Services in 1987.

The discontinuation of these indices, however, does not reflect any marked fall in the relatively high cost of living in the South East. If anything, during the 1980s the regional cost-of-living disparity has increased due to movements in house prices. As Martin and Tyler (1989) demonstrate, by 1987 a sharp divide had emerged between the South East, where the relative cost of living had risen to as much as 135 per cent of the national average and the rest of the country, where a majority of counties had an index of less than 94 per cent and many, a relative cost of living as low as 80 per cent of the national average. This reflects much higher rates of house price inflation in the south of England during the period (see Table 4.8). The regional house price differential in the South East, which had been 25 per cent above the UK average in 1981, peaked

Table 4.8 *Regional house price disparities, 1989*

Average regional house prices (UK average = 100)

	1989
South East	148
East Anglia	119
South West	120
West Midlands	91
East Midlands	92
Wales	83
Yorkshire and Humberside	80
North West	78
North	70
Scotland	64
Northern Ireland	57

Source: Cambridge Econometrics, 1990:15

at 53 per cent above in 1987 (Cambridge Econometrics, 1990; Hamnett, 1988). Moreover, average earnings in the South East have not kept pace with regional housing costs. In mid 1988 house prices in the South East were 2.5 times those of the north, whereas weekly earnings were only 26 per cent greater on average for full-time males.

The relatively high cost of housing in the South East acts as a barrier to the migration of workers from other regions and exacerbates labour shortages. In the late 1980s the relative scarcity of workers in key occupations, including professional, scientific, administrative, technical staff and skilled manual craft workers posed a major problem for both public and private sector employers. Employers have responded by reorganising their remuneration packages, including London allowances, to attract and retain staff in a tightening labour market. Many companies also offer relocation incentives to potential migrants as a means of solving skill shortages. Difficulties in the recruitment and retention of labour in the region have put upward pressure on pay levels. The net effect has been an increase in the magnitude of the South East regional pay differential.

The most aggressive response to labour shortages has come from the financial and business services sector – which provides almost one-quarter of London's employment. Increased competition for staff after the deregulation of financial markets provided the impetus for major increases in London allowances. Moreover, divisions among clearing bank employers over the appropriate level of territorial allowances led to the disintegration of multi-employer pay bargaining in 1987. Subsequently, single-employer negotiations precipitated a 50 per cent rise in inner London allowances to £3,000. Despite an attempt to freeze the allowance, a number of employers (mostly building societies) sought to establish a competitive advantage in 1989 by raising the level above £3,000. Other employers in the financial services sector, including the clearing banks, were forced to follow suit.

The finance sector's utilisation of London allowances to redress labour shortages has had repercussions for both private and public employers. Table 4.9 illustrates the percentage increases in inner London allowances among a representative range of key employers in the South East over the period 1985-9. Despite overall increases, it is apparent that in 1989 – with the exception of IBM, ICI and the major oil companies – relatively few non-finance companies paid their staff inner London allowances in excess of £2,500. The vast majority of public sector allowances are grouped in the range of £1,000 to £2,000 with only the police, electricity supply workers and Civil Aviation Authority staff receiving

Reducing Regional Inequalities

Table 4.9 *Increases in Inner London allowances, 1985-9*

	1985	1989	% Increase
1. Financial services			
Clearing banks	1,845	3,000	63
Bradford and Bingley B Soc	1,461	3,250	122
Nationwide Anglia	1,509	3,000	99
Allied Dunbar	1,605	('86)3,000	87
Prudential	1,483	2,250	52
2. Retail distribution			
Littlewoods	19.05	45.21	137
Asda	12.35	15.00	21
Sainsbury	4.29	23.79	454
3. Oil and chemicals			
ICI	1,584	3,240	104
Shell UK	1,515	2,700	93
4. Other large employers			
IBM (minimum)	636	2,556	301
(maximum)	2,604	4,500	73
British Gas	1,455	2,500	72
British Telecom	1,612	2,100	49
5. Public sector			
British Rail	860	1,295	50
Civil service (non-industrial)	1,365	1,750	28
Electricity supply	1,470	2,250	53
Local authorities	1,317	1,722	31
NHS (administrative)	1,133	1,580	39
(doctors)	839	1,200	43
Police	1,866	2,100	12
Teachers	1,110	1,377	24
Post Office	1,390	1,460	5
Civil Aviation Authority	1,454	2,250	55

Source: IDS Study No 378, January 1987; Study No 400, December 1987; Study No 445, November 1989

allowances in excess of this amount. Employers in retail services have also raised allowances to attract and retain sales staff, but they remain far below those prevailing in the finance sector, with the majority in the range of £500 to £1,500.

Although there have been general – and sometimes dramatic – increases in London allowances in the late 1980s, they rarely represent the real difference in living costs between London and the rest of Britain. On the basis of the January 1989 cost of living figures, one study estimated that the inner London allowance should be £3,353 (Reward Survey, Summer 1989). But, apart from the financial services sector, the majority of London allowances fall considerably below this figure. As the real value of the allowance declines, companies have consolidated the regional differential into basic pay. British Airways, for example, abolished London allowances through consolidation because the payment was no longer considered a realistic measure of the regional cost of living disparity (European Industrial Relations Review, 1989).

Companies traditionally viewed London weighting as a mechanism of cost compensation, with allowances uniformly allocated to all staff within strict geographical boundaries. In the late 1980s, however, the cost compensation rationale underpinning London allowances has been rejected by many employers as an inflationary and ineffective way to alleviate the more acute problem of labour shortages. They have sought instead to target additional payments to particular groups of employees in short supply or to key locations, often on the basis of recruitment and turnover statistics. The finance sector has selectively extended territorial allowances to other locations in the south of England, including Bristol, Southampton and Cambridge and higher allowances in the retail industry are targeted to 'nominated stores', particularly in the West End, rather than to all employees within a geographical boundary.

The selective utilisation of London allowances is symptomatic of a general reorganisation of remuneration packages in the South East, involving the growth of merit and performance-related pay, recruitment supplements and enhanced benefits packages. British Airways, for example, recently awarded pay rises of up to 15 per cent to qualified engineering staff to cut labour turnover rates, which had risen from 2 per cent in 1988 to over 12 per cent in May 1989. These developments are also characteristic of private manufacturing industry. A 1988 survey of 40 engineering companies in the South East revealed that employers were utilising a variety of strategies to remedy serious labour shortages caused by poor standards of training and the high cost of housing (IDS, March 1988). Most firms had given substantial pay rises, extended the salary range and introduced merit payments (particularly for professional grades), but the main emphasis was on targeting key occupations and individuals rather than on blanket regional supplements for all staff.

It is questionable whether imbalances in the South East labour market can be alleviated by the *ad hoc* proliferation of supplements, allowances and fringe benefits. Despite raising London allowances by 37 per cent in 1987, the Alliance and Leicester Building Society found that turnover rates in London were still twice those of the rest of the country (IDS, March 1988). Indeed, in 1967, a National Board of Prices and Incomes Report anticipated the problems associated with the use of the London allowance as a mechanism to ameliorate labour shortages. It argued thus:

> To base London weighting on a scarcity in one particular category of labour would have inflationary consequences throughout an entire pay structure or, alternatively, create serious anomalies. It would also lead to wage increases by other employers without effecting any change in the distribution of labour, ie it would not serve its intended purpose. (NBPI, 1967:13)

In its review of London weighting in the non-industrial civil service, the board recommended that the payment should only compensate employees for the London cost of living differential, primarily travel and housing costs, and take no account of labour market considerations.

The public sector

In the short term, public employers are experiencing intensified competition for staff as firms in the private sector (particularly financial services) use a range of financial allowances and supplements to combat labour shortages in the South East. According to the local authority employer's organisation (LACSAB), the proportion of London authorities suffering shortages of craft workers rose from 30 to 76 per cent over the period 1987-8 and the number registering shortages of secretaries and typists increased by 52 per cent. In 1988, the turnover of civil service clerical staff in London and the South East exceeded 30 per cent (IDS, 1989).

These developments have had a profoundly destabilising impact on public sector payment structures and the system of centralised pay bargaining. As in the private sector, there has been an *ad hoc* expansion of financial incentives, especially for professional staff, including labour market premium payments, performance-related bonuses and improved location allowances. But the government has put additional pressure on public employers to dissolve national pay agreements and to 'regionalise' bargaining structures.

There is little evidence, however, of any perceptible shift to local level or regional bargaining in the public sector, apart from a range of initiatives to deal with the specific problem of labour shortages in the South East. 'Local pay additions' were introduced in the civil service to ease recruitment and retention problems among administrative staff, but these have been used exclusively in London and the South East where turnover rates are highest. After a major industrial dispute in late 1988, the Post Office eventually introduced a systematic set of pay variations to combat labour shortages. Supplementary payments are offered to staff on the basis of unemployment levels (less than 5 per cent in local area) and turnover rates (15 per cent plus), but as in the civil service, these have been exclusively directed to workers in the South East.

In the NHS a range of special pay supplements have been offered to nurses and computer/administrative staff in the London area, but the Nurses Review Body, in particular, has adopted a cautious approach to decentralised bargaining on a regional basis. After agreeing to a pilot scheme in 1989, whereby national rates of pay could be supplemented selectively for recruitment and retention purposes, it warned that such measures could lead to 'uncontrolled pay drift' and that they should not be allowed to become 'a soft option for poor management' (IDS, 1989).

The most intense debate about the future role of national pay agreements has occurred in the local authority sector. Although the vast majority of local authorities are in favour of maintaining national bargaining, many envisaged a shift away from prescriptive agreements towards a nationally negotiated framework conducive to local flexibility and adaptation. In 1989, however, 13 councils in the South East broke away from the national pay agreement to pay more in order to tackle labour shortages. Three-quarters of London boroughs have reviewed their payment systems, with 71 per cent introducing selective market supplements in addition to London weighting by 1988 (Table 4.10). Performance related pay, merit bonuses, and an expansion of 'fringe' benefits have also been used to attract and retain staff. According to the local authority employers' organisation, 69 per cent of authorities in the South East were offering car-leasing schemes in May 1988, compared to 21 per cent in 1987; 24 per cent were offering BUPA schemes in 1988 compared to none in 1987 and the provision of mortgage subsidy schemes had risen from 22 per cent of authorities to 52 per cent.

External labour market pressures, coupled with government exhorta-

Reducing Regional Inequalities

Table 4.10 *Local government labour shortages*

A. *Pay measures used in London 1987 and 1988: % of authorities using:*

	May 1987	May 1988
Revised grading structures	–	90%
Market supplements	59%	71% (£300–£4,500)
Extended scales and career grades	40%	76%
Regrading	25%	57%
Performance pay	22%	38%
Accelerated promotion	–	33%
Retention bonus	–	19%

B. *Pay measures used in South East in 1988: % of authorities using:*

Extended scales and career grades	73%
Market supplements	60%
Performance pay	60%
Regrading	53%
Overtime/premium pay	20%
Accelerated promotion	13%

Source: IDS, 1989:12

tions to decentralise bargaining arrangements, have inexorably driven public sector employers to modify existing pay bargaining structures. But apart from the South East initiatives outlined above, regional or local level bargaining has been eschewed in favour of segmented bargaining on occupational or 'profit centre' lines. In the civil service, for example, national 'single table' bargaining – whereby all unions negotiated as a single body through the council of civil service unions – has been replaced by seven separate national agreements for different occupational groups. As Bailey (1989:39) points out, extensive decentralisation in central government could only occur if there was a corresponding devolution of budgeting control, but the Treasury is resistant to this development. Other moves to decentralised bargaining include British Rail's plan to introduce a system of business-related bargaining units based on five main sectors and the Post Office's proposed shift to 'profit centre' bargaining. In other parts of the public sector, the shift away from national industry-wide negotiations has often been a consequence of organisational restructuring precipitated by preparations for privatisation (eg the water industry, British Steel) or deregulation (eg municipal bus sector).

The present reorganisation of the civil service and the NHS will further fragment existing bargaining arrangements. The government report outlining plans for the creation of free-standing agencies in the civil service, explicitly stated that the influence of national agreements on agency staff would diminish and that 'central responsibility for pay and conditions of service, and the associated negotiations with national trade unions will be progressively and substantially reduced (IDS, 1989:26). In the NHS, 'opting-out' arrangements – whereby hospitals can set themselves up as self-governing trusts – could precipitate individual hospitals ignoring or supplementing national agreements, particularly in London and the South East.

As we have seen, the fragmentation of bargaining arrangements has been partially driven by labour shortages – particularly in the South East – with some employers willing to pay in excess of nationally negotiated rates to attract and retain staff. It is likely, however, that if there were to be widespread introduction of local geographical discretion on pay to national public services with a homogeneous product it would lead to leap-frogging pay claims and loss of managerial control. A partial emphasis on the virtues of regional or localised bargaining is therefore no substitute for a coherent public sector pay policy. In practice the majority of public sector organisations have not introduced regionally differentiated payment systems but have followed the example of large private corporations by decentralising bargaining structures along business or profit unit lines. The success of these ventures will depend to a considerable extent upon how far the new bargaining groups are insulated organisationally from the old arrangements. If the separation of bargaining units is merely an artificial importation from private industry, then informal comparability arrangements may merely fill the vacuum left by the dissolution of national pay agreements.[2]

Summary

We have argued that the growth of selective recruitment supplements, local pay additions and fringe benefits constitutes a short-term and largely *ad hoc* response to intensified competition for labour in the South East. An IDS survey of engineering employers in 1988 revealed that most firms regard labour shortages in the South East as a structural problem, which can only be alleviated by the construction of extra council housing

---

[2] The establishment of review bodies for nurses and teachers is illustrative of a contradictory move towards greater centralisation of bargaining for these two occupations (see Bailey, 1989).

Reducing Regional Inequalities

and a reduction in regional house price disparities (IDS, March 1988). It is probable that the specific problems of the South East region can only be satisfactorily addressed in the long term by an overhaul of housing policy and/or the decentralisation of economic activity to other regions. In this context, Bover *et al* (1989) have stressed that institutional distortions surrounding the owner-occupied housing market, particularly the existence of tax incentives which raise the portfolio returns on owner-occupation relative to other assets, may seriously dislocate the operation of labour markets in the South East through the creation of a 'mobility trap' for labour.

It is clear that high property prices and high rents in the south, combined with increasing shortages of labour, have led to a relocation of economic activity out of London and the South East. Private sector moves have generally been to the South West, East Anglia and the Midlands, whereas most moves into the north of England, Scotland and Northern Ireland have been in the public sector. This process of decentralisation may explain the slowdown in business service employment growth in the South East region — from 8.6 per cent in 1988 to around 6 per cent in 1989 (Cambridge Econometrics, 1990).

Although recruitment and retention difficulties are most severe in the South East, skill shortages have spread to other regions. Furthermore, the evidence indicates that this is not merely a cyclical development, but symptomatic of serious supply side weaknesses. A range of studies have revealed major skill shortages throughout British industry and have highlighted poor training as a major obstacle to the introduction of advanced technologies (eg Steedman and Wagner, 1987). The proportion of national income devoted to education and training in Britain was lower in the late 1980s than in 1979 (Nolan, 1989). Skill shortages can only be successfully addressed in the long term by a radical and sustained reorganisation of training strategies, career structures and childcare provision. The proliferation of local pay supplements and selectively targeted allowances can only constitute a short-term palliative in the context of a national supply side problem. An exclusive reliance on these measures would seriously jeopardise the internal coherence of payment structures in both the public and private sectors.

VII Conclusion

Contrary to current govenment thinking, the long-term trend towards the decentralisation and fragmentation of pay bargaining arrangements has been associated with the narrowing of inter-regional pay differentials (excluding the South East). Moreover, despite extensive deregulation of labour markets in the 1980s, regional pay averages – outside the South East – showed no tendency to diverge. The survey evidence on company pay structures indicates that when firms are freed from national pay agreements, they do not rush to exploit the competitive conditions prevailing in local labour markets. Firms that decentralise their pay bargaining tend to do so according to their product divisions irrespective of where the particular production facilities may be based. The propensity of firms to co-ordinate wage bargaining at company level is a response to long-term organisational changes associated with the widespread devolution of business decision-making structures.

Our analysis suggests that employers pursue proactive pay strategies and that the textbook model of firms as passive wage and price-takers is considerably wide of the mark. This does not, however, imply that wages are completely insensitive to changes in the external labour market or economy. In fact, we have argued that external pressures – labour shortages and the relatively high cost of housing – have exerted a particularly destabilising influence on wage structures in the South East. The important point is that employers have not responded to these external pressures by restructuring their pay bargaining arrangements on regional lines. Employers have dealt with labour shortages and the specific structural problems of the South East on a piecemeal and *ad hoc* basis. The long-term trend in both the public and private sectors is towards the decentralisation of pay bargaining arrangements along business or profit unit lines and not the adoption of geographically differentiated payment systems.

References

Aoki, M (1984), *The Co-operative Game Theory of the Firm*, Oxford University Press, Oxford.

Bailey, R (1989) 'Pay and Industrial Relations in the UK Public Sector'. *Labour*, vol 3, No 2, Autumn, pp31–56.

Bank of England Quarterly Bulletin (1988), 'Regional Labour Markets in Great Britain', August, vol 28, No 3, pp367–75.

Blackaby, DH and Manning, DN (1987), 'Regional Earnings Revisited', *Manchester School of Economic and Social Studies*, pp158–83.

Blanchflower, D, Oswald, A and Garrett, M (1988), *Insider Power in Wage Determination*, Centre for Labour Economics, Discussion Paper No 319, LSE, London.

Booth, A (1989), 'The Bargaining Structure of British Establishments', *British Journal of Industrial Relations*, vol XXVII No 2, July, pp225-34.

Bover, O, Muellbauer, J and Murphy, A (1989), 'Housing, Wages and UK Labour Markets', *Oxford Bulletin of Economics and Statistics*, vol 51, March, No 2, pp97-136.

Brown, W (ed) (1981), *The Changing Contours of British Industrial Relations: A Survey of Manufacturing Industry*, Basil Blackwell, Oxford.

— and Terry, M (1978), 'The Changing Nature of National Wage Agreements', *Scottish Journal of Political Economy*, vol 25, No 2, pp119-33.

— and Wadhwani, S (1990), 'The Economic Effects of Industrial Relations Legislation Since 1979', *National Institute Economic Review*, No 131.

Cahill, J and Ingram, P (1987), *Changes in Working Practices in British Manufacturing Industry in the 1980s: A Study of Employee Concessions Made During Wage Negotiations*, mimeo, CBI, London.

Cambridge Econometrics and the Northern Ireland Economic Research Centre (1990) *Regional Economic Prospects: Analysis and Forecasts to the Year 2000*, January.

Confederation of British Industry (1988), *The Structure and Processes of Pay Determination in the Private Sector: 1979-1986*, CBI, London.

— (1989) *Employment Affairs Report*, November/December, CBI, London.

— (1989) *Pay and Performance 1989-90*, CBI, London.

Craig, C, *et al* (1982), *Labour Market Structure, Industrial Organisation and Low Pay*, Cambridge University Press, Cambridge.

Deaton, D and Beaumont, P (1980), 'The Determinants of Bargaining Structure: Some Large Scale Evidence for Britain', *British Journal of Industrial Relations*, vol XVIII, pp202-16.

Department of Employment (1988), *Employment for the 1990s*, HMSO, London.

European Industrial Relations Review (1989), 191, December.

Gregory, M, Lobban, P and Thomson, A (1987), 'Pay Settlements in Manufacturing Industry, 1979-84', *Oxford Bulletin of Economics and Statistics*, 49, 1.

Hamnett, C (1988), 'Regional Variations in House Prices and House Price Inflation in Britain 1969-88', *The Royal Bank of Scotland Review*, No 159, September, pp29-40.

Incomes Data Services (IDS) (1987), 'London Weighting' Study No 378, January.

— (1987), 'London Allowances', Study No 400, December.

— (1988), 'Labour Market Supplement', No 2, March.

— (1989), 'Negotiating at Industry Level', Study No 434, May.

— (1989), 'London and South-East Allowances', Study No 445, November.

— (1989), 'Public Sector Pay. Review of 1988: Prospects for 1989'.

Industrial Relations Services (IRS) (1989), 'Developments in Multi-Employer

Bargaining: 1, Employment Trends', 440, 23 May.
— (1989), 'Developments in Multi-Employer Bargaining: 2, Employment Trends', 443, 11 July.
— (1989), 'Decentralised Bargaining in Practice: 1, Employment Trends', 454, 19 December.
Lee, K (1990), *Wages, Unemployment and Equilibrium in Regional Labour Markets in the UK*, mimeo, Department of Applied Economics, University of Cambridge.
Mann, K (1989), *Growing Fringes: Hypotheses on the Development of Occupational Welfare*, Armley Publications, Leeds.
Marginson, P, Edwards, PK, Martin, R, Purcell, J and Sisson, K (1988), *Beyond the Workplace: Managing Industrial Relations in the Multi-Establishment Enterprise*, Basil Blackwell, Oxford.
Martin, R and Tyler, P, (1989), *The Impact of Real Wages on Local Disparities in Unemployment*, mimeo, St Catharine's College, University of Cambridge.
Millward, N and Stevens, M (1986), *British Workplace Industrial Relations 1980-84*, Gower, Aldershot.
Moore, B and Rhodes, J (1981), 'The Convergence of Earnings in the Regions of the United Kingdom', in Martin, R (ed) *Regional Wage Inflation and Unemployment* pp46-59.
National Board of Prices and Incomes (1967), *London Weighting in the Non-Industrial Civil Service*, Report No 44, Cmnd. 3436, HMSO, London.
Nolan, P (1989), 'Walking on Water? Performance and Industrial Relations under Thatcher', *Industrial Relations Journal*, 20(2).
OECD (1989), Economic Survey: The United Kingdom.
Purcell, J and Ahlstrand, B (1989), 'Corporate Strategy and the Management of Employee Relations in the Multi-Divisional Company', *British Journal of Industrial Relations*, vol XXVII, No 3, November, pp397-418.
Reward, (1989), London Weighting Survey, Summer.
Robertson, D (1961), *The Economics of Wages and the Distribution of Income*, Macmillan, London.
Sentance, A and Williams, N (1989), 'Skill Shortages: A Major Problem for British Industry?', *CBI Economic Situation Report*, March, pp21-4.
Steedman, H and Wagner, K (1987), 'A Second Look at Productivity, Machinery and Skills in Britain and Germany', *National Institute Economic Review*, November.
Walsh, J (1989), *Structural Change and Industrial Decline: The Case of British Textiles*, unpublished PhD thesis, University of Warwick.

5
*Migration and Regional Policy**

Barry McCormick

Interest in migration, and labour mobility more generally, is greater today than at any time since perhaps the 1930s. This may not be accidental: much of today's concern, as in the earlier period, arises from an historically high level of aggregate unemployment and the unevenness with which it is distributed across regions. Widespread unemployment and the lack of normal new job openings in, for example, the North East and Northern Ireland, has prompted an examination of regional policy which, of course, extends beyond mobility issues. However, while mobility, or lack of it, is inevitably a central part of any policy towards regional adjustment, in much post-war regional analysis that role is not made explicit. The purpose of this chapter is to redress this imbalance by gathering together that part of recent migration analysis which contributes towards understanding the pattern of diversity that exists between UK regions, together with an analysis of policy towards mobility and regional development.

For many post-war analysts of UK regional problems, migration from depressed areas is regarded as a source, rather than a resolution, of problems: by encouraging the loss of more 'skilled' and perhaps more enterprising young individuals, migration is held to create additional problems for depressed regions, and should therefore be discouraged. While the loss of 'skilled' young workers is a good reason for being concerned at the consequences of migration, and might even help to motivate a policy of trying to take 'jobs to the workers', it scarcely justifies the scant attention paid to understanding the pattern and consequences of migration in formulating regional policy. One may be sceptical about

* The paper from which this chapter was drawn was presented at the NEDO Conference on Reducing Regional Inequalities, March 1990. I am grateful to Ian Gordon for detailed comments on an earlier draft and to Jim Taylor, Alistair Ulph and John Muellbauer for helpful suggestions. They are not responsible for the errors that remain.

the current contribution of migration to the moderation of regional inequalities – as I am – while at the same time being concerned that overlooking the consequences of the current structure and level of migration, may have weakened our understanding of why depressed regions stay depressed, and thereby undermined policy-making.

The basic 'regional' economic problem is to allocate firms and households efficiently over the available area. The problem is simply stated but the presence of, for example, sequential decision-making, the overspill effect of policies addressing distortions in other markets, imperfect information and local public goods, combine to make the issue analytically complex. Nevertheless, given the high density of population, and advanced level of economic development in the UK, the regional problem is of pressing significance, and requires a clear policy view.

It is helpful to break down the joint determination of firms' and households' locations by asking how far policy and institutions help to ensure an efficient level of gross (and thereby net) migration between areas, which when combined with natural population change, determines the long-run population in each region. If distortions alter the level or pattern of gross migration, then the allocation of persons to regions will be almost certainly inefficient, even if firms continue to locate so as to minimise costs. It is sometimes not recognised that this remains true even if the distortions reduce only gross flows, leaving net migration unaltered: workers are not perfect substitutes in production – even within OPCS skill categories – and even in the event this could be assumed, reduced levels of gross migration will reduce the welfare of those who would otherwise move for personal reasons.

A natural question around which to begin to arrange our concerns about migration in regional development is the following: does migration presently contribute to or worsen the likelihood of balanced regional growth? If a pessimistic conclusion follows, we must ask why it arises and what can be done.

The remainder of this chapter begins, in section I, with a discussion of the central features of the pattern of regional migration. In section II, the question of how various local labour market circumstances – real wages, job openings, unemployment, etc – influence both the decision to migrate and destination choice will be discussed. While this latter decision is no less crucial than the decision to migrate, the empirical analysis of migration has dwelt primarily on the decision to leave a region; largely, one supposes, because it raises less difficult analytic issues and, when using individual data, can be tackled with a smaller (but necessarily still large) data set. This leads to a discussion in section III of the role of the housing market and the extent to which both (a) council housing and (b) regional differences in the price of owner-occupied housing alter regional mobility patterns. The overall conclusion is reached that migration is not at present acting to reduce in a noticeable way the sharp differences in labour market conditions that presently characterise UK regions.

Section IV considers policy reform, first discussing why migration and the regional allocation of labour may not be optimal under *laissez-faire* competition, and then considers broad policy indications. This is followed by a more detailed discussion of how various *existing* policies have served both to reduce migration artificially below efficient levels and to reduce the flexibility with which the regional labour market responds to shocks. Finally, there will be a discussion of how current policy might be reformed.

Some concluding – rather speculative – comments are offered in Section V concerning how we might try to use our densely-populated regions more efficiently.

I Migration and regional labour markets: stylised facts

This section begins with some crucial stylised facts describing the patterns of regional unemployment – the primary target of regional policy – and labour mobility, which together provide insight into whether migration is contributing to equalising labour market conditions.

Perhaps the simplest way of viewing the relationship between regional unemployment and migration is to assume that shocks change the demand for a region's output relative to the national average, and that a change in relative wages or unemployment results. Provided inputs reallocate quickly to expanding regions, the regional unemployment (or wage) differential will be quickly eliminated; only a serially-correlated sequence of shocks will result in persistent high regional unemployment or low regional wages. If relatively 'depressed' regions do not stay 'depressed' for long, this strengthens the simple view of migration of labour and capital as the efficient means of preserving regional balance, and of regional unemployment differentials as a symptom of disequilibrium. However, if it is the case that labour and/or capital adjust very slowly, or not at all – as it appears in the UK – then the model of persistence, based upon the serial correlation of shocks, may miss

Migration and Regional Policy

important features determining inertia in the regional unemployment pattern or the distribution of wages. Some insight into the need to look beyond the simplest approach comes from examining whether or not regional unemployment patterns are more persistent in Britain than in other countries.

Do depressed regions stay depressed?

Table 5.1 describes the degree of correlation between regional unemployment levels relative to the national average in 1987, with that of other selected years for several countries. (A score close to one indicates that relative unemployment rates in 1987 were closely predicted by relative positions in the relevant year. A score of zero indicates that relative regional unemployment levels in 1987 were not correlated with earlier years. A negative score suggests that relative regional unemployment levels had reversed.)

For all countries except the US, relative regional unemployment in 1987 is well predicted by relative regional unemployment in 1980. However, when viewed over a longer period, other countries demonstrate the flexibility observed in the US. Only in the UK and Japan do the relative regional unemployment levels of two decades ago readily predict recent relative positions – a fact which will comfort only those who had given up hope of finding anything that the UK and Japanese labour markets share in common. In Germany and Italy it would appear that, in the short run, relative regional unemployment is slow to adjust – relative to the US – but, over a 12 to 20-year period, it does indeed produce a

Table 5.1 *Persistence of Regional Unemployment Rates*

	US	UK	Japan	Germany	Italy	Sweden
1980	0.32	0.98	0.95	0.92	0.91	0.82
1975	−0.21	0.92	0.91	0.40	0.87	0.68
1965	0.19	0.75	0.88	0.21	0.19	–

Note:
The figures give correlation of regional unemployment rates in 1987 with those for the years listed.

Source: OECD Employment Outlook, 1989

changing picture of relative unemployment patterns as the fortunes of regions fluctuate.

By comparison with similar European countries – let alone the US – the ability of depressed regions in the UK to bounce back appears comparatively more limited than we might have supposed: the forces bringing about regional unemployment equality appear disturbingly weak. From this we may infer either that the serial correlation of region-specific shocks afflicting depressed areas has been greater in the UK than in other similar countries, or that structural considerations have altered the degree, and perhaps pattern, of UK labour mobility in response to an adverse regional shock. If we adopt the second view then we may proceed to model why unemployment differentials may take long periods to erode, or how structural impediments lead to equilibrium unemployment differentials. In the absence of any reason for supposing that region-specific shocks are more serially correlated in the UK than in Sweden, Germany or Italy (or for that matter in the less similar economies of Japan and the US), my own preference is to proceed to examine the role of structural considerations. However, before we do this, one further matter deserves attention.

A straightforward explanation for persistence is that regional unemployment differences reflect the different labour force composition of a region in terms of participants' age, occupation, marital status, broad industry category, number of dependent children, discrimination against minority workers or the incentives associated with different house tenure patterns. However, when these are controlled for using individual data, the labour force in the 'Celtic Fringe' – Scotland, Wales, the north of England and the North West – is predicated to have an unemployment rate of 75 per cent greater than elsewhere. This compares without about 100 per cent greater on a straight comparison without controls (McCormick, 1988). Thus, most of the difference between regions is not accounted for by labour force composition, and furthermore most of the remaining difference is shown in the form of longer unemployment durations, all else equal, rather than more frequent spells. This pattern has not significantly changed between 1973 and the 1980s. (This will not surprise regional economists, but labour economists seeking to characterise the nature of unemployment among high unemployment groups generally ascribe the difference to higher probabilities of entering unemployment rather than longer durations – for example Johnson and Layard (1986)).

Rather than construct a particular model, first evidence will be described which suggests that the prevailing pattern of UK regional

Migration and Regional Policy

migration and unemployment is not only unlikely to provide the equalising role that is sometimes ascribed, but may well contribute to destabilising the pattern of regional growth. After attempting to explain why this arises, other distortions that might give rise to persistent regional labour market misallocation will then be discussed.

Is there net migration from depressed regional labour markets?

The answer to this question provides one of the most central and as yet poorly-understood features of the British unemployment problem. Until recently the conventional answer has been that there has been net aggregate migration from the depressed regions over the past 40 years and that this migration makes a contribution – albeit a small one – to the equalisation of regional unemployment rates. Measured on a north–south basis, it has been shown that aggregate migration from the depressed areas is rising. For example, Champion (1989) estimates that if 'the south' is defined as comprising the South East, South West, East Anglia, and East Midlands then, between 1971 and 1985, net immigration from the north to the south has risen from about 20,000 per annum to around 60,000 per annum.[1]

Unfortunately, the use of aggregate data, while of general interest, is unhelpful in answering the questions at hand since it obscures a crucial important fact: while it is the manual labour force which experiences a regional unemployment 'problem', it is the non-manual workers who are presently migrating on a net basis from the regions with high aggregate unemployment rates (Hughes and McCormick, 1987). It is noteworthy that this was not always the case. For example, 1981 Census data concerning migration between Scotland and the rest of Britain confirm the LFS data with net immigration to Scotland among semi- and un-skilled manuals – a reversal of the high out-migration levels of 1970/1. The net migration of skilled manuals was still occurring, albeit at half the levels of 1970/1. In contrast, net out-migration among professional and managerial workers remained unchanged in 1980/1 relative to

---

[1] After controlling for the influence of regional characteristics and other variables on aggregate migration flows a number of studies – Creedy (1974), Elias and Molho (1982), Gordon (1985), Hart (1070), Molho (1984), and Pissarides and McMaster (1984) – have concluded that migration will gradually reduce unemployment differentials. The effects of wages and unemployment rates on migration are not so strong and in some studies they have been found to be statistically insignificant, but this only implies that the period of adjustment is likely to be long.

Reducing Regional Inequalities

Table 5.2 *Average net migration and unemployment rates for heads of household 1983-6*

Region	Manual Net in-migration (% pa)	Manual Unemployment (%)	Non-Manual Net in-migration (% pa)	Non-Manual Unemployment (%)
East Anglia	0.546	9.0	0.490	4.3
South West	0.036	9.7	0.625	3.9
South East	-0.251	10.3	-0.127	4.0
East Midlands	0.298	10.6	0.606	4.4
Yorkshire and Humberside	-0.024	15.0	-0.246	5.2
Scotland	0.012	16.2	0.049	5.3
North West	0.033	16.8	0.039	6.6
West Midlands	-0.060	16.9	-0.413	5.7
Wales	0.255	17.1	0.148	5.1
North	0.178	18.1	-0.171	5.5

Source: Authors' calculations based on Labour Force Surveys 1981-86

Notes:
1. The regions are listed in ascending order of the mean regional unemployment rate for manual workers. Thus, if workers are migrating into the low unemployment regions we should expect positive figures in the top rows of the table for net in-migration and negative ones in the bottom rows.
2. The average net migration rates are calculated as an average of the percentages of the population living in the region one year earlier who moved out of the region in the periods ending in April/May of 1981-6 when the Labour Force Surveys took place.

1970/1, whereas heavy net out-migration among junior non-manuals in 1970/1 had ended by 1980/1.[2]

The evidence in Table 5.2 concerns heads of household and is taken from Hughes and McCormick (1989a). It is immediately apparent from Table 5.2 that unemployment is not only low for non-manuals, but also even across regions – the range is 2.7 percentage points. In contrast, unemployment for manuals is both high and uneven – the range is 9.1 percentage points. (This contrast would be even more striking if Northern Ireland were included.) If we distinguish between the labour markets for manuals and non-manuals do we still find a tendancy for net migration from depressed regional labour markets?

To make the implications of Table 5.2 more easily comprehensible, the

[2] I am grateful to Ian Gordon for drawing this to my attention.

regions are listed in ascending order of mean regional unemployment rates for manual workers. There are some obvious surprises: there are net inflows of manual labour into Wales and the North despite these being high unemployment markets; there is a substantial net outflow of manual labour from the South East, despite it being an area of labour shortages. If the economy is divided into north and south, with the East Midlands being the highest unemployment rate region in the south, then we find that manual inflows into the south are just about balanced by outflows. However, for non-manuals there is net inflow into 'the south'. While flows for manuals do not follow expected patterns, those for non-manuals suggest that migration responds to small differences in the buoyancy of regional labour markets.

Not only do these flows suggest that manual worker migration from the relevant depressed markets is presently not occurring, but increases in the ratios of non-manual to manual labour in the south, and reductions of this ratio in the north, may contribute towards undermining the prospects for balanced growth across the regions. While it is entirely possible that it is efficient for migration to be allocating white and blue-collar occupational groups to increasingly separate locations, the social costs of making this assumption without considering at length other less benign explanations – in particular, the potential role of housing and labour market distortions – may be considerable, in no small part because of the consequences of 'the North' being gradually drained of its local provision of industrial leadership.

What was the linkage between regional unemployment and net migration in the 1930s?

At the present time, the relationship between aggregate regional net migration and aggregate unemployment is a fairly weak one as Table 5.2 suggests: with the exception of the South East there are net inflows into the remainder of 'the south'. However, there are also substantial net inflows into Wales, with the North West and Scotland being more or less in balance. To what extent has there been a weakening since the Second World War of the tendency for workers to leave 'depressed' regions?

The early volumes of *Oxford Economic Papers* include a number of interesting articles by Makower *et al* exploring UK migration in the inter-war years. One of their figures is reproduced here as Figure 1(a) and (b). These show that at both county and regional level, there existed a remarkably close positive relationship of net out-migration from a region and its relative unemployment, that has been eroded in recent decades,

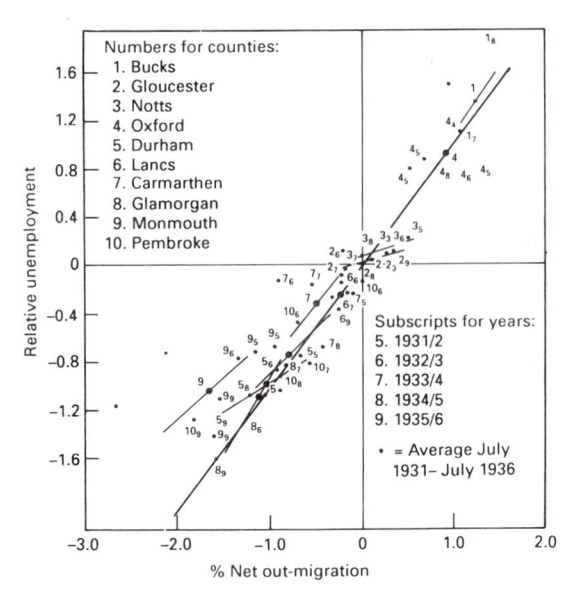

Reducing Regional Inequalities

Figure 5.1a *Scatter diagram of migration and relative unemployment for ten selected counties, July 1931–July 1936*

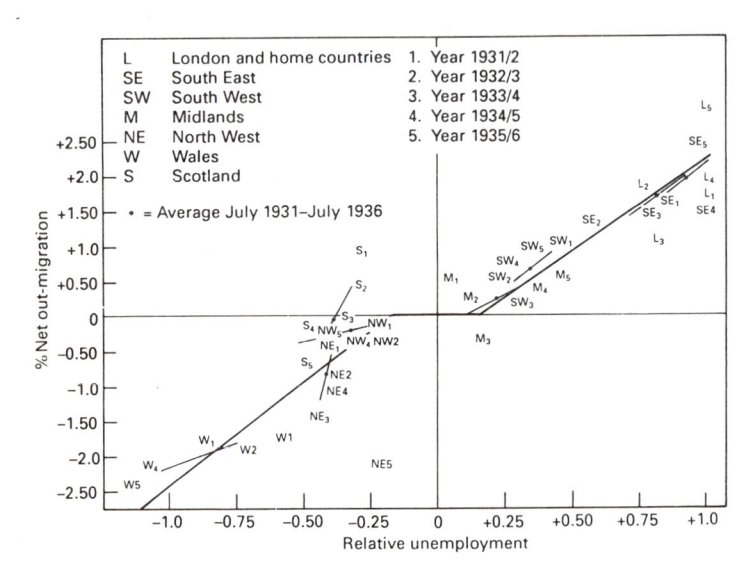

Figure 5.1b *Scatter diagram of migration and relative unemployment by divisions*

Source: Makower, Marschak and Robinson (1938–40)

224

Migration and Regional Policy

and the scale of net in/out-migration has dramatically declined since the 1930s, so that even if the slope of the regression line were unchanged, the regions would be clustered more closely around zero net-migration. Studies by Ian Gordon (1985, 1987) suggest that migrational adjustment absorbed a large proportion (around 80 per cent) of a local demand shock within three years – or one year in the case of Greater London – until the 1960s, but that this is no longer true. While pinning down the date of the onset of the inflexibility of the location of labour to labour market slack is likely to prove difficult, Table 5.2 suggests a clue that the problem lies especially with changes that altered the behaviour of the manual labour market.

Finally, we may note in passing that with regard to the 1930s data, the investigators report a sharp fall in gross migration after 1931 when unemployment rose, which is paralleled by the decline of migration in the 1980s.

Low migration among manual workers

It is now well known that migration rates among manual workers are low – about 1 head of household in 200 migrates between regions each year. Most of this migration is *not* for 'job-related reasons', in fact only about 1 in 1,000 migrates for 'job-related reasons'. In comparison, among non-manuals, 1 in 100 migrates for job-related reasons. One might ask how far this difference is true elsewhere. Among non-manuals in the US the rate of inter-state job-related migration is broadly similar to that for inter-regional job-related migration in the UK. However, for US *manuals* the rate of inter-state job-related migration is 18 per 1,000 – that is, 18 times the rate of inter-regional job-related migration for British *manuals*. (For details see Hughes and McCormick (1987)). This leads to the following question: why is job-related migration over a given boundary so pronouncedly low for British manuals, when for non-manuals the rates in both countries are broadly similar? Since manual workers exercise a predominant influence on unemployment in both countries, could the high geographic mobility of US manual workers be the key to explaining why depressed regions in the US do not stay depressed, as was shown in Table 5.1?

So far in this section, four interrelated facets of the pattern of migration and regional unemployment in Britain have been described. It is now time to analyse the factors that underpin these relationships. First, the extent to which the buoyancy of local labour markets and the individual experience of unemployment exercise separate impacts on the likelihood that an individual will migrate will be examined in greater detail and an unsolved puzzle concerning the influence of local unemployment on out-migration rates will be discussed. Then this will be related to how labour market conditions influence the choice of destination. Second, how housing markets influence migration will be examined, and in particular, the ways in which council housing and regional differentials in the price of owner-occupied housing together act to influence the overall level and pattern of migration over the life-cycle.

II Assessing how regional labour market circumstances influence the likelihood of out-migration and the location choice

While there has now been extensive study in the US of the out-migration decision using reliable individual data, and a few papers published concerning European countries, there are only one or two studies of destination choice and its economic determinants. This is important since without an understanding of how economic factors influence where migrants go, we cannot provide a complete picture of how net migration is related to local labour market conditions. Since migration is rare and migrants have many potential destinations, a useful analysis of destination choice requires very large samples of data.

This section will draw heavily on Hughes and McCormick (1989a), which studies migration in the combined Labour Force Surveys (LFS) from 1981 to 1986. This provides a very large database since each LFS is a 0.5 per cent random sample of UK households.

The analysis of out-migration decisions shows that, all else being equal, being unemployed significantly increases the probability of migrating, which is consistent with prior expectations and both US and Dutch evidence (see Van Dijk *et al* (1989) and Da Vanzo (1978)). For all socio-economic groups, being unemployed one year earlier approximately doubles the probability of migrating, (from, on average, 1 per cent for a head of household to about 2 per cent).

While the effect of individual unemployment on migration is as expected, the effect for regional unemployment is not. Regional unemployment rates – defined appropriately using the LFS for the socio-economic grouping of the individual being studied – have a significant *negative* correlation with the likelihood of individual out-migration, for all but professional and managerial workers. The impact of high regional unemployment reducing the out-migration rates is especially strong for skilled manual workers, being twice that for other affected groups. (The regional vacancy rate mirrors that of the regional unemployment rate.) Being unemployed increases migration rates of all socio-economic groups by equal proportions, and by an amount which is the same for residents of both depressed and prosperous regions.

The result for the role of regional unemployment is perhaps more surprising than for vacancies. Vacancy data relate to registrations at Job Centres and have recently been questioned since rates are frequently higher in depressed regions than elsewhere – see Jackman and Kan (1988). While *à priori* surprising, the failure of higher local unemployment to encourage out-migration is well established in US data – see, for example, Da Vanzo (1978) and Greenwood (1975). For Holland, Van Dijk *et al* (1989) also find parallel results to those uncovered by Hughes and myself – higher local unemployment *reduces* out-migration. In our own work we have found a significant counter-intuitive relationship for British data from the 1970s and the 1980s and have come to believe that it is empirically robust. Why might this be? One conjecture is that higher local unemployment may discourage migration if employees anticipate that migration may be disappointing, so that return migration may occur. Higher local unemployment reduces the probability that return migrants can find a job as attractive as their former one and, by worsening the down-side risk associated with migration, reduces the resulting expected gain.

In the same study we also estimate the sensitivity of out-migration to relative local wage rates for the socio-economic group to which the individual belongs. We find that the response differs between occupational groups: higher real wages reduces the out-migration for all manual worker categories, whereas the relationship is weaker for non-manual workers. This may be because the manual occupational groups are more homogeneous than the non-manual ones so that the regional real wage differentials that we calculate are more accurate for manual workers than for non-manual workers, for whom the estimated differentials may reflect the concentration of superior professional/managerial jobs in certain regions. Pissarides and Wadsworth (1989) report a common wage elasticity of migration for all workers which is negative and significant.

Overall, however, the decision to migrate is not substantially affected by living in a region with high relative wages, even among manual workers. This accords with the first component of a broad view in US thinking, in which it is held that economic circumstances have little impact in determining the likelihood of migration, but a far greater one in the choice of destination (for example, Lowry (1966)). How far is this second claim true for the UK?

Economic influences upon destination choice

The one study of destination choice using UK national level individual data is Hughes and McCormick (1989a). This uncovers a strong quadratic relationship between the likelihood of choosing a region and the distance between the origin and destination region, which is similar in nature to that found in various US studies using aggregate data. Of greater interest is that real wages in a destination area have a conventional positive influence on in-migration in our various experiments, and is consistently significant: migrants are attracted to regions with high real wages. We included the job turnover rate as a measure of job availability – within a region for a migrant's socio-economic category – and found this to have a substantial influence on the willingness of migrants to choose a region. (For an analysis using aggregate rate of turnover in US migration behaviour see Fields (1976).) It seems to capture the role of the relative tightness or slackness of a region's labour market in attracting or repelling migrants more satisfactorily than either the unemployment or the vacancy rate. After the inclusion of the regional dummies, the coefficient on regional unemployment is positive rather than negative and so, in spite of its near significance, we decided to omit it from the final model. Having examined both out- and in-migration we can now answer the following question.

Does an increase in unemployment or a reduction in local wages stimulate net out-migration?

We have found above that the unemployed are much more likely to migrate than are the employed, whereas an increase in regional unemployment leads to a downward shift in all migration probabilities. The net effect of higher unemployment in a region on migration out of the region will therefore depend upon the relative importance of the higher level of migration due to the increase in the number of the unemployed versus the reduction in the migration rate for those who remain employed. In order to calculate the overall effect of changes in regional unemployment on migration, Hughes and McCormick have simulated their model for the sample of heads of households drawn from the 1986 LFS. The figures show that the net effect of an increase in unemployment

Migration and Regional Policy

rates in one region relative to the national average will depress outflow rates. Inflow rates, on the other hand, are hardly affected. Thus, net migration from the region is reduced, *ceteris paribus*, by higher local unemployment. The reasons for this remain unresolved.

Real wage adjustments offer a more conventional effect, with higher relative real wages leading to net inflows.

This leads to an interesting conclusion: the extent to which migration is likely to stabilise the consequences of an adverse regional shock depend on the extent to which the resulting excess labour supply prompts local real wage reductions. Only if real wages are flexible to local excess demand is there much likelihood of inter-regional stabilising flows in the manual labour market. In other words, pockets of high local unemployment which leave local real wages unaffected, are likely to persist.

Finally, we might note two related analyses of regional adjustment. First, Burridge and Gordon (1981) have argued that the failure of migration to eliminate unemployment differentials arises from the 'structural' component to unemployment, which they suggest gives rise to only a slight motive for migration, rather than demand deficient unemployment. There is not scope here to pursue this argument in detail, but it does raise interesting issues. Second, across county boroughs (urban areas) between 1966-71, employment growth subsequently failed to reduce local unemployment rates – perhaps due to in-migration or long-distance commuting – whereas employment contractions do significantly increase local unemployment (McCormick, 1983). Thus, new jobs in growing urban areas may have been filled by non-residents, while the redundant local workers in contracting areas did not migrate out. This is again consistent with the view that location choice is sensitive to local conditions, whereas out-migration is sluggish and perhaps insensitive.

III Why is there low mobility among British manual workers?

There are four factors that may contribute towards explaining the low mobility of manual workers in the UK relative to the US: taxation arrangements, education levels, wage setting and the extent of unionisation, and the nature of the housing system. Of these, the evidence points to the arrangements for tenant relocation that have been adopted by the public housing system as being the most significant contributor. It is a system which features little co-ordination between local authorities to

Reducing Regional Inequalities

meet tenants' requirements. The announced arrangements towards privatisation could well reduce mobility still further, since the private landlords will have no incentive to accept migrating tenants at subsidised rents when the sitting tenant quits. It is quite remarkable that the huge investment in public housing has been accompanied by measures to facilitate mobility that no politician of any party has, to my knowledge, expressed satisfaction with, and yet none in a series of govenments has seriously tried to reform. One can only conclude that it reflects the political interests of both major parties that manual workers in Britain are virtually immobile, and that unemployed manual workers in distressed labour markets should not be given more than nominal access to southern rental housing. The consequences for the efficiency of UK labour markets, and growth without inflation, are as considerable as the disinterest our political leadership has shown in removing them. In this author's view there are few more saddening examples of the consequences of the two major player political game that has been played in the UK.

Taxation

There presently exists a sharp and arbitrary distinction between the tax treatment of migration expenses according to whether or not these are incurred during a move sponsored by an employer. Employers are able to set such expenses against corporate revenue with corresponding tax savings; independent migrants, on the other hand, are not able to set expenses against income taxation (in contrast to the US). About half of migration is with a continuing employer, but this is overwhelmingly among non-manual workers. In this way, the tax system subsidises migration among non-manuals. This bias is scarcely justifiable – not least because of the need for labour mobility among manual workers to help equalise the circumstances in UK regional labour markets. Furthermore, in some cases, the use of generous bridging loans and refurnishing loans and allowances have become a tax-efficient way of compensating employees.

Education levels

The normal school-leaving age in the US is 18, as compared with 16 in the UK. While the extent of 'drop-outs' and college-going complicates the picture, it remains the case that the manual labour force has experienced more years of formal full-time education in the US and individual migration rates are strongly increased by formal education. In an attempt to allow for this, Hughes and McCormick (1987) simulated their model of UK migration and estimated that with US levels of formal education, manual worker migration rates would be increased by about 20 per cent.

Trade unions

Trade unionism is much more widespread in the UK than in the US, which also experiences higher migration. This may partially reflect an interesting two-way interrelationship between these variables. First, geographic immobility may have influenced the size of the union sector. To adopt Hirschman's terminology, trade unions provide a means for the expression of 'voice' when an issue of concern to the workforce arises. Without representation workers may alternatively respond to the source of concern by seeking work elsewhere – the 'exit' option. When mobility is costly, workers have an increased incentive to exercise the 'voice' option and to join and be active in unions. Adopting a more traditional argument, where labour turnover is high and substitute labour can easily move into the area, the difficulties confronting a labour organiser are greater and, if a union is formed, it will have more difficulty enforcing its preferences. Thus, the high unionisation rate in the UK may partially result from the factors contributing to geographic immobility, particularly among manual workers. Furthermore, the choices of unions (for example, whether to strike or not) and the adversarial character of management–union relations – the traditional explanation for low UK productivity – may also reflect the costliness of exercising the 'exit' option for many British workers.

Second, a large union sector will presumably result in a larger share of manual workers earning above a readily-available alternative wage. The existence of 'surpluses' in jobs will lead to a diminished incentive to migrate between regions and, the greater the proportion of workers earning a surplus, the lower are the migration rates. The highly competitive nature of the US job market may have reduced the existence of such surpluses and thus the costliness of job turnover to an individual worker, although there is as yet no empirical estimates of its importance.

Housing market issues

It is helpful to summarise the ways in which geographic mobility is influenced by housing policy by beginning with a description of how these policies have determined the evolution of tenure patterns. Three

Reducing Regional Inequalities

housing policies are of prime importance to the geographic tenure patterns: the decision to construct – densely in certain regions – large amounts of public rental housing; the adoption of rent control; and tax relief to encourage owner-occupation. These policies have combined to reduce the private rental sector to a comparatively tiny size – about 8 per cent of households – so that the ability of rental accommodation to serve as a supply of short-term accommodation has been dramatically reduced. These housing policies have almost certainly had a deep impact on the functioning of the labour market, and in ways which we are only slowly coming to recognise.

Council housing

About one-quarter of UK families live in council housing – down from one-third 15 years ago – and a little under one-half of manual workers are accommodated in the public sector. The contraction of the private rental sector has amplified the weaknesses in the operation of the council housing system, which itself discourages migration by its tenants. Having little access to private rental housing, council tenants wishing to migrate who are unable or unwilling to purchase a dwelling must obtain council housing in their destination region. To locate suitable accommodation council tenants are expected to rely upon council house exchanges/transfers; if there are unequal numbers of households seeking to enter and leave an area some workers will be 'rationed' out of an exchange. Such an imbalance is inevitable given the asymmetry of the UK labour market: the largest stocks of council housing are located in regions of high unemployment, from where workers are likely to want to migrate.

The use of tenure as an explanatory variable in migration decisions is complicated by the fact that, for some individuals, the decision to migrate and the choice of tenure are jointly determined. There is not the space here to discuss the problems raised by the resulting selectivity bias – more than to note that the comparisons between owner-occupiers and council tenants may, if anything, *overstate* the migration rates of council and private tenants relative to owner-occupiers. (Those interested in this are referred to Hughes and McCormick (1987).)

The available evidence (Hughes and McCormick, 1981, 1987; Pissarides and Wadsworth, 1989) suggests that council tenants have migration rates about 20 to 25 per cent those of owner-occupiers, all else held equal (including age, industry, occupation, education, region, etc). Against this evidence one might argue that council tenants are a self-selected group who are less likely to move even after factors such as age, education, industry, occupation and region are allowed for. It is noteworthy, therefore, that when total house movement is studied, making no distinction according to whether a regional boundary is crossed, council tenants have a movement rate which is about 60 per cent higher, *ceteris paribus*, than for owner-occupiers. Thus, within local areas, the council system seems to work well. It is no less striking that if *intended* migration is studied, council tenants are no less likely than owner-occupiers to wish to migrate, merely less successful in fulfilling their intentions. The difficulties arise where house exchanges are sought *between* labour market areas so that the cost of collecting information may be high, and there may be a substantial imbalance in the numbers of intended movers in each direction.

Regional house price differentials and migration

Land prices are perhaps the major source of variation in both production and living costs between different geographic areas in the UK. Long-run differentials are therefore important in discouraging land intensive activities from locating where the demand for land is relatively high: for example land intensive manufacturing industry will locate away from areas in which gains from the agglomeration in service activities arise. Distortions in the supply of land, deflecting from its use of highest value (including recreational) will alter the long-run relative price of land and the allocation of work and housing throughout the economy.

While there has yet to be detailed econometric study of the role of *long-run* land or house price decisions on the location decisions of UK firms and households, there has recently arisen an interesting analysis of how short-run (or cyclical) fluctuations in relative regional house prices affect the efficiency of the labour market.

The innovative work of Bover *et al* (1989) argues that to understand wages and unemployment it is important to allow for the role of relative regional house prices on regional labour mismatch, and that government policy shocks in financial markets may influence the portfolio demand for housing predominantly in the south, creating cycles in the ratio of 'south' to 'north' house prices. Thus, quite apart from the conventional view that long-run average house price differentials influence migration, the Oxford group has stressed the labour market inefficiencies that may result from *cyclical fluctuations* in regional house prices. In particular, they stress how this may result in the net migration flows to the south having a cyclical pattern, the volatility of which may be increased by tax relief to mortgage interest payments, thereby increasing the argument for the abolition of such relief. Finally, it is conjectured that year-to-year fluctuations in annual migration to the South East may be sufficiently great as to bring about fluctuations of labour shortages in the South, thereby influencing the cyclical pattern of aggregate wage determination.

A more conventional argument would stress product market shocks as the major source of relative regional price and quantity adjustments. Thus, the initial shock is to product markets which changes the regional pattern of labour demand, which in turn leads to regional wage and/or unemployment adjustment, and these finally change local house prices. This is not to argue that house price changes do not influence the incentives to migrate, simply that shocks in the financial and thus the housing market are not the primary *cause*, but the *consequence* of adjustment in the product/labour markets.

Independently of whether the shocks prompting the regional allocation of resources arise primarily in financial/housing or in product markets, if it is recognised that immigrants to the buoyant labour market regions are relatively young, then there exists a further interpretation of the link between house prices and the labour supply.

As the buoyant region expands, house prices and earnings rise while unemployment falls. This draws in young immigrants who will tend to displace those who enjoy only a marginal gain from living in the increasingly high cost of living, high wage region – the retired or those approaching retirement. This migratory process reduces the average age and increases the labour supplied by residents in the buoyant region, so one should hesitate therefore, to infer from a rising relative price of housing in a region that the labour supply is being 'choked off'. However, the extent to which this is empirically important remains an open question.

The empirical work that Hughes and McCormick (1989b) have undertaken on behalf of the Treasury suggests that short-run cycles in relative regional house prices primarily influence migration through their effect on out-migration rates for those over 54 years and, to a lesser extent, those aged 35–54. However, this effect is not estimated to be very large in terms of the total amount of migration affected. For this reason year-by-year relative regional house prices may not be capturing year-by-year fluctuations in labour market mismatch when entered, with a short lag, into an aggregate wage equation (Bover *et al*, 1989). Even so, the distortion of migratory patterns by these short-run regional house price fluctuations may still impose substantial labour market inefficiency. Finally, we may note that the view that *long-run* differences in mean regional house prices exercise an influence on migration remains plausible.

IV Policy analysis

The absence of a commonly agreed model which explains the regional pattern of unemployment, wages and migratory flows, and the prevailing difference of views concerning the crucial market failures and distortions that require correction, inevitably makes regional policy prescriptions a difficult, some might say impossible, terrain. Nevertheless, while there is no substitute for a fully articulated analysis of these issues, the regional problem is too significant to be neglected until this research is done.[3]

How, then, might we usefully proceed in the space available? First, we need a keen sense of why, even if markets in labour and capital were perfectly mobile, the competitive allocation of labour between UK regions is unlikely to be that which a wise government might arrange. Having educated our intuitions concerning potential market failures in an economy with mobile factors, and considered how we might 'patch' this up, we may then turn to the very significant forces that provide barriers to efficient (not 'maximum'!) levels of labour migration – particularly of manual labour – in the UK. In other words, we need to imagine an optimal set of policies once artificial barriers to regional factor mobility are eliminated, and to combine this with a strategy of sweeping away these restrictions.

Let us begin with a frictionless two-region view of the UK economy in which labour and capital are perfectly mobile and firms are competitive. Why might labour and capital not be allocated between regions in a welfare-maximising fashion, given the UK institutional policy context? The source of distortions would appear to mainly arise in the markets for labour and land. Before detailing the arguments it is worth noting that deficiencies in the competitive allocation are, in general, appropriately corrected by intervening in the distorted market itself. Thus, it follows from our preliminary diagnosis that first best policy intervention is likely to primarily revise policy in the markets for labour and land. It is useful to begin with the market for land.

The supply of land for development

There is no practical alternative in a densely-populated western country

---

[3] There is not the space here to review the contribution of earlier policies. For a useful discussion see Green *et al* (1986).

to land-use controls of some type, if only because the use to which land is put on a neighbouring plot is of vital importance to one's own utility. Thus, a basic assumption in the theory that competitive behaviour leads to an efficient outcome is transgressed: other people's choices in a locational context directly affect one's own utility. However, efficient land-use rules must allow for the 'rise and fall' of regional comparative advantage – we cannot rigidify the location of production and housing without great cost. These rules must also aim to preserve the 'quality' of the environment – with the well-known facets of congestion, pollution and beauty that are of concern. This leads to a basic question.

Is it likely that current institutional arrangements encourage communities to 'internalise' these concerns and provide approximately the socially-optimal amount of land for development?

The answer appears to be that there exists systematic and, over the past two decades, increasing political pressure to under-supply land for housing development. This arises because of the absence of a mechanism which enables the major beneficiaries from development (land-owners and the Treasury – thus, the general taxpayer) to compensate the 'losers' in any particular project. The 'losers' are local residents and voters who, most noticeably in the past 20 years, understandably oppose new housing projects: the additional housing reduces local amenities and increases the supply of competing assets. If part of the surplus from the project was allocated to compensating the local residents – and there are various ways of doing this – then an efficient allocation of land to housing use would emerge. Since the highest demand for housing in the past two decades – as reflected in price – has been in the south, it is there that the under-allocations of land-use to housing are likely to be greatest. I am certainly not suggesting here a dramatic unrolling of a 'concrete carpet', but simply that by failing to sensitively develop in those areas where marginal housing projects are of slight environmental concern, we are likely to reinforce the long-run trend towards higher house prices in the South East, to prompt an inefficient allocation of production to the South East, and to incur substantial costs to the whole economy. As one symptom of this, during the 1980s there has been more or less zero net-migration into the South East (including Greater London), which, given current unemployment patterns and the approach of 1992, appears unlikely to be optimal.

If land development arrangements have limited the supply of housing and raised the cost of living in the 'south', a similar consequence is likely to follow from the various policies that have contracted the provision of private rental accommodation. Whereas in high-density areas of various other OECD countries good quality, private rental apartment-style accommodation – suitable for small professional families – is readily available, this is not so for Britain. This under-provision of apartment buildings will affect mostly the high land value areas of the country – notably the South East and Central London – reducing the amount of accommodation supplied per acre of developed land. Until the subsidies on other forms of tenure are reduced, it appears unlikely that simply removing restrictions on the working of the private rental market will sufficiently increase the demand for such accommodation (Whitehead and Kleinman, 1989). To move towards an optimal distribution of labour between high and low-density areas of the country, and the construction of purpose-built private apartment blocks in the south, a withdrawal of government from its involvement in subsidising certain types of housing tenure – both owner-occupied and council tenant – appears no less desirable than the ending of rent controls.

If government could be persuaded to remove the distortions which influence both the supply of land and the distribution of accommodation between tenures, would this bring about the broad circumstances in which there is an optimal allocation of labour to the south? There would remain, it is believed, at least two further major distortions. First, presently, restrictions on land supply for housing development are the primary means to control population growth in the South East and, if these are relaxed, there are likely to arise congestion externalities. Second, in the labour market, pay for manual labour – and certain white-collar occupations – is largely inflexible between regions. On the latter, we need to carefully examine the cost-benefits of current inflexibility, but on the prevailing evidence it would appear a potentially crucial obstacle to reaching an optimum allocation – seriously distorting the supply of manual labour away from the South East. This will be considerably amplified by Nicholas Ridley's reforms aimed at reducing the subsidy content in council rents, which is especially large in the South East.

Congestion

If restrictions on population growth in the South East are eased, this will increase congestion externalities in the region. These should be tackled with taxes carefully directed at whichever market the congestion or pollution arises. We have not yet gone very far along the road of 'green' taxation, and this is important. However, in so far as appropriate taxation proves impractical for certain types of congestion, and in the event that excess persons in the 'south' remains the source of the congestion, a simple incremental local tax for residents in the south deserves careful consideration. Rather than rely upon the arbitrary implications of restricting the supply and nature of accommodation to limit population in the south, the differential taxation of southern residents would discourage those who place only a marginal valuation on being in the south from locating in the region. Alternatively, a land value tax would provide one avenue not only to meet congestion but, with the rise and fall of comparative advantage and congestion pressure, also to provide individuals with some insurance against adverse changes in regional comparative advantage.

If, as appears likely, a regionally based poll or land tax is regarded as politically unacceptable then, as a second-best device, the government could consider other ways of using subsidies/taxes to marginally reallocate population from the South East. One such possibility will be taken up in the final section.

Regional policy in capital and output markets

Thus far we have examined leading distortions that might result in a regional misallocation of resources, despite there being perfect labour mobility. Some readers may be disturbed by a neglect of possible distortions in the markets for capital and output. For example, it could be argued that high-tech industries based in the south are likely to be ignorant of the opportunities for capital investment in the north. This is an argument for subsidised information – presumably a crucial function of the development boards. Alternatively, it is argued that the costs of socially-excessive unemployment in the depressed region can be reduced by capital market intervention. However, this is generally an argument for correcting the distortions that bring about excessive unemployment: to correct distortions in market x with policies in market y needs to be justified in terms of constraints on policy instruments in market x. It is not clear that such arguments are justified. In summary, I am not convinced that there exist strong arguments for regional subsidies or other intervention in the capital and output markets. The relatively low price of land, labour and access to workplace in the less prosperous areas will, by themselves, serve to attract capital. This would be reinforced with labour market policies directed at correcting the distortions responsible for socially excessive unemployment. This approach, perhaps, is recognised in the passing of that collection of policies – Industrial Development Certificates, and various location controls (Distribution of Industry Act 1958) – aimed at constraining the movement of capital to depressed areas.

Public goods

Before turning to barriers to migration, there is the difficult question of the optimal regional supply of public goods. At an equilibrium distribution of labour between regions there will be an appropriate level of public goods – roads, public buildings, parks, etc – together with a range of potentially private goods that require a minimum-sized area to warrant their presence – orchestras, opera halls, high-quality football clubs, etc. If the government understates the optimal population of a town then public goods are under-supplied, so that certain private goods are also likely to be under-supplied, and the resulting population will be less than the optimum, but consistent with the public good. This, of course, poses a difficult problem for governments allocating scarce resources, and can be used to justify in a second-best context populations in excess of the first-best optimum in a declining area. While there is clearly scope for poor decisions in this area, public good provision is crucial and there is little evidence that the north and west would fail to benefit from a redirection of expenditures from the South East.

On balance it appears that there are likely to be greater net costs of too small a 'north' – largely because of the difficulties of taxing congestion – and that public goods in the depressed areas should err on the side of being too much rather than too little. There is a second, more subtle, source of public concern about public goods in 'the north' that relates to migration patterns. For the foreseeable future it appears likely that the non-manual workers will leave 'the north' at a greater net rate than manuals – largely as a result of artificial considerations which affect barriers to mobility and are discussed below. This may well have seriously contributed towards undermining balanced regional growth so that, until the long-run effects of these barriers are unpacked, then there is a 'second-best' case for being especially concerned to support incremental public goods expenditure in the north that non-manuals in particular may value. Thus, if non-manuals value highly health care, quality of schools, leisure facilities, etc when choosing where to live, then it appears especially important that the standards that are offered are carefully maintained. A policy of carefully monitoring – and publicly advertising – standards in these depressed areas, could pay substantial dividends towards balancing skills and, thereby, regional development.

Policy reform and barriers to regional mobility

While it is important to be clear why competitive behaviour may require policy intervention in order for it to provide the optimal allocation of

Reducing Regional Inequalities

labour and capital between regions, we must also recognise, and gradually eliminate, the battery of policies which have artificially reduced migration among certain segments of the working population.

Harmonisation of tax treatment of migration expenses

I am unaware of any serious argument by which we may justify allowing firms to set the mobility expenses of employees against corporation tax, but to exclude the migration expenses of unsponsored migrants of working age as an income tax deduction. This bias against unsponsored migration places the greatest burden on manual workers, which is where the largest inequalities in regional unemployment exist and, thus, *ceteris paribus*, where the greatest need for migration arises. The US tax code allows migration expenses to be tax-deductible and UK reform in this matter would not appear to be exceptionable.

How might we proceed? From an equity viewpoint the neatest arrangement would be to disallow corporations from setting mobility expenses against taxation, but to allow all individuals to set migration expenses, below an upper limit, against income tax when moving over 30 miles (say). Thus, firms would find it tax-efficient to compensate employees for migrating on the job with a higher (possibly lump sum) salary, provided corporation tax was not less than the combined marginal rates of income tax and national insurance contributions.

At present migration levels it would not be unreasonable to assume that, each year, in the order of 200,000 heads of household migrate between regions and would be eligible for such a tax allowance. Of these, about 30 per cent have expenses which are already set against corporate taxation. This leaves about 140,000 regional migrants eligible for tax relief against migration expenditures and, assuming an average migration cost of £3,000 and a standard rate of income tax of 25 per cent, this implies an annual cost to the Exchequer in the order of £105 million per annum. This would not appear an excessive sum for a reform which may not only substantially improve labour market efficiency and raise the numbers in employment, but would particularly assist and encourage those who are ineligible for support by an employer, and thus are likely to be in a manual (and thus low income) category.

Privatisation of 'council housing' and regional mobility

The short-run consequences of privatisation for mobility are likely to be mildly unfavourable: purchasing tenants are often obliged to return discounts if they sell their property within a few years, and house sales have been heavy in the South East, thereby probably increasing the excess demand for north-south tenant migration. In the longer run the effects are more favourable, given the evidence concerning the relative migration rates of council tenants.

Our primary concern, however, should be with the consequences of 'privatisation' for the mobility of the larger portion who remain tenants – the 'residual' potentially comprising as great as 20 per cent of all households. Under present plans for the council housing stock, privatisation is likely to result in a patchwork of small landlords – either voluntary housing associations or private – administering the housing stock at controlled rents, so long as the sitting tenant continues in occupation. As new tenants will be expected to pay rents at closer to market levels, the total costs to migrating by means of exchange will be much greater. Equally, landlords have no incentives voluntarily to join exchange schemes, where incoming tenants can enjoy the low rents offered to the sitting tenant. The only arrangements for exchange likely to emerge are those contracted for initially when the housing is sold by the local authority. Except for local arrangements within a given housing association scheme, I am unaware that such provision is being arranged. If this is not altered then we may anticipate the continued and even accentuated rigidity of location among newly privatised 'council' tenants.

Migration, pay flexibility and Nicholas Ridley's rent subsidy reforms

During the 1980s, real council rent levels have risen considerably with regional differences in rent levels becoming more marked. This will emerge as migration from the South East to council housing elsewhere – as Mr Ridley intended – although perhaps not on the anticipated scale. While Ridley was probably wise to undertake rent reform – provided the rental increase is indeed gradual, as planned – the additional need for *pay flexibility* to accompany this substantial deviation from uniform regional living costs appears important if the South East is not to become increasingly constrained by labour shortages. It is not hard to envisage how regional pay flexibility in certain markets, coupled with inflexibility in others, will result in increasingly serious economic misallocations and social distress.

Reserving council relets for the 'depressed' region unemployed

A policy requiring local authorities in prosperous areas to increase from 1 per cent to, say, 5 per cent the council relets reserved for migrants could have a dramatic impact on mobility and upon both job prospects for

Reducing Regional Inequalities

those unemployed in depressed areas, and the allocation of manual workers between regions. If the unemployed by redundancy were given priority to these relets in 'southern markets' – again, hardly an exceptionable proposal – this would further improve the efficiency of labour market adjustment, and help to offset the tendency towards regional imbalance in the manual and non-manual labour force.

V Efficiency and densely populated regions

This chapter began by arguing that British regions stay depressed for exceptionally long periods, by international comparisons, and that by primarily reducing the mobility of manual workers, various policies may contribute to, or even accentuate, this tendency.

Not only do mobility-reducing policies reduce gross migration and thus the efficiency with which workers are matched to jobs, but, by being applied primarily to manual workers, contribute to uneven unemployment, inequality, and imbalanced regional growth.

If these controls upon migration are removed and policies towards the allocation of land to housing in 'the south' are reformed as discussed above in section III – will this not lead to over-congestion in 'the south'? How is the brake to be applied to the allocation of population to the more productive south, if not by inefficient quantity controls in the land market?

One plausible answer is that prices, rather than the present system of quantity controls, could play a larger role in allocating households and firms to the regions. Regional differences in house prices, wages and various congestion taxes, would provide firms and households with a more accurate opportunity cost of their choice of locations than current arrangements. The chief concern that many would express with this approach is that for various practical reasons, congestion taxes would be set below their optimal level, and that 'the south' will become unduly crowded. If this is so, and if an increased local tax on residents in 'the south' were politically unacceptable by itself, then there remain policies which could resolve the problem. A small local property tax could be levied in 'the south' to finance migration from 'the south'. Thus, southern residents might view themselves as contributing to a fund to reduce congestion, from which they benefit. The subsidy might be framed as an exemption from stamp duty when purchasing a property elsewhere, or in terms of generous provision for tax allowances.

If the poll tax/migrant subsidy rates were to imply a negative expected value for migrants to 'the south', then the impact effect of the scheme would be to encourage net migration from 'the south', and lower population congestion. In due course the lower population and house prices in 'the south', together with higher wages would restore migration balance. This approach could be given a further twist.

If those approaching retirement are more sensitive than younger workers to non-labour market migration incentives – as the response to house prices suggests – then the dead-weight loss from the migrant subsidy may well be less if the subsidy were to apply only to those over a certain age (say 55 years).

Thus, a large migration subsidy to those over about 55 years might be the least costly way of securing a given level of net migration from 'the south', while at the same time having the potentially advantageous effect of reducing both the mean age in 'the south' and the congestion resulting from allowing a shift in the regional allocation of employment towards the south and east. If the costs of distorting the economy by raising tax revenue were very low then we might proceed with substantial migration subsidies to reduce congestion to levels at which it was not regarded as socially costly. The argument is given in rather a bare outline, but there is evidence that those near retirement migrate with a view to financial advantage (Hughes and McCormick, 1989b). A policy which increased out-migration from the South East by 10,000 long-standing southern households per annum, for those with heads of households over 55 years – financed by a tax on southern residents – would in a steady state allow in the order of an extra 200,000 people of working age to live in the South East (including Greater London), with no increase in population. At present about 30,000 persons over 55 leave the South East each year. If a subsidy of £3,000 would achieve the extra migration, then the total cost would be £120 million. The dead-weight costs of raising the taxes to finance this sum may easily be less than the efficiency gained from increasing employment in the South East without altering population.

This chapter has focused on the role of migration in the formulation of regional policy, but it is important not to close before emphasising the crucial nature of both distortions in various urban markets, and imperfections in the market for retraining in formulating policies which will assist depressed regional labour markets.

Bibliography

Bover, O, Muellbauer, J and Murphy, A (1989), 'Housing, Wages, UK Labour Markets', *Oxford Bulletin of Economics and Statistics*, March 1990.

Burridge, P and Gordon, I (1981), 'Unemployment in British Metropolitan Areas', *Oxford Economic Papers, vol 133*, pp274-97.

Champion, A (1989), 'Internal Migration and the Spatial Distribution of Population', in Joshi, H (ed), *The Changing Population of Britain*, Basil Blackwell, Oxford.

Creedy, J (1974), 'Inter-regional Mobility: A Cross-section Analysis', *Scottish Journal of Political Economy*, vol 21, pp41-53.

DaVanzo, J (1978), 'Does Unemployment Affect Migration? - Evidence from Micro Data', *Review of Economics and Statistics*, vol 60, pp504-14.

Elias, P and Molho, I (1982), 'Regional Labour Supply: An Economic/ Demographic Model', *Population Change and Regional Labour Markets*, OPCS Occasional Paper 28, HMSO, London.

Fields, G (1976), 'Labour Force Migration, Unemployment and Job Turnover', *Review of Economics and Statistics*, vol 58, pp407-15.

Gordon, I (1985), 'The Cyclical Interaction Between Regional Migration, Employment and Unemployment: A Time Series Analysis for Scotland', *Scottish Journal of Political Economy*, vol 32, pp135-58.

Gordon, IR (1987), Evaluating the Effects of Employment Changes on Local Unemployment, *Regional Studies*, vol 22, pp135-47.

Green, AE, Owen, DW, Champion, AG, Goddard, JB and Coombes, MG (1986), 'What Contribution Can Labour Migration Make to Reducing Unemployment?', *Unemployment and Labour Market Policies*, Joint Studies in Public Policy, 12, Gower, Aldershot.

Greenwood, MJ (1975), 'Research on Internal Migration in the US: A Survey', *Journal of Economic Literature*, vol 13, No 2, pp397-433.

Hart, RA (1970), A Model of Inter-regional Migration in England and Wales, *Regional Studies*, vol 4, pp 279-96.

Hughes, GA and McCormick, B (1981), 'Do Council Housing Policies Reduce Migration Between Regions?', *Economic Journal*, vol 91, pp919-37.

— (1985), 'Migration Intentions in the UK: Which Households Want to Migrate and Which Succeed?', *Economic Journal*, vol 95, Conference Supplement, pp76-95.

— (1987), 'Housing Markets, Unemployment and Labour Market Flexibility in the UK', *European Economic Review*, pp615-45.

— (1989a), 'Is Migration in the 1980s Narrowing the North-South Divide?', University of Southampton Discussion Paper No 8913.

— (1989b), 'Do Regional Differences in House Prices Influence the Patter of Migration?', mimeo.

Jackman, R and Kan, B (1988), 'Structural Unemployment: A Reply', *Oxford Bulletin of Economics and Statistics*, vol 50, pp83-7.

Johnson, GE and Layard, R (1986), 'The Natural Rate of Unemployment: Explanation and Policy', *Handbook of Labour Economics*, vol 2, North Holland Press, Amsterdam.

Lowry, I (1966), *Migration and Metropolitan Growth*, Chandler, San Francisco.

Makower, H Marschak, J and Robinson, HW (1938-40), 'Studies in the Mobility of Labour', *Oxford Economic Papers*.

McCormick, B (1983), 'Housing and Unemployment in Great Britain', *Oxford Economics Papers*, vol 35, pp283-305.

— (1988), 'Unstable Job Attachment and the Changing Structure of British Unemployment', *Economic Journal*, supplement, pp132-47.

Molho, I (1984), 'A Dynamic Model of Inter-regional Migration Flows in Great Britain', *Journal of Regional Science*, vol 24, pp317-37.

Pissarides, CA and McMaster, I (1984), 'Regional Migration, Wages and Unemployment: Empirical Evidence and Implications for Policy', mimeo, Centre for Labour Economics Discussion Paper 204, LSE.

— and Wadsworth, J (1989), 'Unemployment and Inter-regional Mobility of Labour', *Economic Journal*, September 1989, pp739-55.

Schlottman, AM and Herzog, HW (1981), 'Employment Status and the Decision to Migrate', *Review of Economics and Statistics*, vol 63, pp590-8.

Van Dijk, J, Folmer, H, Herzog, H and Schlottman, AM (1989), 'Labour Market Institutions and the Efficiency of Inter-regional Migration', in *Housing and the National Economy*, ed J Ermisch, NIESR: Gower, Aldershot.

Whitehead, C and Kleinman, M (1989), 'The Viability of the Privately Rented Market', in *Housing and the National Economy*, ed J Ermisch, NIESR: Gower, Aldershot.

6
Regional Economic Disparities: The Role of Housing

John Muellbauer and Anthony Murphy

I Introduction

In this chapter we consider the role of housing in contributing to regional economic disparities. The discussion draws on and summarises our joint work in a number of areas – the interaction between UK housing and labour markets; the determinants of national and regional house prices; and the determinants of migration to and from the South East region. Here is a short summary of the chapter.

Housing tenure, labour mobility and regional mismatch

It is widely recognised that the UK housing tenure structure inhibits labour mobility. The lack of a free rented sector, as a result of controls on rents and security of tenure, and the restrictions on the mobility of council tenants have long been seen as handicaps reducing the efficiency of British labour markets by writers such as Hughes and McCormick (1981, 1985, 1987) and Minford *et al* (1987, 1988). Incentives for owner occupation, as well as direct controls, account for the decline of the private rented sector. Holmans (1987) is of the opinion that the former factors are the more important.

At the regional level, the low rate of labour mobility due to the structure of housing tenure increases mismatch, since the various regional labour markets are partly segmented. As a result, regions with high unemployment and few vacancies, reflecting excess supply, coexist with regions with low unemployment and many vacancies, reflecting excess demand. This wide dispersion of excess demands across regions increases aggregate wage pressure, since regional wages are relatively unresponsive to excess supply but not excess demand. It also increases aggregate unemployment, for a given level of vacancies, and adds to the persistence of regional unemployment differentials.

Owner-occupied house prices

The literature on the relationship between housing tenure and labour mobility tends to focus on the council and private rented sectors. Labour mobility in the owner-occupied sector is low by international standards, but is not considered to be the main problem. However, the 'free' market in owner-occupied houses is extremely distorted which also has implications for labour mobility and mismatch.

Our research has examined the effects of owner-occupied house prices, as well as the structure of housing tenure, on aggregate wages and unemployment. We find that the regional dispersion of house prices also significantly contributes to mismatch in the short to medium term, adding to aggregate wage pressure and aggregate unemployment via regional mobility trap and cost of living effects. This line of research is relatively novel.

Regional house price differentials and aggregate wages

Everyone knows, of course, that higher earnings quickly drive up house prices. But it has not been realised that there is also a reverse effect, with a delay of between one and three years. In Bover *et al* (1989) we model real wages and show that both the average UK house price to wage (HP/W) ratio and the South East/UK difference in this ratio, as well as the structure of housing tenure, help to explain real wages in later years.

Why focus on the South East region and on the ratio of house prices to income? We focus on the South East region since this region has a high rate of activity, high wages and low unemployment relative to the UK as a whole, and particularly so during the 1980s. We use the house price to wage ratio to capture house price effects for the simple reason that this ratio is what really matters to people.

Interpretation of regional house price effects

Much controversy still surrounds the exact mechanisms by which these house price effects operate on wages. The house price/wage ratio in the South East, relative to the rest of the economy, rose from 1983 to 1988, reaching unprecedented levels in 1988. We suggest that this helped to crowd out workers, especially manual workers, from the South East, adding to wage pressure there. But the extra workers in the more depressed parts of the economy did not exert a corresponding reduction in wage pressure. Thus, overall wage pressure increased.

Variations in net migration are, of course, small relative to the mismatch between vacancies and unemployment. However, this does not imply that variations in regional mobility due to the structure of housing tenure necessarily have little or no impact on wage pressure. Both theory and empirical evidence suggest that changes in both unemployment and mismatch are more important than stocks of these in generating wage pressure. In addition, it is plausible that the workers whose mobility is restricted by high regional house price differentials are the more skilled and highly paid ones who are of greater importance in determining the pattern of pay increases.

There are also direct regional cost of living effects from house prices and mortgage costs on wage demands. Many young workers with mortgages find that 40 per cent or more of their outgoings – more than twice the derived average share of housing expenditure obtained in recent Family Expenditure Surveys – are swallowed up by mortgage repayments. House prices are by far the major regionally variable component in the cost of living.

Further evidence

We obtain further confirmation for the regional mobility trap associated with both the regional dispersion of house prices, or more strictly the difference in the HP/W ratio between the South East and the UK, and the structure of housing tenure in our models for the relationship between unemployment and vacancies, and for migration to and from the South East region. Of course, before we draw any policy conclusions from these results, we must examine the determinants of regional house prices.

Determinants of regional house price differentials

In the short run, the supply of housing is fixed and in the medium run the supply of housing in the South East is less elastic than elsewhere, presumably because planning controls bite deeper in the South East. This implies that both positive national demand (or supply) shocks, as well as regionally differentiated shocks favouring the South East, will increase the difference in the HP/W ratio between the South East and the UK as a whole. The short-run jump in the regional difference in the HP/W ratio is to be expected. Over time the rise in this ratio is eroded as housing supply increases, wages adjust and both firms and jobs relocate between regions. This is a plausible neo-classical adjustment process.

However, we find little evidence that the house price effects in our wage equation are merely proxying unobservable demand shocks. The house prices effects enter with quite long lags and are robust to various attempts to construct other proxies for demand shocks. In addition, the unemployment/vacancy and migration equation results are consistent with either regional mobility trap or regional cost of living interpretations. The results are not consistent with a regional demand shocks interpretation of the house price effects.

Moreover, we argue that, in the short to medium term, both the national HP/W ratio and the regional difference in the same ratio overshoot excessively, resulting in costly dislocation, job mismatch and inflationary pressure. Here we must distinguish between 'efficient market' overshooting and 'inefficient market' overshooting due to extrapolative expectations or speculative bubbles. We must also distinguish between the demand for housing services and the demand for housing as an asset. The real return on owner-occupied housing is considerably greater on average than the returns on all other common forms of saving. The tax incentives favouring owner occupation contribute to this high rate of return. We argue that, due to extrapolative expectations and speculative frenzy, both national house prices and regional house price differentials exhibit inefficient market overshooting. The resulting instability of the housing market amplifies rather than dampens down shocks to the economy.

We find evidence for this in our research on national and regional house prices. Although South East house price differentials do respond to local demand shocks, we find that this effect is relatively small. The major factors driving the South East house price differential are the same as those driving the national HP/W ratio. Among other standard factors, we find strong 'frenzy' type effects as well as evidence of extrapolative house price expectations. The implication is that, as house prices accelerate, participants in the housing market are made even more fearful of missing out on the rise or of being locked out of owner occupation and never again being able to join as house prices rise out of their reach. Combined with extrapolative expectations, there is strong evidence here of the powerful instability of the housing market.

The structure of this chapter

In section II we briefly review national and regional trends in the structure of housing tenure and in house prices. We also discuss how to construct an index of mobility based on the structure of housing tenure.

Reducing Regional Inequalities

We consider the effects of the structure of housing tenure and house prices on aggregate wages and unemployment in section III. We also outline our view of the interaction between housing and labour markets. We look at regional migration to and from the South East in section IV. The determinants of national and regional house prices are examined in section V and we conclude in section VI by outlining some policy options.

II Trends in the structure of housing tenure and house prices

The structure of housing tenure

Although the structure of housing tenure varies widely between regions, the changes over time are remarkably similar as shown in Table 6.1. More

Table 6.1 *The regional pattern of housing tenure*

	Owner occupied '71 %	'80 %	'89 %	Council rented '71 %	'80 %	'89 %	Other rented '71 %	'80 %	'89 %
North	42	46	59	39	40	30	19	12	11
Yorkshire and Humberside	50	55	65	32	32	25	18	12	10
North West	53	59	68	28	30	23	17	10	9
East Midlands	53	58	70	28	29	20	19	13	10
West Midlands	52	57	67	34	33	25	14	10	8
East Anglia	53	59	70	26	26	18	21	15	12
South East	53	57	69	25	27	20	22	16	11
South West	59	64	73	22	22	15	19	14	12
Wales	56	60	71	28	29	20	16	12	9
Scotland	31	36	47	52	54	44	17	10	9
Northern Ireland	47	48	–	34	41	–	19	11	–
UK	50	55	66	31	32	24	19	13	10

Note: Other rented includes rented from housing associations

Sources: Regional Trends and Housing and Construction Statistics

detailed time series for the UK and the South East region are set out in Table 6.2. It is clear that changes in the national pattern of housing tenure are considerably more important than changes in the inter-regional tenure pattern when considering regional labour mobility.

Table 6.2 shows that the UK share of owner-occupied housing has risen continuously since the mid 1960s from 47 per cent in 1966 to 66 per cent in 1989. The share of council housing rose from 29 per cent in 1966 and peaked at 32 per cent in the late 1970s. It then fell back to 24 per cent in 1989. However, the share of other rented housing, which includes housing association lets, has declined from 19 per cent to 10 per cent since the mid 1960s. The decline in this sector since the 1950s is even more dramatic. In addition, the composition of the private rented sector is far from uniform. For example, there are furnished and unfurnished lets. Unfurnished lets may be controlled or regulated.

Table 6.2 *Trends in housing tenure structure*

	Owner occupied %	Council rented %	Other rented %
UK			
1966	47	29	24 (–)
1971	50	31	19 (–)
1976	53	32	15 (–)
1981	57	31	12 (2)
1986	63	27	10 (3)
1989	66	24	10 (3)
South East			
1966	50	24	26 (–)
1971	53	25	22 (–)
1976	55	27	18 (–)
1981	59	26	15 (3)
1986	65	23	12 (3)
1989	69	20	11 (3)

Notes: Housing associations are included in the other rented sector. Data for the share of houses rented from housing associations for more recent years are shown in brackets

Source: Regional Trends and Housing and Construction Statistics

Minford *et al* (1987) provide a detailed discussion of the composition of the private rented sector. The implications of these divergent housing tenure trends for labour mobility will now be considered and a summary index of labour mobility will be constructed.

The structure of housing tenure and mobility

There is a very substantial body of research on housing tenure and mobility reviewed, for example, by Minford *et al* (1987). These studies suggest that council house tenants are by far the least mobile. Owner occupiers are less mobile than private renters, although we know that the mobility of this group has declined significantly as a result of the controls in the 1965 and 1974 Rent Acts. Within the private rented sector, furnished tenants are the most mobile.

These findings are robust even when controlling for socio-economic characteristics. Some of the observed immobility of council tenants is due to unobservable sample selection bias, since council housing is housing of last resort. However, we have little doubt that only a part can be explained in these terms since there is ample evidence that, given the desire to migrate, frustrated migrants are particularly prevalent among council tenants.

In Bover *et al* (1989) we derive a mobility index based on the structure of housing tenure. We find that this index is a useful summary measure and helps to explain aggregate wages and unemployment. However, the index is necessarily a little complicated since the private rented sector is far from uniform and we need to take account of the declines in mobility that resulted from the Rent Acts 1965 and 1974. The Rent Act 1965 introduced 'fair' rents for unfurnished property while the Rent Act 1974 aimed to extend the same set of regulations to furnished property without a resident landlord.

Our mobility index is a weighted sum of the predicted rates of migration associated with different tenure groups, based in part on the results of Hughes and McCormick (1981, 1987), where the weights are the proportions of households in each tenure group. We distinguish between private furnished and unfurnished tenants and disaggregate the latter group into those with controlled rents, regulated or 'fair' rents and the remainder consisting of those whose rents are neither controlled nor regulated. We allow the Rent Acts 1965 and 1974 to have spillover effects reducing mobility more generally in the unfurnished and furnished sectors respectively. Some of these effects are estimated.

In our mobility index we standardise on owner occupiers by setting their mobility rate equal to 1. Our index then incorporates a mobility rate of 0.3 for council tenants, 3.2 falling to 1.7 post 1974 for furnished tenants, 0.55 for controlled unfurnished tenants, 1 for regulated unfurnished tenants and 3.2 falling to 1.7 post 1965 for other unfurnished tenants.

Our mobility index displays a sharp rise in the early 1960s due to the decline of the controlled unfurnished sector and the rise in owner occupation. The index peaks in the mid 1960s and then falls sharply until the early 1970s as the Rent Act 1965 bites. The mobility index displays a slight downward trend throughout the 1970s as the positive effects of the rise in owner occupation are offset by the falling share of private rented housing combined with a reduced rate of mobility as a result of the Rent Act 1974. This slight downward trend continues until the early 1980s when the mobility index begins to rise again as the share of council housing declines.

There are very real difficulties in measuring mobility so there is plenty of scope to use other approaches than the one we have used. Minford *et al* (1987) argue for what is in many respects a more sophisticated mobility index than ours. They attempt to estimate the gap between actual and free market rents in the different tenure groups and assume that relative migration rates are not merely proportional but equal to the ratios of actual to free market rents in the different tenure sectors. Though there is cross-section support for their approach, the regional mobility indices do not help to explain changes in regional unemployment rates over time.

The regional pattern of house prices

Regional house price differentials for selected years are set out in Table 6.3. The data are derived from the 5 per cent sample survey of building society mortgages. They are not adjusted for differences in mix of housing and are subject to a number of qualifications. Holmans (1990) discusses these qualifications in detail. He also provides a full discussion of the regional pattern of house prices. Nevertheless, the data in Table 6.3 do provide a reasonable indication of the pattern of regional house prices. The ranking of regions in terms of house prices evolves slowly although actual differentials may change by large amounts. Note the relative decline of house prices in Wales, Scotland and Northern Ireland and the relative rise in the South East, East Anglia and the South West.

The ratio of house prices to incomes is what really matters to individuals. In Figure 6.1 we chart mix adjusted house prices in the South

Reducing Regional Inequalities

Table 6.3 *Regional house prices differentials (UK average = 100)*

	1969	1974	1979	1984	1989
North	80.0	76.8	77.5	77.7	68.1
Yorkshire and Humberside	74.1	75.4	75.3	76.8	76.2
North West	84.5	80.9	84.8	83.9	76.8
East Midlands	81.7	83.6	79.9	83.8	90.1
West Midlands	93.7	93.3	92.8	85.9	90.8
East Anglia	92.6	100.1	92.7	92.2	117.8
Greater London	133.5	135.2	129.5	135.2	150.2
Rest of South East	124.8	126.9	123.8	128.3	148.8
South West	96.9	105.6	102.9	105.2	122.2
Wales	89.8	85.5	85.6	81.3	78.4
Scotland	99.3	88.9	87.2	88.9	64.5
Northern Ireland	84.9	79.3	109.5	73.7	55.2

Notes: Average house price per building society mortgage survey

Sources: Table B.1, Holmans (1990) and Housing and Construction Statistics

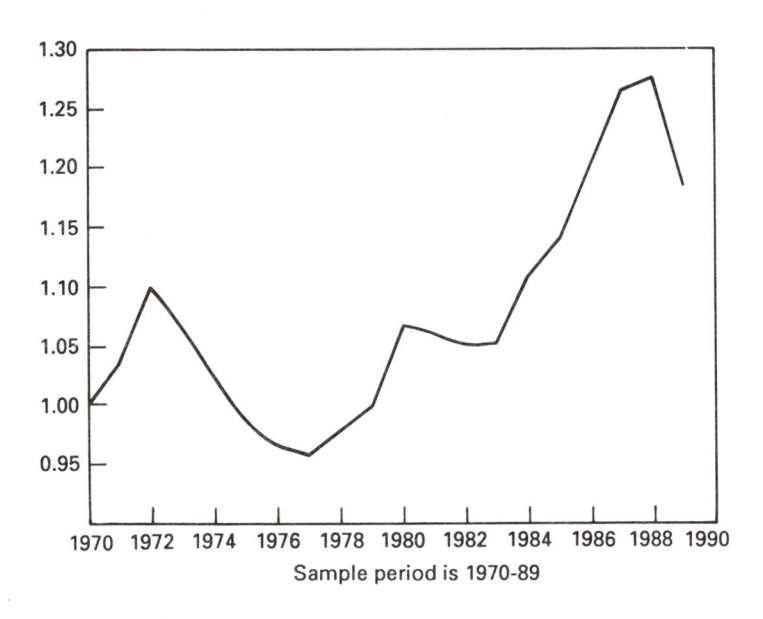

Source: Department of the Environment mix-adjusted house price index

Figure 6.1 *The ratio of South East to UK mix-adjusted house prices (all dwellings excluding sales not at market prices; DoE)*

Regional Economic Disparities: The Role of Housing

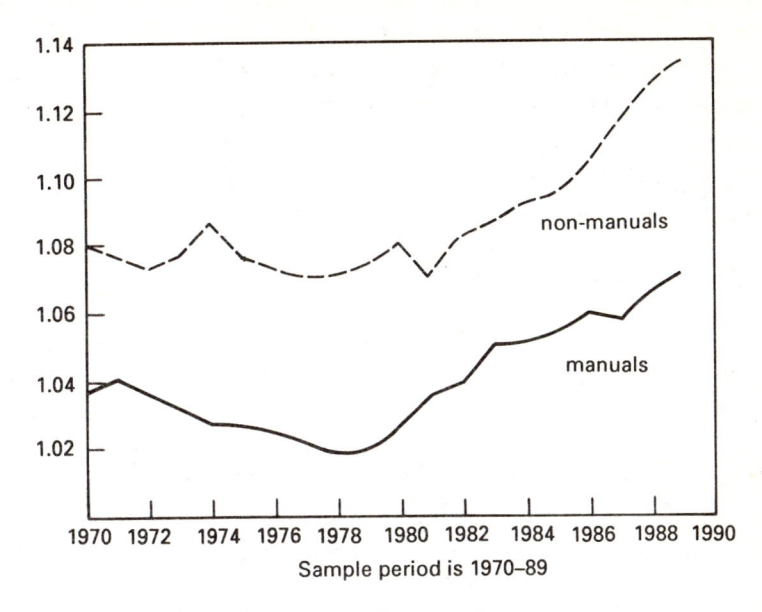

Source: Department of Employment New Earnings Surveys

Figure 6.2 *The ratio of South East to UK male earnings (full-time adults whose pay was not affected by absence; NES)*

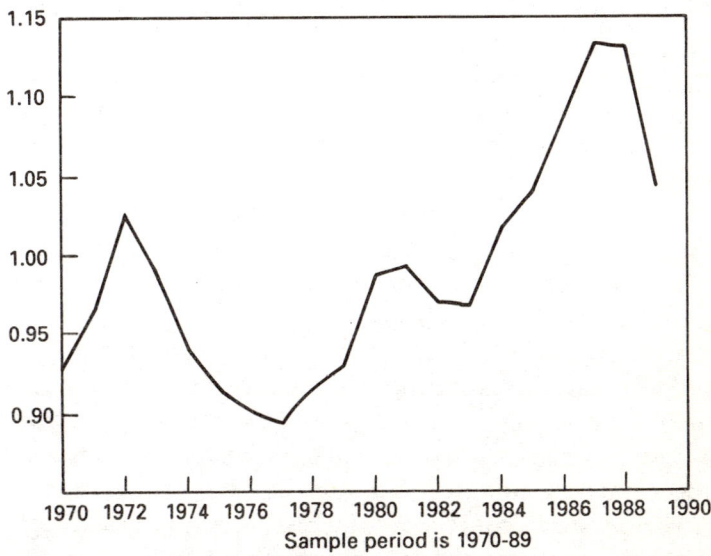

Sources: Department of the Environment mix-adjusted house price index
Department of Employment New Earnings Surveys

Figure 6.3 *The ratio of HP/W in the South East to HP/W in the UK (HP = mix-adjusted house prices, W = non-manual earnings)*

Reducing Regional Inequalities

East relative to the UK. Regional differentials in manual and non-manual male earnings are charted in Figure 6.2. In Figure 6.3 the regional difference in the HP/W ratio is charted. Mix-adjusted house prices are used for HP and non-manual male earnings for W. The record rise in the regional difference in the HP/W ratio in recent years is very apparent. If the regional difference in the HP/W ratio is weighted by the proportion of owner occupiers, which is an appropriate weighting on theoretical grounds, an even sharper rise is obtained.

The regional difference in the HP/W ratio also widened during the 'Barber' boom in the early 1970s. The larger rise in the regional HP/W differential during the 'Lawson' boom in part reflects the more favourable demand shocks experienced by the South East. It also reflects an additional factor, financial liberalisation, the exogenous shock par excellence of the 1980s. Since planning controls are more restrictive in the South East, the rise in aggregate demand as a result of financial liberalisation drove up the South East house price to wage differential.

III House prices, wages and the trade-off between unemployment and vacancies

In Bover *et al* (1989), we examine the interaction between the labour and housing markets in the UK, integrating the owner-occupied sector more fully than in previous research. We explore the effects of this interaction on the behaviour of aggregate wages and on the relationship between aggregate unemployment and unfilled vacancies, which reflects mismatch between job slots and job seekers. Our analysis confirms the importance of housing markets in explaining labour market behaviour. In particular, the lagged regional difference in the house price to wage or HP/W ratio is very significant in both the aggregate wage and unemployment/vacancy equations.

Our analysis emphasises the importance of segmented regional labour markets in which labour is relatively immobile, in part as a result of the operation of the housing market. As a result, one sector or region of the national labour market may experience high unemployment while there are, at the same time, unfilled vacancies or excess demand in another sector or region. Economic theory suggests that given such segmented markets, aggregate wage behaviour is influenced not only by aggregate excess demand, but also by its sectoral dispersion or mismatch. Hansen (1970) develops this type of model.

House prices and wage pressure

The starting point is the wage equations developed by Layard and Nickell (1986), which are now widely used. Other expositions of their approach include Layard and Nickell (1988) and Nickell (1987). Their work explains employment, wages and the price level, based on a theoretical framework of imperfect competition in the product market with some normal cost mark-up pricing. In the labour market, employers and unions bargain over wages while employers set employment. Wages adjusted for the trend in productivity depend upon both 'wage push' and 'wedge' factors. Both sets of factors are found to be empirically important.

The wage push factors include the level of and change in log unemployment with an adjustment for the share of the long-term unemployed, the ratio of unemployment benefits to wages or the 'replacement ratio', union power proxied by union density and a measure of the change in labour market mismatch proxied by the absolute change of the employment share in industry and construction. The wedge factors consist of direct and indirect taxes and weighted relative import prices. These factors determine the size of the wedge between the real product and consumption wages.

Annual data from 1958 to 1986 are used to estimate an equation explaining the behaviour of the aggregate product wage adjusted for the trend in productivity. As expected, both push factors and wedge factors, to a lesser extent, help to explain wage pressure. However, we also find significant mobility or housing tenure and house prices effects. To many, these housing effects are unexpected.

On mobility we find that wage pressure is related to our housing tenure based index of labour mobility. However, we find that variations in the lagged regional HP/W difference between the South East and the UK have been quantitatively more important. Our measure of this variable operates as a moving average with an average lag of two and one half years. There are three possible interpretations of this effect. It may be picking up regional mobility effects, regional cost of living effects and/or regional labour demand shocks. However, our work on the relationship between aggregate unemployment and vacancies, regional migration to and from the South East and the determinants of South East house prices supports the mobility and cost of living interpretations and not primarily the regional demand shocks one. Average house prices appear to be a significant part of the wedge between producer and consumer prices. In our equation we use the national HP/W ratio which

Reducing Regional Inequalities

enters as a moving average with a two-year lag. This suggests that the cost of living effects of house prices take a long time to feed through into wage pressure.

We find important roles for both the level and change in unemployment, although the deviation in the wage from trend productivity is more responsive to the latter. This is consistent with hysteresis as in Nickell (1987) and with most accounts of insider-outsider theory. Given our sectoral labour markets framework, it follows that if changes in excess demand are important then so will be changes in mismatch, or more precisely, the sectoral dispersion of excess demand. This is precisely what is found.

We only find a temporary effect from relative import prices and no effect from other components of the wedge. Union power adds significantly to wage pressure. There is evidence of negative feedbacks which we interpret as reflecting the lagged response of wages to price shocks and incomes policy catch-up effects. International competitiveness also has a small effect on wages.

The trade-off between unemployment and vacancies

One of the consequences of restricted labour mobility is increased mismatch in the labour market, and this should be reflected in outward shifts in the aggregate unemployment/vacancy or U/V locus, that is, more unemployment for given vacancies or vice versa. We therefore use annual data from 1958 to 1986 to estimate a second equation, in which unemployment depends on vacancies as well as on the 'replacement ratio' capturing search unemployment and the proportion of new entrants in the labour force. The housing market adds to mismatch by restricting regional mobility. We introduce housing market influences using the same lagged moving average of the regional difference in the HP/W ratio and through the housing tenure based index of mobility used in the wage equation.

We obtain a stable unemployment/vacancy equation with very significant housing effects. As expected, increased mobility due to a favourable change in the structure of housing tenure or reduced regional HP/W differentials reduce unemployment at a given level of vacancies.

The interaction between owner-occupied housing and labour markets

In view of these results it may be helpful to sketch out our view of the interaction between housing and labour markets. Our view, which was formulated a few years ago, seems rather familiar and less contentious now with the benefit of hindsight. We recognised the implications of the house price boom for consumption and the balance of payments, but we did not anticipate the amount by which the government would raise interest rates to dampen demand. Our exante views are presented below.

In the long run higher house prices in the South East benefit the unemployed elsewhere in the UK – they are 'the best friend of the unemployed man in Liverpool' as Patrick Minford has quipped. We acknowledge the long-run allocative role of house prices in the owner-occupied housing market. Higher South East house (and land) prices not only create an incentive to expand the supply of housing in the South East, but also incentives for firms and households to locate elsewhere.

Our research suggests, however, that the institutional distortions associated with owner-occupied housing, combined with tight planning controls, introduce important dynamic distortions into the housing market, with consequent damaging short to medium-term effects on national and regional labour markets. Two major tax incentives favour owner-occupied housing relative to other financial assets or rented accommodation. These are mortgage interest tax relief and the absence of capital gains tax on one's principal residence. These distortions, which have been added to by the abolition of domestic rates, artificially raise the portfolio returns on owner-occupied housing relative to other assets, with profound implications in economic upswings, especially when these are accompanied by rapid growth in financial liquidity.

In such upswings the response of house prices to the growth of income and liquidity results in high own rates of return on owner-occupied housing, which further stimulates housing demand. Even if other factors did not lead to higher economic growth in the South East, higher national housing demand tends not only to raise the national HP/W ratio, but also to widen the South East's ratio relative to that of the rest of the UK, because housing supply is less elastic in the South East.

The rise in the regional difference in the HP/W ratio leads to a mobility trap. As the relative appreciation of house prices increases, households in the South East are initially more reluctant to move to other areas: they would miss out on any further relative appreciation and may thus be unable to move back to the South East at a later date. Households outside the South East become increasingly unable to bridge the gap in house prices and so are less inclined to move.

As the regional differential in the house price to wage ratio approaches a peak, outward migration from the South East increases. At the same time, the credit constraint for potential migrants to the South East reaches a maximum. Also, by this time, additional new housing in the South East will have been built. This situation cannot persist and speculative expectations are reversed: the result is a rapid fall, such as in 1973 to 1975, of the South East's premium in the regional HP/W differential. The peak and the early part of this post-peak phase are likely to be particularly uncomfortable for firms in the South East trying to hold on to or hire workers and, unless labour demand in the South East slackens, is likely to be associated with strong wage pressure there. For example, 1973 saw record out-migration from the South East, with further large outflows in 1974 and 1975.

This process eventually leads firms and workers to locate outside the South East and so relieves unemployment in other regions. In the short run, however, this process imposes significant costs. Wage increases in the South East, quickly followed by even larger price increases there, can give workers in the South East an incentive to leave and be relatively ineffective in attracting new workers. Firms may therefore have to bear the brunt of the resource reallocation shifts engendered by this interaction of housing and labour markets.

Another element in the overshooting of excess demand for labour in the South East arises from the regional multiplier effects, especially in services and retailing, that arise from the South East's consumer boom that itself is fuelled by a South East house price boom.

IV Regional migration to and from the South East

In Muellbauer and Murphy (1988, 1990b) we set out to model regional migration to and from the South East region as a direct check on the implicit story regarding segmented regional labour markets and the regional mobility trap in Bover *et al* (1989). As a background we note the persistence of large regional differences in unemployment rates as well as participation rates. The wider regional differences in the 1980s are not explained by conventional measures of mismatch constructed from regional employment, unemployment and vacancy data. See Jackman and Roper (1987) and Jackman *et al* (1990), for example. Of course, there are very real measurement problems with the vacancy data.

Nevertheless, the persistence of large regional unemployment differentials indicates that both the regional wage adjustment and migration responses are insufficient. We know that regional wages are relatively responsive to excess demand but much less responsive to excess supply. Thus, the puzzle is why in the 1980s high wages, low unemployment and a general boom in activity in the South East coincided with significant net out-migration, in particular, net out-migration of manual workers. Does

Regional Economic Disparities: The Role of Housing

the dramatic rise in the regional difference in the HP/W ratio in the South East help to explain this puzzle?

Our model and some data

In our research, we use 19 years of aggregate time series data from 1971 to 1989, derived from NHS central register data on patient moves, to model gross and net regional migration flows to the South East. One disadvantage of this data is that an age disaggregation is only available since 1982. However, this is the only available time series data available with a reasonable number of observations. To date, the only pooled time series cross-section study of migration we are aware of, Hughes and McCormick (1990), is based on only four years' Labour Force Survey data, which may be too short a period to pin down cyclical house price and other effects. Their data period is also one with little variation in migration and regional house price to wage differentials.

We estimate equations for the gross migration flows in addition to the net migration flow. Given the limitations of the data – small sample size and very limited availability of disaggregate data – this provides an important test as our model of migration implies testable restrictions on

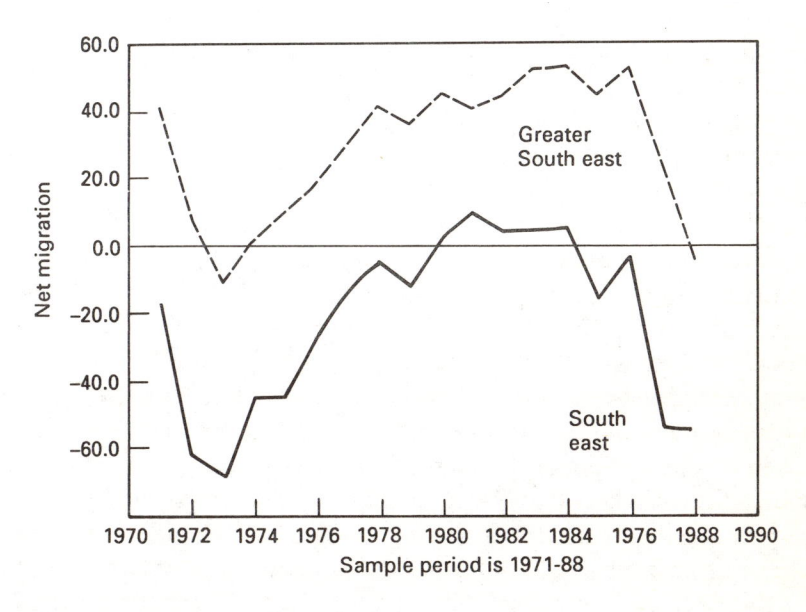

Source: Population Trends (OPCS)

Figure 6.4 *Net migration from South East and 'Greater' South East regions*

the gross flow equations. For reasons of economy, however, we will only discuss our results of estimating the net migration equation. Net migration from the South East region is shown in Figure 6.4. Recall that regional differentials in mix-adjusted house prices, manual and non-manual male earnings, and the HP/W ratio are shown in Figures 6.1 to 6.3 respectively.

Results for net migration equation

In our net migration equation we find significant roles for the regional difference in the HP/W ratio, the exact same variable we use to explain aggregate wages and unemployment, as well as the regional difference in wages. The estimated coefficients suggest a weighting of 2:3 on house prices relative to wages. This weighting is considerably higher than the weighting of housing in the retail price index or the average share of housing in the Family Expenditure Survey. The relative sizes of the house price and wage effects are readily rationalised. *Ceteris paribus*, the young have higher propensities to migrate and housing is a higher share of their outgoings. In addition, they have a lower equity stake in their houses so that bridging the gap between high and low-price regions is a major problem. These arguments apply to both long and short-term differences in regional HP/W ratios. However, the regional mobility trap is greatest in the short run when the housing market is gripped by speculative frenzy.

Recall that the lagged regional difference in the HP/W ratio appears in both the aggregate UK wage and unemployment/vacancy equations which we discussed in the previous section. There are three possible explanations for the presence of the regional difference of the HP/W ratio in our wage equation – first, as an indicator of local or national demand shocks since house prices are 'jump' variables; second, as an indicator of regional cost of living differences, the main regional difference being the cost of owner-occupied housing; third, as an indicator of a short-term regional mobility trap. Our migration equation results are consistent with the second and third interpretations.

Other factors influencing net migration to the South East are the relative labour market strength of the South East – captured by relative employment growth and the male unemployment rate differential there – international migration which tends to crowd out other potential migrants since it is relatively concentrated in the South East, our housing tenure based mobility index which we use in our aggregate wage and unemployment equations, demography since migration rates decline with age generally, the change in the nominal interest rate and the general level of unemployment which both serve to depress mobility generally.

Regional house price differentials, the regional mobility trap and wage pressure.

Some critics argue that relatively high house prices in the South East may not crowd out that many workers in the short term. For example, when house prices are relatively high, more retired people or workers near retirement may move to areas with lower house prices. In addition, some workers may trade off commuting costs for housing costs by also moving to low house price areas. Thus, it is possible that the strong role of the regional difference in the HP/W ratio may be reflecting residential moves to a greater extent than labour market moves.

However, this line of argument suggests that the lagged regional difference in the HP/W ratio should not show up significantly in our aggregate wage and unemployment equations. In addition, the age breakdown of migration flows, which is available from 1982 onwards, does not reveal any significant shift in the age composition of the flows. We suggest that long-distance commuting is still not that common and tends not to be permanent. Finally we estimate our migration model for a Greater South East area, consisting of East Anglia, the South East and the South West, and obtain very similar results. As Figure 6.4 shows, the pattern of net migration from the two overlapping areas is very similar. The use of this 'Greater South East' area serves to eliminate from our data the vast majority of residential moves which are not matched by labour market moves.

Other studies of regional migration

In Muellbauer and Murphy (1990b) we survey a number of studies of regional migration. Significant wage and house prices effects, when included, have been found in various studies. However, recent studies of migration have produced many contradictory results.

For example, Hughes and McCormick (1990) pool four years' Labour Force Survey data from 1983 to 1986. They find that house prices and wages have little effect on regional migration, being unemployed a year ago encourages migration, high regional unemployment discourages migration, and high regional vacancies encourage migration. They do find stronger house price effects in more recent work. However, we find their other results rather strange. Over half of all migrants are in employment so we would expect to pick up some wage effects. Certainly in time series data significant wage and correctly signed regional unemployment and vacancy rate effects are the norm. However, Jackman and Savouri (1990), using NHS central register data from 1975 to 1987 and a job-matching theoretical framework to model the pattern of regional migration, find an unexplained perverse wage effect. They also find a weak house price effect and significant regional unemployment and vacancy effects. Using two years' Labour Force Survey data, Pissarides and Wadsworth (1989) find significant wage effects and some weak effects from regional unemployment and vacancy rates. Clearly there is considerable scope for more work, using a variety of models and data, in the area of regional migration.

V The determinants of house prices

In this section we discuss our research into the determinants of national and South East house prices. In practice we model the ratio of house prices to wages or HP/W since this is what really matters to people. Since many of the same factors drive both the national and South East HP/W ratios, we must discuss our results on the determinants of UK house prices at some length. We also discuss why there is only limited evidence of the instability of the housing market in the period before financial liberalisation. Finally we look at the determinants of South East house prices.

UK house prices

Our model of the national house price to wage ratio builds upon the research of Hendry (1984), but with somewhat different results. In the long run, we find that rising real incomes, a rising ratio of liquid assets to income, an increased proportion of the population in the key house-buying ages, an increased proportion of the labour force in non-manual occupations, lower real tax-adjusted interest rates and cutbacks in public and quasi-public sector house building, but not the sale of council houses, drive UK house prices. We also find that people tend to expect their recent experience of house price changes to continue, which suggests that they form extrapolative expectations.

In the short run we find that house prices increase relative to wages if there is an increase in the outstanding mortgage stock relative to income or a reduction in the rate of equity withdrawal relative to income. These advances and withdrawals are of course endogenous and have themselves to be explained. More importantly, there is also strong evidence for the 'frenzy' effect uncovered by Hendry which suggests that, as house prices accelerate, potential home owners fear that they will either miss out on the expected capital gain or else be unable to afford the higher expected future house prices.

The combination of extrapolative expectations and speculative frenzy on the demand side lead to instability in the housing market and 'inefficient' overshooting of house prices. This instability is exacerbated by the lags in the supply of new houses, which tends to peak after the housing bubble bursts and then collapses. Land speculation also plays a part.

Evidence of housing market instability in earlier periods

An obvious question is why there is only limited evidence of the instability of the housing market in the 1960s and 1970s. The answer is simple. The instability of house prices was generally kept in check by a system of rationing of mortgages and other consumer credit. In the 1970s, when the real cost of mortgage borrowing was negative between 1971 and 1981, rationing had to do double duty by also keeping borrowing in check. The classic previous example of instability in the housing market was the 1971 to 1973 house price boom. House prices almost doubled and increased by 60 per cent in real terms, falling again in real terms by almost one-half over the following four years. A proper understanding of that period is made difficult by the 1973 inflation shock in oil and raw material prices and by overly crude monetarist interpretations. During the 1971 to 1973 period the extent of mortgage rationing was significantly reduced as banks increased their share of net mortgage advances. They subsequently cut back their share to its previous level.

The system of mortgage and consumer credit rationing was abandoned as financial liberalisation was introduced in the 1980s. Some of the key stages in this process were the abolition of exchange controls in 1979, the lifting in 1980 of the 'corset' which had restricted the development of bank lending and borrowing, the strong re-entry of the banks into the mortgage market in 1981, securing over one-third of net mortgage advances in 1982, and finally, the Building Societies Act 1986. Among other things, building societies now have access to wholesale money markets and can extend unrestricted consumer credit as long as it is backed by housing collateral. Thus, in the 1980s financial markets were liberalised while the existing major distortions in the housing market were maintained. As a result the latent instability of the housing market was unleashed.

South East house prices

Our explanation of the regional difference in the HP/W ratio between the South East and the UK involves many of the same aggregate demand factors that drive the national HP/W ratio as well as some labour market variables that reflect regional differences in economic activity. This differential labour demand shock effect explains only a small part of the variance in the regional HP/W differential, despite extensive searching for other variables that would enhance the size of this effect. This suggests to us that the lower supply elasticity of land in the South East accounts for much of the rise in the regional house price to wage differential when aggregate demand for housing in the UK increases.

VI Policy options

In this chapter we argue that both the structure of housing tenure and the regional dispersion of owner-occupied house prices contribute to regional economic disparities. They increase mismatch and thereby add to both aggregate wage pressure and unemployment. These factors operate via the regional mobility trap and regional cost of living effects. In the longer term the regional dispersion of house prices reflects fundamentals and fulfils an important allocative role. However, we argue that in the short to medium term, the housing market is unstable and that inefficient overshooting of house prices occurs.

Identifying the problems

In relation to housing tenure and labour mobility, the problems are the lack of a free rented sector and the restrictions on the mobility of council tenants. The Housing Act 1980 did not result in an expansion of the private rented sector. Instead, this sector continued to contract. The sale of council houses has increased mobility, but the cutback in council house building has helped to drive up house prices. The Housing Act 1988 provides for rent decontrol and reduces security of tenure. However, the resulting increase in the supply of private rented housing has been small, while decontrolled rents have inevitably followed house prices up, increasing rapidly relative to incomes. We fully agree with Holmans' (1987) authoritative assessment that the long-run decline of the private rented sector has as much to do with tax reliefs on owner occupation as with rent and tenure controls.

Turning to owner-occupied housing, we note the British obsession with providing incentives for owner occupation. Mortgage interest tax relief and zero capital gains tax on one's main residence stimulate the portfolio demand for housing, in other words the demand for housing as an asset as opposed to the demand for housing services. On the supply side, planning controls are very tight, especially in the South East. As a result, house price to income ratios are higher than they otherwise would be and the housing market is unstable and prone to inefficient overshooting. Financial liberalisation throughout the 1980s, combined with a surge in economic activity in the later half of the 1980s, revealed the extent of this instability. The abolition of domestic rates and its replacement by the poll tax also played a part. Financial liberalisation is a worldwide phenomenon. Though a framework of regulation to ensure capital adequacy and limit systemic instability needs to be maintained, financial liberalisation cannot and should not be reversed. The same cannot be said of the distortions in the housing market.

The options

On the supply side, our general view is that it would be highly desirable if the supply elasticities for housing were higher than at present. During overshooting episodes supply elasticities appear to be very low or even perverse. Achieving high supply elasticities involves a more rational system of land use planning which takes better account of the costs of restricting planning. Indeed, some critics have argued that the system of land-use planning should be abolished altogether, but externalities are surely large here. Evans (1988) has argued the case for drastic reform. One simple proposal is to relax planning restrictions in the case of rented housing or housing for low-income families.

In addition, local authorities or other quasi-public institutions holding residential building land should be encouraged by a system of sticks and carrots to release it for development by housing associations or by local authorities themselves. More generally, public or quasi-public provision of housing should be increased and certainly not be cut back any further.

On the demand side, the tax treatment of owner-occupied housing needs reforming. Mortgage interest tax relief could be restructured to make it self-financing or, at least, less costly in tax revenue lost. A national property tax, ideally in the form of a tax on the market value of residential land, could be introduced. These reforms would compensate for the distortions of the tax and planning system and the abolition of domestic rates. They would prevent the overshooting of house price to income ratios and reduce the portfolio demand for housing. The private provision of rented accommodation would be encouraged since, with the current levels of house price to income ratios, there is little scope for landlords to make good economic returns from renting property at affordable rents.

At present, mortgage interest tax relief is an extremely inefficient subsidy at meeting its overt aim, which is to help first-time buyers over the early phase of the life-cycle when households are most cash-constrained. The bigger demand for housing land and for credit which results, makes house price to income ratios and interest rates higher than they would be in the absence of the tax subsidy. First-time and recent home buyers would benefit more if the subsidy were available only for the first eight to ten years.

A residential property tax should aim to raise about as much revenue as domestic rates. However, it would differ from rates in a number of important respects. First, it would be a tax on residential land and land zoned for residential use. Second it would be indexed annually to local land price indices. Third, it would be collected by the Inland Revenue with the rates of tax being set nationally at uniform levels. Such a tax would have important macroeconomic benefits by preventing the overshooting of house price to income ratios. The tax revenue could be used in constructive ways to improve the supply side of the economy by, for example, reducing business taxes or stimulating training.

A tax on residential land values would also have major microeconomic benefits as a form of congestion tax, a regional policy that would work through the market rather than through bureaucracy, and as a 'betterment' charge, as a way of recouping to the public purse part of the benefits accruing to landowners of public investment projects.

In practice, a national residential property tax may be unlikely. However, if there is a return to a local property tax funding local services and replacing the poll tax, then it would improve greatly on domestic rates if it were assessed on current market values instead of imputed rents and if it were indexed annually. A more sophisticated system of rebates would be required. The retired with low cash incomes but large housing wealth could be permitted to build up debt with the local authority to be settled from their estate.

There is also a good case for the reform of lending practices. The combination of financial liberalisation and the distortions on both the demand and supply sides of housing and land markets account for the recent overshooting of house prices. Since financial liberalisation makes asset prices more volatile, the capital adequacy ratios, that is the levels at which lenders can lend relative to their capital base, applied by financial regulators to financial institutions, should become more stringent. In particular, given the volatility of house prices recently experienced in the UK and experienced in parts of the US and the Netherlands for example, the risk weighting of mortgages should be increased. As a result a smaller proportion of house values would be advanced as mortgages. This would both protect depositors and investors in financial institutions and protect mortgage borrowers from the risk of dispossession. Capital adequacy reforms would be a permanent change to compensate for the permanent changes induced by financial liberalisation and should not be seen as a temporary form of credit control.

Further details of many of these policy options are set out in Muellbauer (1990). Other policy options are set out in McCormick (1990). Minford *et al* (1988) do not appear to accept that the tax treatment of owner-occupied housing has any implications for the supply of private rented housing. They correctly argue that the owner-occupied housing market is a 'free' one where prices adjust to clear the market. However, they incorrectly equate this 'free' market with an undistorted market. Planning controls and tax incentives are important distortions. They also have major spillover effects on the market in private rented housing.

Of course, reform of the tax treatment of owner-occupied housing is desirable for a number of reasons not involving the link between house prices, tenure and regional mobility. For example, on the demand side there is the link between high housing wealth to income ratios, the fall in UK personal savings and recent record balance of payment deficits. Muellbauer and Murphy (1989, 1990a) discuss this link.

Some concluding remarks

We do not argue that the housing market is 'the tail that wags the dog'. Distortions in the housing market contribute to regional economic disparities but they are not the major cause of them. Problems in the national housing market do, however, have important regional consequences. We suggest that it is easier to devise acceptable policies to reduce or remove problems in the housing market than policies to increase regional wage flexibility, for example, or deal with the other problems contributing to regional economic disparities.

Bibliography

Bover, O, Muellbauer, J and Murphy, A (1989) 'Housing, Wages and UK Labour Markets', *Oxford Bulletin of Economics and Statistics*, 51, pp97-162 (with discussion).

Evans, A (1988), *No Room, No Room!*, Institute for Economic Affairs, London.

Hansen, B (1970), 'Excess Demand, Unemployment, Vacancies and Wages', *Quarterly Journal of Economics*, 83, pp1-23.

Hendry, D (1984) 'Econometric Modelling of House Prices in the UK', in Hendry, D and Wallis, K (eds), *Econometrics and Quantitative Economics*, Basil Blackwell, Oxford.

Holmans, A (1987), *Housing Policy in Britain: A History*, Croom Helm, London.

— (1990), 'House Prices: Changes Through Time at National and Sub-National Level', Government Economic Service Working Paper No 100.

Hughes, G and McCormick, B (1981) 'Do Council Housing Policies Reduce Migration between Regions?', *Economic Journal*, 91, pp919-37.

— (1985), 'Migration Intentions in the UK: Which Households Want to Migrate and Which Succeed', *Economic Journal Conference Supplement*, 95, pp76-95.

— (1981), 'Housing Markets, Unemployment and Labour Market Flexibility in the UK', *European Economic Review* pp615-45.

— (1990), 'Do Regional Differences in House Prices Influence the Pattern of Migration?', mimeo.

Jackman, R and Roper, S (1987), 'Structural Unemployment', *Oxford Bulletin of Economics and Statistics*, 49, pp9-36.

— Layard, R and Savouri, S (1990), 'Labour Market Mismatch: A Framework for Thought', London School of Economics Centre for Economic Performance Discussion Paper No 1.

— and Savouri, S (1990) 'An Analysis of Migration Based on the Hiring Function', London School of Economics Centre for Economic Performance Working Paper No 14.

Layard, R and Nickell, S (1986), 'Unemployment in Britain', *Economica*, 33 supp, S121-170.

— 'The Performance of the British Labour Market', in Dornbusch, R and Layard, R (eds), *The Performance of the British Economy*, Oxford University Press, Oxford.

McCormick, B 'Migration and Regional Policy', *Chapter 5 of this volume*.

Minford, P, Peel, M and Ashton, P (1987) *The Housing Morass*, Institute for Economic Affairs, London.

— (1988), 'The Effects of Housing Distortions on Unemployment', *Oxford Economic Papers*, 40, pp322-45.

Muellbauer, J (1990), *The Great British Housing Disaster and Economic Policy* Institute for Public Policy Research, London.

— and Murphy, A (1988), 'House Prices and Migration: Economic and Investment Implications', Shearson, Lehman and Hutton Securities research report.

— (1989), 'Why has UK Personal Saving Collapsed?', Credit Suisse First Boston Research Report.

— (1990a), 'Is the UK Balance of Payments Sustainable?', forthcoming *Economic Policy*.
— (1990b), 'Housing and Regional Migration To and From the South East', revised mimeo.
Nickell, S (1987), 'Why is Wage Inflation in Britain so High?', *Oxford Bulletin of Economics and Statistics*, 49, pp103-28.
Pissarides, C and Wadsworth, J (1989), 'Unemployment and the Inter-regional Mobility of Labour', *Economic Journal*, 99, pp 739-55.

7
*Regional Economic Development in the UK: The Case of the Northern Region of England**

John B Goddard and Alfred T Thwaites

I The development of the Northern Region of England

Introduction

The purpose of this chapter is to examine the changing nature of regional economic development problems in the UK by reference to one particular region. By adopting a region-specific approach the chapter goes beyond regional industrial policy which aims to influence the location of private sector investment within the UK to an examination of a broad range of often interrelated factors that influence economic growth and change within a geographically defined area. In considering the smallest and most peripheral region within England and one which has had the longest history of public intervention to arrest economic decline, the chapter examines a 'worst case' situation for public policy, a case unencumbered by the special political and institutional factors which have shaped events in Scotland, Wales and Northern Ireland.

A regional development perspective which stresses the long-term and interrelated nature of regional problems raises the thorny question of regional definition. This is an issue that has not been confronted seriously by regional industrial policy. Under this regime the map of assisted areas has changed frequently in response to shifting patterns of unemployment and changing economic and political philosophies. Thus the aggregation of travel-to-work areas that has defined the assisted areas in a particular period has never coincided with any administrative or functional regions.

Like the more numerous travel-to-work areas the eight standard

---

* Paper presented to the National Economic Development Office Conference on Regional Policy, Lumley Castle, County Durham, March 1990.

Regional Economic Development: The Northern Region of England

regions of England are principally statistical reporting units; since the abolition of Regional Economic Planning Councils and the Regional Planning Boards of Whitehall Departments in 1979, the regions have limited formal meaning in the governance of England. The Northern Region is composed of the four northern-most counties of England, namely Northumberland, Cumbria, Tyne and Wear, Durham and Cleveland with a combined population of three million, which is $\frac{1}{18}$th of the English total. There are some regional bodies with responsibilities for a similar area (for health and the arts and privatised 'regional' utilities for gas, electricity and water); three Whitehall Departments (Trade and Industry, Employment and Environment) do have regional offices in Newcastle covering similar territories and there is a regional promotion agency supported by local authorities, the CBI, the TUC and DTI (the Northern Development Company). Nevertheless, the main thrust of area economic development policy in the 1980s has shifted to the sub-regional scale, particularly urban areas. In so far as any regional planning remains (ie co-ordinated public action over a defined territory), it occurs in the region as a requirement of the Community support framework of the European Regional Development Fund and covers the two sub-regions of Tyne and Wear/south-east Northumberland and Cleveland/County Durham. Central government co-ordination in the region takes place through City Action Teams for the two urban areas of Teesside and Tyneside which both have their own Urban Development Corporations. Tyneside, Wearside and Teesside each have their own business-led urban regeneration teams. Last but not least, enterprise training and development are to be organised in the future through six local Training and Enterprise Councils covering different parts of the region.

The current emphasis on local as opposed to regional economic development is in part justified by history and geography. Within the Northern Region there are striking contrasts in the trajectory of economic development between Tyneside and Teesside. Tyneside has a long urban tradition based on trade in coal, a tradition which predates the industrial revolution; subsequently it was a centre of industrial innovation through most of the nineteenth century and first witnessed large-scale industrial decline in the 1930s. In contrast, Teesside was entirely a creature of the later industrial revolution and, with a small hiccup in the 1930s, continued to expand particularly during the 1960s on the basis of heavy capital investment in chemicals and steel. It experienced a sharp downturn in its economic fortunes in the 1970s and remains dominated by one employer, ICI. Unlike Tyneside it has not benefited from the services-oriented role of a regional capital. The east and west coasts of the

Reducing Regional Inequalities

region both have dispersed patterns of urban and industrial development, again often based on a dominant employer.

There are equally strong contrasts between the region's rural areas with high levels of prosperity in parts of Northumberland and Cumbria, but considerable deprivation in parts of West Durham.

Local contrasts in unemployment rates are found in all of the regions of England and Wales, but what is equally apparent is the rising level of unemployment from the centre to the periphery of England and Wales. While there are local variations, there are obvious north-south disparities which need to be considered on a broader scale. Notwithstanding internal differences, the Northern Region includes areas with shared if different experiences of growth, decline and revival and which all stand in a similar peripheral relation to the prosperous centre of the UK based in the south.

The evolution of the northern region economy

The regional problem first came to national public attention in the 1930s when large parts of areas like the North failed to recover from the great depression. During the 1930s the region did not participate in the growth of lighter consumer-based industries. Much of this growth derived from inward investment to London and the South East as US companies began to invest overseas for the first time. It was regional disparities, not just the absolute levels of local unemployment, which prompted the Jarrow Marches.

Like most major recessions, the 1929-32 period saw a massive rationalisation of excess industrial capacity in the North in sectors such as steel, shipbuilding and chemicals. Unemployment was eventually reduced after 1936, not by this rationalisation or by diversification into new factories opened in the first publicly supported industrial estates on the Team Valley in Gateshead, but rather by rearmament. Rearmament increased demand in the older industries using traditional methods of production. The need to diversify was once again avoided as it had been in the period of rearmament up to the First World War.

The rationalisations of the 1930s established an emerging trend which has come to dominate the economic development of the North, namely the loss of local control of industry as headquarters of the emerging national companies created through the merger of formerly regionally based enterprises moved out of the region. Parallel developments also took place in the service sector as regional banks, the press and the media centralised in London, downgrading the role of Newcastle as an office centre for regional firms.

Regional Economic Development: The Northern Region of England

The post World War Two period of expansion, based on mass production and consumption, transformed the economic and social structure of the North, dramatically intensifying a tendency that had begun to emerge in the inter-war period. Traditional industries continued to decline and more control passed outside the region. The region became a 'branch plant economy' through the external acquisition of indigenous firms and the establishment of new manufacturing plants from outside the region which were seeking economies in production on greenfield sites. Thus, by 1978, 73 per cent of manufacturing employment in the region was in externally owned firms (Smith, 1979, p425). New investment was attracted to the region by State subsidies (advanced factories in New Towns plus capital grants) and plentiful labour released from traditional industries and through growing female activity rates (Hudson, 1985, p229). Technical innovation in transport and communications, as well as new methods of mass production, underpinned this regional transformation. The development of the motorway network enabled the dispersal of production into this peripheral region while the widespread diffusion of the telephone enabled this production to be controlled from London (Goddard, 1976). Advances in communication therefore facilitated the translation of the emerging functional division of labour between conception and execution into a spatial division of labour (Massey, 1984).

During this period central government came to play a dominant role in both the production and consumption sectors in the region (Hudson, 1986). Many traditional industries (coal, steel, shipbuilding) were transferred to State ownership and headquarters removed to London. Central government also played a major role in rebuilding the infrastructure of the region – new towns, roads, the Tyneside Metro, etc. As a dominant supplier of industrial premises through bodies like English Estates and the New Town Corporations, the State became the prime provider of industrial premises; by offering low rents to attract new firms it subsequently deterred private sector property investment.

As the central government increased its role in regulating the national economy – for example, through encouraging industrial concentration, through direct and indirect support for research and development and through public purchasing, companies found it increasingly necessary to transfer their headquarters to London which was the focus of this State regulatory activity. The growth of collective wage bargaining through national sector based unions was a further centralising influence. However, as strategic State functions were centralised, routine activities like the processing of social security payments were dispersed to the

North, with the Department of Health and Social Security administrative offices becoming the largest employer on Tyneside (reaching a peak of over 10,000 employees in the late 1970s).

The State also became a significant source of finance for industry through regional industrial policy – regional based investment banking functions having already centralised on London. Until 1980 no venture capital organisations existed within the region.

In terms of consumption, local government grew in importance as an employer and provider of services, especially housing. Nationally agreed wages in the public sector, driven up by London standards, coupled with low private ownership of housing (the main focus for national savings) produced a high propensity to spend. As the regional capital, Newcastle was transformed from being a centre for producer services in the nineteenth century to primarily a centre for consumer services.

In many ways the institutional structure of the region came to be dominated by collectivist modes of action. Local government was dominated by the Labour Party and perceived itself as largely concerned with wealth redistribution (social welfare); it was increasingly dependent on transfer payments from central govenment subventions (the rate support grant). Trade unions came to occupy a strong bargaining position in most industries, with close connections to local authorities; many trade unionists were co-opted on to regional bodies. In contrast, with its high level of external control, much of the business community was not actively involved in community affairs and was bereft of leadership; branch factory managers came and went on an average two-year cycle and lived in executive housing 'ghettos'. The region's two universities, one originally created by local industrialists to service local links, became national institutions funded by central government and recruiting students nationwide. They were in the region but not of it. The old order of paternal capitalists was replaced by a new order of State institutions on which the working class was equally dependent – although many of the region's old capitalists did occupy key positions in these State institutions. Altogether, the region became more dependent on London both economically and politically.

From about the mid 1970s it became clear that this form of organising and regulating the economy both nationally and regionally could not be sustained. The IMF intervention in 1976 forced a major reduction in regional industrial assistance on the government. In the 1979-83 recession and its aftermath, many of the region's older industries witnessed another sharp decline with the closure of steel works (Consett, Hartlepool, Workington), and shipyards (on the Tyne, Tees and Wear).

Regional Economic Development: The Northern Region of England

Table 7.1 The North in recession, 1979-83

	North	South East
% change in output	-15.7	-6.1
% change in employment	-29.8	-17.5
Redundancies per 1,000 employees (1979-82)	64.3	28.1
GDP per head as % UK average		
1979	92.0	115.3
1988	89.6	121.5

Source: R Martin and B Rowthorn (1986)

Coal-mining also witnessed further contractions in another wave of pit closures following the 1984/5 strike. Many branch factories established with regional policy support proved vulnerable to corporate restructuring prompted by the recession. There was a steady exodus of many British based multinationals like Courtaulds, Dunlop, Plessey, TI, RHM and WE & HO Wills. The recession halted the long-term growth of the service sector, particularly due to public expenditure restraint.

As a result of these changes the number of employees in employment fell by 185,000 between 1978 and 1983, a decline of 14.9 per cent, and regional unemployment increased by 100,000 to 225,000 or from 9.3 per cent to 16.3 per cent. While the recession clearly affected all regions, the North and other industrial regions suffered much more than the South East. For example, between 1979 and 1982 redundancies per 1,000 employees in the North were three times those in the South East. Between 1979 and 1983 manufacturing output in the region declined by 15.7 per cent, compared with a decline of 6.1 per cent in the South East region (Table 7.1). After a period of convergence in the post-war period the recession once again opened up the north-south divide. To what extent has the gap been narrowed in the recovery period?

Economic recovery in the North since 1983

Since 1983 there has been some recovery in the region, but there are still fewer jobs than in 1981 and 140,000 fewer than in 1978. To set it in its

Reducing Regional Inequalities

Table 7.2 *Employment change in the North in recovery*

	North	GB
% employment change 1983-9	5.2	9.7
% self-employed		
1981	7.6	9.8
1988	10.3	13.4

Source: Employment Gazette 4/84, 7/84, 4/89, 2/90

context, the rate of employment increase in the region since 1983 has been approximately half the national average. Self-employment, which in 1981 stood at 7.6 per cent of the workforce compared to 9.2 per cent nationally, has also increased to 10.3 per cent. However, the gap with the national average has widened from 2.2 to 3.1 percentage points (Table 7.2).

Unemployment has also declined in the region and slightly faster than the national average with the gap in rates falling from 4.8 percentage points in 1983 to 4.1 percentage points in 1989. This may be partly attributable to a reduction in labour supply through net migration of working age population (a net loss of 2.5 per cent between 1983 and 1989), which has contributed to a fall in the working age population of 0.8 percentage points (compared to a national increase of 2.9 percentage points) and to a smaller increase in participation rates (+0.2 percentage points) than the national average (+1.7 percentage points). Taken together these factors have meant that more of the jobs created in the region have been available for the unemployed (Table 7.3).

Given that much of the increase in employment was in part-time jobs for women, it is also not surprising that male unemployment in the region relative to the national average has in fact increased (from 46 per cent above the national average in 1985 to 58 per cent above in 1989).

The majority of other key indicators of regional prosperity suggest that the North has witnessed an economic recovery, but at a slower rate than in the nation as a whole. Between 1983 and 1989, GDP increased by 19.2 per cent, compared with 26.5 per cent for the UK. House prices in the North increased by 83.2 per cent over the period, but this increase was 15.6 percentage points below the national average; although there was some convergence in house prices between regions from the end of

Regional Economic Development: The Northern Region of England

Table 7.3 *Unemployment change, labour market participation and migration, 1983-8*

	North	GB
Unemployment % point change	-4.4	-4.0
Net migration %	-2.5	0.2
% participation rate working age population		
1983	69.8	75.3
1989	70.3	77.0
Relative male unemployment rate		
1985	146.0	100.0
1989	158.0	100.0

Source: Employment Gazette 7/84, 2/90
Population Trends, 1989

1988, this made little impression on the gap that had opened out during the previous three years (Table 7.4).

Rising prosperity has also not significantly filtered down to the deprived sections of the region's population, with the proportion of household income derived from supplementary benefits increasing between 1980/1 and 1986/7 (up from 16.2 per cent to 16.9 per cent, compared with a slight fall nationally from 12.6 per cent to 12.5 per cent). Although average male manual wages remain around the national

Table 7.4 *Indicators of regional prosperity, 1983-9*

	North	UK
% change in GDP	19.2%	26.5%
% change in house prices	83.2%	98.8%

Source: Regional Trends, 1990
Halifax Building Society (private data)

Reducing Regional Inequalities

Table 7.5 *Changes in regional income distribution and wages*

	North	GB
% household income from supplementary benfit		
1980/1	16.2	12.6
1986/7	16.9	12.5
Male manual wages		
1983	102.3	100.0
1989	100.9	100.0
Male non-manual wages		
1983	95.1	100.0
1989	88.6	100.0
% male full-time workers receiving over £250 pw (1988)	30.3	36.5
Female full-time manual wages		
1983	98.1	100.0
1989	96.8	100.0
Female full-time non-manual wages		
1983	94.5	100.0
1989	88.6	100.0

Source: New Earnings Survey, 1984, 1989

average, the gap in non-manual male wage levels has widened from 4.9 per cent below the national level to 11.4 per cent below. Thus, only 30.3 per cent of males in full-time employment in the region earned over £250.00 per week in 1988 compared with 36.5 per cent nationally. The gap in female earnings is equally pronounced and widening (Table 7.5).

The only indicator in which the region out-performs the nation is in investment in manufacturing industry. In 1981 net capital expenditure per job in manufacturing in the region was 19.5 per cent above the average; by 1986 it had increased to 29.2 per cent. However, this did not lead to an improvement in value added per worker which fell from 6.0 per cent above the average to 1.3 per cent below over the same period (Table 7.6).

To conclude this statistical portrait it is clear that the Northern Region

Table 7.6 *Investment in manufacturing industry, 1981-6*

		North	UK
Net capital expenditure per capita	1981	119.5	100.0
	1986	129.2	100.0
Value added per capita	1981	106.0	100.0
	1986	98.7	100.0

Source: Census of Production, 1981, 1986

has lagged behind in the strong national economic recovery post 1983. If comparisons had been made with the South East then an even poorer performance would be apparent indicating that the north–south divide, which is a long-established feature of the British economy, opened up once again in the last recession and has widened further during recovery.

II Dimensions of the regional problem in the North

Interpretations of the region's problems

The recent trends that have been described can be subjected to a number of interpretations and related policy responses. The first is that the problems of the North are primarily the product of the imperfect functioning of the market; if policy allows divergences in various costs between the region and more prosperous parts of the country to increase, appropriate adjustments in investment will take place, which will inevitably lead to sustained economic growth in the region. Some supply-side interventions to boost entrepreneurship and bring under-utilised land into production may be necessary, but these are essentially local rather than regional initiatives.

The second interpretation is that market forces inevitably pull economic growth to more prosperous areas and consequently significant co-ordinated public intervention is necessary at the regional scale through investment in training, infrastructure and financial assistance to attract new industry and redress the balance of advantage. Proponents of this interpretation also acknowledge that some attention to the barriers to entrepreneurship arising from the history of the region will also be necessary.

Both of these interpretations, if to varying degrees, see the solution to the region's problem as residing within the region and the wider framework of opportunities arising from short-run macroeconomic cycles which can stimulate the diversion of investment from more prosperous to less prosperous regions.

The third interpretation, or more correctly set of interpretations, would consider the North's problems in terms of long-run changes in the structure of industry and commerce and the way economic activity can be regulated; changes which are apparent throughout the industrialised world. From this viewpoint the period of active regional policy succeeded in attracting mobile industrial investment to the North because the prevailing form and organisation of production made it a suitable location for relatively labour-intensive assembly operations. The past ten years have, however, seen fundamental changes in the organisation of industry and commerce associated with the emergence of new high technology production industries, the use of new information and communications technology in production and in the management of economic activity, new divisions of labour between large and small firms, the industrialisation of the service sector and last but not least the globalisation of enterprise. These changes have been associated with new ways of managing the interface between enterprises and the State, particularly in relation to the provision of industrial infrastructure, with greater emphasis on the regulation of the market rather than direct intervention. This has chiefly taken the form of the fragmentation of public intervention between single function regulatory bodies and to a multiplicity of local agencies which cut across the hierarchically ordered and territorially integrated structure of central, regional and local government of previous eras. The implications of these tendencies in terms of policies which seek to redress regional imbalance are as yet unclear.

The question then for the North is how well is it placed given its industrial and institutional inheritance and its peripheral location within Britain and Europe to withstand the threats and exploit the opportunities of a new industrial and regulatory order? What actors and policies need to be mobilised at the local, regional, national and indeed European level to assist the region to make the necessary structural adjustments?

The next section of the chapter considers some of the evidence on industrial trends in the region, focusing on the industrial dynamics highlighted by the last group of interpretations. Some of the evidence is also relevant to understanding for example the opportunities for relocation of investment between regions and for local entrepreneurship which is relevant to the other interpretations.

The discussion is largely based on the analysis of statistical data. These data usually come in the form of averages for the region and can therefore conceal quite different developments in individual enterprises. For example, some firms in the region may be at the 'leading edge' of some of the transformations referred to above, notwithstanding the generally adverse situation in the region reflected in the average figures. As such leading behaviour could act as a role model and focus for policy which seeks to achieve a demonstration effect, some individual cases are introduced into each component of the following discussion.

Regional industrial structure

One traditional explanation of poor performance points to the region's overspecialisation in declining industrial sectors. The policy response has been to achieve diversification by the attraction of new industries from elsewhere. While this might have been relevant in the past, successive rounds of rationalisation coupled with the establishment of new firms from elsewhere has served to bring the sectoral structure of the North's economy closely into line with the national average (Table 7.7). Shift-share analysis of regional manufacturing employment change over the period 1952-79 undertaken by Fothergill and Gudgin (1983) has indicated that industrial structure was relatively unimportant in account-

Table 7.7 *The industrial structure of the North, 1987*

	North	UK
Agriculture, utilities construction	10.4	8.3
Mining, metals, minerals, chemicals	8.2	4.7
Engineering	11.6	12.1
Other manufacturing	10.5	10.5
Transport and communication	5.4	6.6
Business services	7.1	11.4
Private personal services	24.2	26.0
Public services	22.6	20.4

Source: Employment Gazette, November 1987

Reducing Regional Inequalities

Table 7.8 *Shift-share analysis of manufacturing employment change in the Northern Region, 1952-79*

1952 Manufacturing Employment (000s)	409
% actual change	+7.8
National component[1]	-7.8
Structural component[2]	-4.4
Differential shift[3]	+20.0

Notes:
1. The national component is the change that would have occured if total manufacturing employment in the region had grown at the same rate as total manufacturing employment in the country as a whole.
2. The structural component is the change relative to the country as a whole that can be attributed to the region's particular mix or industries. This is calculated as the change which would have occurred if each industry in the area had grown at the national rate for that industry, less the national component.
3. The differential shift is the difference between the expected change and the actual change in the regions.

Source: Fothergill and Gudgin, 1983

ing for the region's employment performance - the large positive differential shift indicates that other factors were influential (Table 7.8).

Other commentators, recognising the inadequacy of such analyses based on broad sectors of the economy, have pointed to the national shift in industrial structure towards 'high technology' sub-sectors. During the active period of regional policy, the North clearly succeeded in attracting a considerable amount of production in what at the time were perceived to be high technology sectors such as electronics. But, in reality, much of this was routine assembly operations of electro-mechanical products at the end of their product life-cycles. A great deal of this capacity was eliminated during the 1976-83 recessionary years as new products were introduced elsewhere (Williams and Charles, 1986, p29). Between 1971 and 1981 employment in electronics industries in the North declined by 27.0 per cent and, as Table 7.9 shows, many of the major producers who had moved in the previous 20 years closed or shed large amounts of labour.

Taking a wider definition of those sectors with a high research intensity Begg and Cameron (1989, p361) have identified 12 high technology industries for the UK. In 1984 the North had an above-average share of national employment in only four of these sectors:

Table 7.9 *Some notable job losses in the North East electronic industry*

Firm	Sector	Location	Known peak	Job loss	Reason
Timex	consumer	Washington	850	850	closure 1980
Burroughs	computer	Cramington	200	200	closure 1981
Commodore	computer	Eaglescliffe	213	213	closure 1979
Mullard	components	Thornaby	266	266	closure 1976
Rank Xerox		Shiremoor	200	200	closure late 1970s
Thorn Radio Valves	components	Sunderland	2,000	2,000	closure 1971-4
Corning Electrosil	components	Sunderland	500+	350+	contraction to 150
Philips	components	Washington	950+	450+	contraction to 500
STC	components	Aycliffe	700	700	closure 1985
Allen Bradley	components	Jarrow	1,100	750	contraction to 350
Ferrograph	consumer	South Shields	327	327	closure 1978
GEC	telecommunications	North Shields	417	417	closure
GEC	telecommunications	Hartlepool	4,174	3,974	contraction to c200
GEC	telecommunications	Middlesbrough	1,772	1,772	closure
GEC	telecommunications	Aycliffe	2,200	800	contraction to 1400
Rediffusion	consumer	Billingham	1,317	1,317	closure 1985
Mullard	consumer	Durham	1,400	400	contraction to 1000
Thorn EMI	consumer	Sunderland	1,700	1,550	contraction to 150
Plessey	telecommunications	Sunderland/S.Sheilds	6,000	6,000	closure 1970s-1984

Source: Various sources in the public domain (eg press cuttings, directories)

Reducing Regional Inequalities

Table 7.10 *High technology industries in the North*
Location quotients GB = 100

	1981	1984
		Above Average
Resins, plastic and rubber	243.2	534.3
Basic electrical machinery	130.3	142.3
Active components	120.7	133.6
Pharmaceuticals	130.2	121.9
		Below Average
Telecommunications equipment	89.6	85.8
Measuring instruments	78.4	78.6
Telecommunications	62.6	65.7
Research and development	42.6	59.3
Other instruments	53.0	56.5
Computing services	24.3	26.6
Aerospace	8.2	7.1
Office machinery	12.6	6.9

Source: Begg and Cameron, 1988

resins, plastics and rubber; pharmaceuticals; basic electrical machinery; and active components (although it had improved its position in a further three sectors over the previous three years). Not surprisingly, Cameron and Begg's analysis showed that it is the South East which is pre-eminent in all but 2 of the 12 sectors (Table 7.10).

Such an emphasis on high technology does assume that changing industrial structure occurs only at this aggregate level of analysis. Equally significant is the process of industrial diversification at the level of the enterprise through industrial innovation and in the introduction of new and improved products. Firms in traditional industries can equally renew and transform old products by the incorporation of new technology. To what extent has this been happening in the North?

The Science Policy Research Unit (SPRU) has assembled a database on the introduction of significant innovations across the spectrum of British manufacturing industry since 1945. These data suggest that the North's share of innovations has fallen through successive decades from 5.9 per cent to 4.5 per cent of the total. A detailed regional analysis of these data

Regional Economic Development: The Northern Region of England

Table 7.11 *Shift-share analysis of innovations, 1945-79 (per unit of net output)*

	1945-63	1964-79
Actual	7.41	3.22
Differential[1]	2.00	-1.85
Structural[2]	-7.20	-5.07
Residual	9.20	3.23

1. Difference in comparison with national average
2. Difference due to mix of industries

Source: Harris, 1988

by Harris (1988, p361) has indicated a statistically significant decrease in the north's share of innovations over the period in two out of five groups of industries (amalgamated from the most to the least innovative sectors). Part of the explanation of this poor performance is the North's specialisation in industries which nationally recorded a low level of innovation; however, comparing the periods 1945-63 and 1964-9, Harris's analysis suggests that in the second period the North performed worse than would be expected from its industrial structure as indicated by net output per sector (Table 7.11).

More recent data gathered by Northcott on the incorporation of the microprocessor into products in British industry confirms the poor performance of the North. Based on a survey of 1,200 establishments in Britain carried out in 1986, Northcott (1986) found that only 6 per cent of establishments in the North had introduced microprocessors, compared with 19 per cent in the South East and 21 per cent in East Anglia. A more detailed study by Alderman *et al* (1989) of technological innovation in establishments in nine engineering industries also found that only 16.1 per cent in the North had adopted microprocessors by 1986, compared with 39.4 per cent in the West Midlands (Table 7.12). Moreover, 42.0 per cent of firms in the North regarded microprocessors as 'not applicable' to their products, compared to 26.3 per cent in the West Midlands, perhaps indicative of a lower level of technical sophistication of products in the region.

Industrial diversification occurs not only by the introduction of 'significant' innovations and the incorporation of microelectronics into products, but by the incremental improvement of existing products. A

Reducing Regional Inequalities

Table 7.12 *Adoption of microprocessors in engineering establishments*

%	Non adopter	Adopted 1981/86	Adopted pre 81	Not applicable
North	42.2	6.2	9.9	42.0
South East	43.2	16.1	18.7	21.9
West Midlands	34.3	21.2	18.2	26.3
North West	42.0	14.5	15.9	27.5

Source: Alderman, Davies and Thwaites, 1987

now somewhat dated study of the introduction of new and improved products into three industries – metal-working machine tools, scientific and industrial instruments and radio and electronic components over the period 1973-7 by Thwaites *et al* (1981) discovered a notably poor innovation record in the North, particularly among single-site firms. While multi-site firms, chiefly the branches of national and multinational companies had a much better record of innovation, the survey results suggest that two-thirds of these were the result of research and

Table 7.13 *Technology transfer into the north 1973-7*

	North	South East
% all establishments with developing product innovation in-house	53.0	73.0
% multi-site establishments with in-house product innovation	45.0	71.0

Note: Sample of 110 establishments in metal working machine tools, radio and electronic components and scientific instruments

Source: Thwaites *et al*, 1981

development effort elsewhere and transferred to the North for production (Table 7.13).

Notwithstanding this overall poor record, a number of small firms in the North have successfully introduced new products into the market as a basis for growth. Integrated Micro Products (IMP) at Consett for example is a 1980s start-up, founded by a researcher from the Open University. The company has developed a series of multi-user microcomputers, mainly for research and educational markets, using the newest designs of microprocessors from US and Japanese semiconductor firms. IMP has been able to grow rapidly within its market niche, and most recently acquiring a competitor based in Silicon Valley, California.

Some software firms have also been successful, notably Sagesoft with high volume business packages and Quality Software Products (QSP) with IBM mainframe accounting packages. The latter initially borrowed time on customers' mainframes to develop software, until they could afford to purchase their own. Subsequently, they have established a joint venture with IBM, whereby IBM will market the company's products throughout the UK allowing QSP to double its employment of 200 over the next two years.

Diversification by innovation in small firms in the region is therefore possible: the essential problem is insufficient numbers to create a critical mass that can transform the environment.

Occupational structure

Although industrial structure in aggregate does not seem to work against the North, these findings do support the idea that a micro-level northern industry, particularly indigenous industry as a whole, is failing to diversify through product innovation. What does characterise the high technology and innovative enterprise is the high proportion of employment in research and development occupations. As the pace of technological change quickens, the importance of R&D is increased in all industries. While regional assistance may have attracted production to the North in a wide range of sectors, these generally did not include R&D functions. As is well known, industrial research and development in the UK is heavily concentrated in the South East. Even in sectors like pharmaceuticals, which have a strong presence in the North, research activity is not well represented. A study by Alderman and Thwaites (1987) of mechanical engineering sectors experiencing rapid technological change shows that in spite of some improvement in the period 1981–6, the proportion of establishments without on-site R&D in the

Reducing Regional Inequalities

Table 7.14 *Research and development in the North, 1981-6*

	North	UK
% of establishments with No R&D		
1981	54.3	36.4
1986	50.6	35.5
% of establishments with R&D but none full-time		
1981	13.6	16.1
1986	7.6	15.7
% of establishments with full-time R&D		
1981	32.1	47.5
1986	41.8	48.7

Source: Alderman and Thwaites, 1987

North remained well below the national average (Table 7.14).

This under-representation of R&D in the occupational structure of northern manufacturing industry is symptomatic of a wider orientation towards skilled manual work and away from managerial and professional occupations. Skilled manual workers represented 24.3 per cent of the workforce in 1981, compared with only 16.1 per cent in the south. At the sub-regional scale a comparison of Tyne and Wear and Berkshire using occupational data from the 1981 census undertaken by Howells and Green (1986, p83) clearly indicates an over-representation of lower grade occupations in the former county and higher grade occupations in the latter (Table 7.15). Within manufacturing industry, the proportion of employment in managerial and professional occupations in 1981 was 15.0 per cent below the national average and 34.0 per cent below the figure for the South East.

Although on aggregate measures of R&D, the northern region performs poorly, there are none the less a number of major R&D units in the region, some of international status. Thus, for example, ICI Wilton Technical Centre on Teesside employs around 1,000 people on materials R&D. British Gas has two major research stations north of Newcastle, Sterling Winthrop has its European Research Centre at

Table 7.15 *Occupational structure, Tyne and Wear and Berkshire, 1981 (Location quotients GB = 100)*

	Tyne and Wear	Berkshire
Managerial and admin	79	107
Higher level service	85	119
Higher level industrial	89	162
Lower level service and supervisory	109	102
Craft/foremen	122	89
Lower level industrial	101	80

Source: Howells and Green, 1986

Alnwick in Northumberland and British Steel has centred its process research near its Redcar facility. Other local companies, such as NEI and Swan Hunter, although apparently in traditional industries, maintain considerable R&D effort, and there is also a range of smaller companies, particularly in engineering and IT sectors, devoting a considerable proportion of turnover to R&D. Within these research facilities a number of significant innovations have been developed in recent years, some recognised by Queen's Awards for Technological Achievement, a recent example of which is the pipeline inspection vehicle of British Gas, a unique product combining advanced engineering design with electronics, and requiring a major data processing and interrogation facility to process output.

Within all of these firms there is a recognition that quality R&D can be undertaken in the North, the only problem being the difficulty of recruiting staff from the south. However, in interviews, most R&D employers in the region report that graduates can be recruited from elsewhere in the North and Scotland, and any problems of recruitment are counterbalanced by low labour turnover rates. Recent anecdotal evidence appears to suggest that the lower overheads, proximity to production and lower labour turnover is currently leading a number of firms to decentralise more R&D to their plants in the North.

New and small firms

The occupational structure of employment in the North goes a long way to accounting for its relatively poor record in new firm formation and

Reducing Regional Inequalities

Table 7.16 *Net change in business registrations, 1980-6*

	Numbers (000)		Rate (per 000 employees 1981)	
	North	UK	North	UK
Production industries	1.5	29.8	3.7	4.3
Finance, property and product services	1.5	49.1	7.1	10.6
Other services	2.0	99.5	11.1	25.4
All industries	7.5	284.4	6.7	13.1

Source: VAT statistics
Employment Gazette, April 1985

subsequent growth and the difficulty it has faced in the indigenous economic development. Research by Storey and Johnson (1987) has clearly demonstrated a higher survival and faster growth rate among new businesses established by entrepreneurs with a managerial background. The regional significance of this factor has been demonstrated by Coombes and Raybould (1989) using VAT registration data. These data indicate that while the number of new businesses in the North has increased, the rate of increase is half the national average (Table 7.16). This difference is largely attributable to low rates of registration in the fast growth business and private service sectors. However, the crude VAT registrations data do not take account of the high turnover of small businesses. Coombes and Raybould have therefore constructed a growth index which incorporates birth and survival rates for VAT registered firms by county for the period 1983-5, which reveals that all the counties in the North East are in the bottom quartile according to this index.

In modelling these variations Coombes and Raybould have discovered that the proportion of the workforce that are managers is the most significant factor accounting for variations in the growth index in every single year of their study period (Table 7.17). Analysis of the residuals from the model do however show certain counties in the North performing better than might be expected from the characteristics of the local entrepreneurial environment.

Regional Economic Development: The Northern Region of England

Table 7.17 *Factors accounting for inter-county variations in local enterprise activity potential, 1980–5*

	Rank order model
% workforce that are managers	1
% households owner-occupied	2
Index of peripherality	3
% population change	4
% households with more than one car	not significant
% households that are dual income, no children	not significant

Note: % age explanation = 85%
Source: Coombes and Raybould, 1989

The VAT statistics do not directly provide information on employment growth in new and small firms. However, a sample survey comparison of new starts in Teesside and Reading over the period 1980–5 by Johnson (1989) does reveal a pronounced difference between the two areas, with expanding Reading firms creating jobs at almost twice the rate achieved in Teesside (Table 7.18). Johnson's findings on the uptake of new technology also confirm those of Thwaites *et al.* In his sample, new firms in Reading recorded a higher use of new technology in all areas of business activity than is the case on Teesside.

There are, of course, many individual exceptions to this overall pattern among new and small firms in the region. Some of the fastest growth businesses in the region are in the small firm sector and are not necessarily new foundations. For example, J Barbour and Sons is a long-established family firm manufacturing a range of waterproof clothing which has appeal in a range of industrial and fashion markets. While the clothing sector is frequently classified as a traditional industry, the firm is far from traditional in its approach to business operations. In addition, it has invested in the latest computer-controlled equipment to help in the production of its products. An indication of its performance is reflected in the number of employees in the firm which grew from 150 in 1981 to 350 in 1987.

Although there are exceptions, the North has a long way to go if it is to attain the levels of economic growth associated with new and small business growth that has been achieved in more prosperous parts of the

Reducing Regional Inequalities

Table 7.18 *New firms in Teesside and Reading, 1980-5*

A *Employment*	Teesside	Reading
Survivors: Job gain per firm	+3.0	+5.6
Job loss per firm	-2.3	-1.3
New firms: Job gain per firm	+5.5	+8.7

B *% adopting of new technology*

	Limited adopters		Significant adopters	
	Teesside	Reading	Teesside	Reading
Office processes	10	33	8	17
Production processes	21	37	11	7
Corporation	27	37	3	11
Storage and distribution	7	33	7	10

Source: Johnson, 1987

country. The national shift of economic activity to smaller firms is in part the result of the externalisation of functions, especially services from large enterprises; there is little indication that this process is a potent force in the North, given that the service demands of branch establishments are usually met from corporate headquarters.

Peripherality

A characteristic of many small businesses is their initial reliance on the local market. Low levels of domestic demand and high levels of non-local ownership of industry have inevitably meant that small businesses in the North have had to look to distant national and international markets in contrast to their counterparts in the South East who have had a prosperous market on the doorstep. Thus, 22.0 per cent of the small firms in Johnson's sample in Cleveland served an entirely national market, compared with only 4.0 per cent in Reading. This difference also reflects the higher proportion of the Cleveland sample in the manufacturing sector. (It should also be noted from Table 7.17 that an index of peripherality is the second most important factor in Coombes and

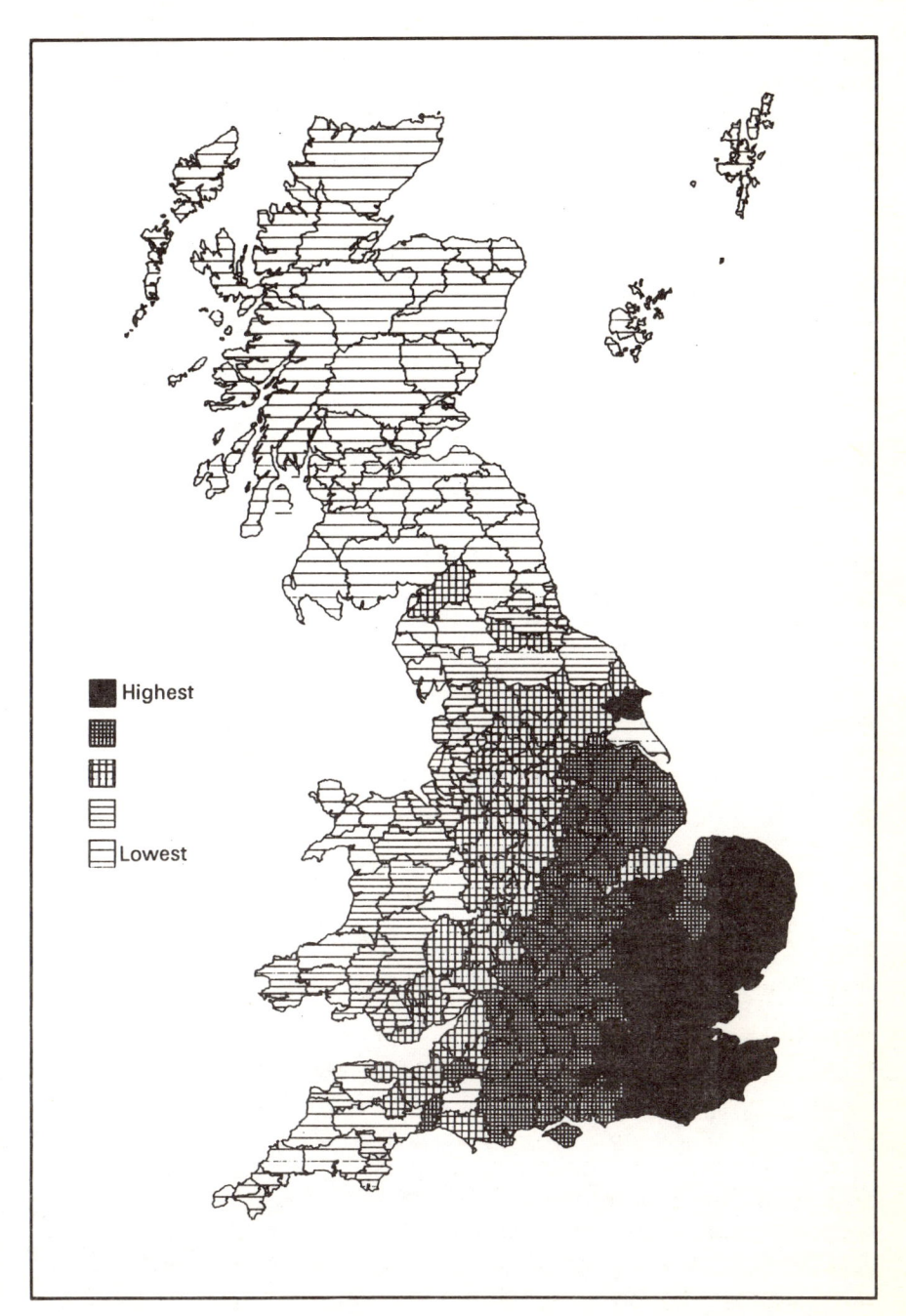

Regional Economic Development: The Northern Region of England

Figure 7.1 *Change in market access (regional market assumptions used)*

Raybould's analysis of small firm growth and survival.)

The peripherality of the northern region with respect to the UK and European markets is difficult to deny. An analysis of market potential undertaken by Owen and Coombes (1986) has shown the extent of the peripherality problem in the UK in the context of the European market. The recent work by Coombes *et al* (1990) in this context, taking full account of the opening of the Channel Tunnel and other non-fixed links, shows the extent to which the North can expect to see a much smaller increase in its accessibility to the European market. Figure 7.1 records the changes in an index of peripherality for travel-to-work areas arising from the opening of the Channel Tunnel: the index is based on travel distance and the distribution of economic activity (employment) within Britain and Europe.

These regional differences within Britain clearly feed through into higher expenditures on communication in the North. A recent survey of a matched sample of manufacturing establishments in four regions undertaken for the DTI by Diamond and Spence (1989) reveals that transport and communications expenditure in the North represents 7.1 per cent of total operating costs, compared with the average of 6.6 per cent for the four regions (Table 7.19). Lowest costs were recorded in the

Table 7.19 *Expenditure on transport and telecommunications, 1987*

	% total operating costs
North East	7.1
North West	6.6
West Midlands	4.2
South East	7.5
Average	6.6

Source: Diamond and Spence, 1989

West Midlands with a figure of 4.2 per cent. Surprisingly, the survey reveals the highest costs for firms in the South East. A breakdown of the data (Table 7.20) suggests that this is chiefly attributable to higher expenditure on personal communications (business cars, air passenger transport and telecommunications). Notwithstanding their peripheral

Table 7.20 *Average components of expenditure (£000s) on transport and communication*

	North	South
A *People and information*		
Business cars	14.5	52.3
BR passenger costs	2.8	4.8
Air passenger costs	12.6	70.3
Telecommunication costs	13.9	31.1
B *Goods*		
Commercial vehicles (numbers)	1.9	6.7
Bought in road services	106.7	167.7
BR freight	6.7	5.4
Other freight	13.4	276.5
Sea freight	35.7	155.9

Source: Diamond and Spence, 1989

location, firms in the North spend less on keeping in contact, a reflection of the under-representation of information-based non-production activities that has already been referred to.

To what extent can firms in the North use new information and communications technology to reduce these peripherality problems? A key development here is the convergence of telecommunications and computing technologies in the form of computer networks which can be used to connect workplaces within and between organisations. Such networks are increasingly central to a wide range of industrial and commercial processes and can make a significant contribution to industrial competitiveness, particularly for firms in peripheral locations.

A recent survey of firms in the northern region undertaken on behalf of the Northern Development Company's Telecommunications Sector Working Group has revealed a rapid increase in the use of advanced telecommunications in the region. Between 1985 and 1988 an 85.0 per cent increase in telecommunications usage was recorded in large multi-site organisations and a 35.0 per cent increase in small independent firms (Bianchi *et al*, 1989). This contrast between the two types of organisations in the region confirms the findings from the Department of Employment's Workplace Industrial Relations Survey for 1984 which revealed that only 5.0 per cent of single-site workplaces in Britain were

Reducing Regional Inequalities

Table 7.21 *Computer facilities, 1984*

% of establishments with:	North	London	UK
No facilities	54.0	29.4	41.5
On-site facilities	26.4	34.1	32.7
Network link	19.6	36.5	25.8

Source: Workplace Industrial Relations Survey, 1984

linked to a computer network. This survey also revealed that only 19.6 per cent of workplaces in the North had a computer link compared with 25.8 per cent nationally and 36.5 per cent in London (Table 7.21). London is therefore the hub of UK computer networks reflecting the concentration of corporate control functions in the capital.

Research in CURDS (1989) has sought to update these figures by surveying the use (or non-use) of computer networks in organisations based in the North East, the North West, East Anglia and the City of London and linking the use of networks to economic performance. Table 7.22 reveals a significant difference between networked and non-networked firms in employment turnover and profitability growth. This

Table 7.22 *Computer networks and company performance, 1985-7*

	North East	North West	East Anglia	City of London
A *Average employment change %*				
Networked	36.3	19.5	15.4	21.2
Non-networked	9.1	0.1	4.6	-4.6
B *Average growth in turnover %*				
Networked	73.5	42.9	48.9	104.4
Non-networked	27.6	18.9	25.6	3.8
C *Average growth in profitability %*				
Networked	17.5	4.1	9.1	16.9
Non-networked	0.7	0.6	1.4	-3.4

Source: CURDS, 1989

difference is particularly marked in the Northern Region. Multivariate analysis shows that after allowing for size differences, the use of a computer network is a significant factor accounting for inter-organisational differences in economic performance according to all three measures and that this effect is most pronounced in the North East. One inference that can be drawn from this analysis is that the firms that are adopting computer networks in the North East have been able to differentiate themselves from other enterprises in the region by effectively linking themselves to sources of information, customers and suppliers outside the area.

Although not in the northern region, a case study of a 'northern' firm based in Doncaster undertaken by CURDS (Williams *et al*, 1988) serves to illustrate how a computer network can be used to overcome the problems of peripherality and assist with adjustment to the changing geography of markets.

John Carr plc is a manufacturer of doors, windows and related components for the building industry. In its early development, the firm concentrated on being the least cost producer in locally defined markets, chiefly for local authority housing. Subsequent expansion involved duplication of production in different areas. The use of information and communication technologies (ICTs) at this stage of the firm's development was restricted to finance and cost control functions. The network was used to reinforce the strategic objective of being the least cost producer.

Over a period of three years, the company's locally differentiated markets collapsed, and the firm was left with an organisation suited to securing its growth in market conditions that no longer prevailed. The orientation of the firm was towards local authority markets in northern England and to the manufacture on a ten-week order-to-delivery cycle. The new market conditions manifested themselves in a shift to the private sector and DIY markets in southern England and the manufacture of product on a three-day order-to-delivery cycle.

The response of the firm to these structural changes in its market were interwoven with the development of its computer network. The firm had to bring about the spatial reorganisation and integration of its production, as well as its sales and marketing functions, in order to achieve the requirements of a three-day order-to-delivery cycle. This reorganisation was centred upon a new information strategy capable of assembling, analysing and integrating information externally and assimilating it with information on its own internal operations. The computer network was central to this restructuring of the firm.

Reducing Regional Inequalities

In terms of production, the firm reorganised its manufacturing plants so that each had prime responsibility for a particular product range, so as to achieve economies of scale in production. The branch plants despatched their output to warehouses which were responsible for deliveries to the market. The sales function was also centralised and its geographical focus reoriented towards the market in southern England.

The integration of these spatially fragmented functions of manufacturing, distribution and sales is now managed through the computer network. A hierarchical relationship has been established, whereby the activities of the company are determined by the actual uptake of product in the market. The company has also sought to align its sales activities with several of the large merchant chains through the development of inter-corporate networking. In so doing it has attempted to use its computer network to exert greater influence on the market and to secure for itself the distribution of product into new geographically defined markets.

Finally, the spatial restructuring of the company has occurred within its existing locations. Through the use of computer networked information flows the firm has managed to reassert its territorial claims in markets in southern England without new investment in this high cost area. In short, the development of computer networking has allowed the spatial reorganisation of the company and ensured the integration of different functions, as well as facilitating its entry into new markets.

Service industries

The use of information and communications technology can clearly be a means of providing competitive advantage for manufacturing enterprise based in the North. What about the business and financial services sector, where the transfer of information is an even more central activity and where there has been a widespread 'industrialisation' of transactional activities in response to competitive pressures (eg deregulation) and the opportunities provided by information and communications technology? Although now incorporated into regional industrial policy, business and professional services have not been its primary focus. Nevertheless, as Table 7.23 shows, banking, finance and insurance employment increased in the North by 26,000 or 49.1 per cent between 1978 and 1989, while other parts of the service sector experienced decline or smaller increases. The bulk of this employment is within Newcastle. (Some of Newcastle's growth may have been at the expense of

Table 7.23 *Service employment in the north 1978-89*

Thousands

	1978	1983	1988	1989
Wholesale distribution, hotels, catering	101	91	98	102
Retail distribution	118	102	106	107
Transport and communication	65	58	53	53
Banking, insurance, finance	53	63	74	79
Public administration and defence	99	87	102	97
Education, health, other services	214	227	263	250

Source: *Employment Gazette*, November 1989

smaller office centres in the region, as large enterprises based in the city increased their share of regional service markets – for example the 20 largest accountancy practices in the UK accounted for 86 per cent of the audit market in the Northern Region in 1986.)

Leyshon and Thrift (1989) have undertaken a detailed analysis of the growth of financial and business services in provincial centres. According to their figures, Newcastle ranked ninth in 1984 in terms of employment in this sector. Its chief competitors were Leeds (which had more than twice the commercial office floor space) and Edinburgh (ranked fifth). Leyshon *et al* attribute the growth of provincial business activity outside London to three factors: the growth of local demand; the expansion of multi-locational firms from London; and the dispersal of back offices from the capital. They suggest that local demand can arise from sources such as the externalisation of services by locally based businesses, the large volume of insolvency business arising from the high rate of business failure in the regions, the need for services to support local takeover and merger activity and the large volume of public sector business generated in the regions as a result of the drive to greater cost-effectiveness and accountability. Rising personal incomes may also be a factor.

The lack of interest of City institutions which have become increasingly involved with international markets has created opportunities for local businesses in provincial cities. In the case of the North East there has been a local entrepreneurial response, but as the VAT registration data reveals (*vide* Table 7.17), the rate of establishment of new businesses in this sector in the region is well below the national average. A more significant part of the expansion in Newcastle to meet rising demand has therefore come from national and international financial service businesses, particularly accounting practices, establishing or expanding local offices. Between 1975-6 and 1985-6 the number of offices of the 20 largest accounting, commercial, property and investment banks located in Newcastle increased from 12 to 15. Nevertheless, this was a smaller increase than was recorded in Leeds or Edinburgh (plus 6 and plus 11 respectively).

Leyshon *et al* (*op cit*) attempt to measure the extent to which local business and consumer demand is met by local supply by means of a regression analysis of county GDP and gross/net personal salaries against financial and business service employment. Although there must be some reservations about the data and this form of analysis, the residuals do suggest that the expansion of financial and producer services in the North has not kept pace with demand (Table 7.24). The explanation for this may be the effect of competition from Leeds and Edinburgh and the fact that much of the GDP growth in the region is attributable to manufacturing enterprises with headquarters outside the area.

It would be wrong to equate the expansion of business and financial services in Newcastle solely with local or regional demand. One large accountancy practice in the city has over 200 staff and forms part of a network of UK and international offices, each with its own specialisms; operating within the network, the Newcastle office serves national and international, as well as regional clients. This structure means that staff can not only earn income from outside the region, but also mobilise different expertise from outside in order to provide the best possible advice for local clients.

The third factor accounting for the growth of financial and business services outside London, the relocation of back offices, has clearly not benefited the North. Of the 115 major office relocations out of London recorded by surveyors Jones, Lang and Wootton over the period 1979-86, only 20 per cent were in the financial and producer services sector; moreover 78 per cent of all moves were confined to the South East region. Analysis of rent and clerical wage savings for organisations considering a relocation suggests that the maximum benefits can be found in office centres in the East Midlands (eg Leicester, Nottingham and Derby). Beyond this distance extra communications costs absorb many of the savings. The current limited supply of office space in Newcastle, itself a product of low office rents in the past, is a further problem. The rapid expansion of office supply in London, associated

Regional Economic Development: The Northern Region of England

Table 7.24 *Residual levels of financial and producer service and employment not 'explained' by local levels of 'consumer demand' and 'business demand', 1984*

Centre	Residual financial and producer services employment
	No
Edinburgh	10,748
Bristol	6,957
Leeds	5,526
Norwich	2,136
Brighton	1,967
Reading	1,826
Birmingham	1,278
Glasgow	1,273
Southampton	847
Northampton	302
Cardiff	261
Bournemouth	−752
Bradford	−925
Sheffield	−1,301
Liverpool	−1,548
Coventry	−1,699
Aberdeen	−2,652
Manchester	−2,765
Leicester	−3,726
Newcastle	−4,116
Southend	−4,227
Nottingham	−4,268
Oxford	−5,092

Source: Leyshon and Thrift, 1989

with Canary Wharf and Kings Cross, could discourage further investment in more remote provincial cities like Newcastle, notwithstanding rising rental values there.

The communication constraints referred to above do not apply to heavily computerised back office jobs. For example, British Telecom has successfully relocated large parts of its London directory enquiries

Reducing Regional Inequalities

service to the region. A large part of the national social security system also continues to be run from Newcastle's single largest employer, the DSS at Longbenton. However, it is doubtful whether the region will gain large numbers of additional Whitehall back office jobs. The agency structure for central government presaged in Sir Robin Ibbs's report, *The Next Steps* (1988), will encourage departments to act and behave more like private sector organisations. Greater flexibility in pay may enable agencies to tackle pay differentials between the public and private sector and remain within London or, if they relocate, to follow the optimal pattern of short to medium-distance moves established by the private sector. Agencies are also likely to avoid a labour market already

Table 7.25 *Employment main locations of non-industrial civil service staff, 1979-87*

Administrative district	Absolute	Change 1979-87 % Change	Total employees
London	-19,013	-14.1	116,012
Newcastle	-4,055	-23.1	13,481
Edinburgh	-1,333	-9.6	12,557
Glasgow	+58	+0.5	11,429
Cardiff	-1,419	-13.7	8,904
Birmingham	-349	-4.0	8,410
Plymouth	-393	-5.2	7,162
Manchester	-2,339	-25.1	6,979
Liverpool	-866	-11.5	6,634
Leeds	-1,061	-13.9	6,592
Sefton	+1,098	+21.7	6,159
Swansea	-687	-10.5	5,879
Bristol	-578	-9.1	5,804
Bath	-838	-13.5	5,369
Blackpool	-49	-1.0	4,921
Sheffield	+1,760	+56.9	4,855
Portsmouth	-2,206	-32.6	4,552
Durham	+568	+14.9	4,375
Southampton	-843	-16.8	4,161
Nottingham	-1,352	-24.6	4,141
Southend	-874	-17.4	4,069

Source: Civil Service Statistics
Marshall, JN (1990)

dominated by one large government employer, particularly one with a reputation for militancy.

A major expansion of civil service jobs from within the region is also highly unlikely. The North already has the largest concentration of non-industrial civil servants outside London and has also recorded the highest net decline over the past ten years (Marshall, op. cit.) (Table 7.25). The National Audit Office (1989) attributes more than two-thirds of the net loss of 152,000 civil service jobs nationally over the period 1979-88 to improved efficiency/streamlining and much of this is associated with computerisation. The DSS computerisation and the use of the government's data network may be associated with the 'migration' of jobs away from 'the back office' in Newcastle to local offices throughout the country. In an increasingly computerised 'network economy', large labour-intensive office factories in both government and the private sector are unlikely to be a major source of employment for the North, particularly as English-speaking developing countries with satellite links compete successfully for routine data entry tasks.

Mobile manufacturing investment

If the possibilities of office relocation to the North are limited, what are the opportunities within the manufacturing sector in terms of mobile investment? Since the DTI have ceased publication of their records of openings and closures it is difficult to assess the extent and significance of inter-regional mobility of manufacturing industry within the UK. The withdrawal of UK multinationals from the North in the 1979-83 recession has already been noted. More recent anecdotal evidence suggests that much of the recent expansion of manufacturing in the UK as a whole has taken the form of *in situ* growth. One possible explanation for this is that rapid rates of technological change and shorter product life-cycles are necessitating a closer link than in the past between research and production: the establishment of large-scale mass-production activities on greenfield sites manufacturing goods with an anticipated long-life span is increasingly uncommon. The removal of intra-UK job transfers for eligibility for regional industrial assistance may be another factor contributing to a decline in mobility.

The same conclusions would not apply to mobile investment from overseas. The region has been very successful in capturing new greenfield investment as Japanese and other Far Eastern manufacturing firms have sought a platform within Europe prior to 1992. The North now has 22 Japanese and Far Eastern manufacturers with a total employment of

Reducing Regional Inequalities

Table 7.26 Components of change analysis of foreign-owned manufacturing establishments in the northern region, 1978-89[1]

Country/area	Stock in 1978	New openings	Acquisitions from UK ownership	Expansions	Contractions	Unchanged	Divestments to UK ownership	Closures	Net change	Stock 1989
(i) No of plants										
USA	114	21	8	24	38	2	14	34	-19	95
Europe	52	32	17	21	16	0	3	13	+33	85
EEC	[34]	[11]	[5]	[11]	[11]	[0]	[2]	[10]	[+4]	[38]
non-EEC	[18]	[21]	[12]	[10]	[5]	[0]	[1]	[3]	[+29]	[47]
Far East	1	19	1	2[2]	0	0	0	0	+20	21
Other[3]	12	1	6	2	4	1	3	2	+2	14
TOTALS	179	73	32	49	58	3	20	49	+36	215
(ii) Associated employment										
USA	39,597	2,245	2,641	1,406	9,380	0	3,333	9,121	-15,542	24,055
Europe	11,053	3,862	6,846	1,602	2,051	0	279	1,138	+8,842	19,895
EEC	[8,239]	[911]	[1,313]	[830]	[1,627]	[0]	[220]	[1,015]	[+192]	[8,431]
non-EEC	[2,814]	[2,951]	[5,533]	[772]	[424]	[0]	[59]	[123]	[+8,650]	[11,464]
Far East	287	4,355	494	343	0	0	0	0	+5,262	5,549
Other[3]	2,761	17	2,050	18	120	0	226	710	+1,029	3,790
TOTALS	53,698	10,479	12,031	3,369	11,551	0	3,838	10,969	-409	53,289
Percentage change over 1978	—	+19.5%	+22.4%	+6.3%	-21.5%	0	-7.1%	-20.4%	-0.8%	

[1] End year data refers in some cases to 1988. The northern region comprises the counties of Northumberland, Tyne and Wear, Durham, Cleveland and Cumbria
[2] Expansion includes one US firm transferred to Japanese ownership
[3] Includes Canada, Australia and South Africa

Source: Compiled using authors' northern region FOS database, Smith and Stone (1989)

approximately 6,000 in 1989. Altogether 10,500 jobs were created between 1978 and 1989 by 73 new overseas plants in the region.

A detailed components of change analysis undertaken by Smith and Stone (1989) has indicated that these gains were offset by the loss of 11,000 jobs in the closure of 49 foreign plants and employment contraction in the survivors over the same period (Table 7.26). The overseas sectors did however increase their share of the North's manufacturing employment to around 20 per cent of the total by 1989, chiefly through the acquisition of 32 British companies, a transfer of ownership involving some 12,000 jobs.

In net terms then, Japanese inward investment has compensated for American disinvestment. Over the 1980s 34 out of a total of 114 US manufacturing plants closed in the region (eg Caterpillar, Timex, Cummings and Proctor & Gamble). US companies have also been transferred to UK ownership through acquisition of US parents (eg Elmwood Sensors) or local management buyouts (eg Allen Bradley). Acquisition by European firms from outside the EC (the most notable being the Nestlé takeover of Rowntree) has also contributed to an increase in foreign ownership in the region. Taking all forms of investment together, Europe clearly replaced the US as the principal component of new investment in the region in the 1980s. Altogether these changes have meant that foreign manufacturing employment has been maintained at its 1978 levels, while total manufacturing employment has declined by 37.2 per cent. The strong performance of manufacturing industry in the region during the recovery period can therefore largely be attributed to foreign companies.

It is too early to assess the impact of this latest round of inward investment on the region. Developments at Nissan are surrounded by controversy. In the earlier stages, until production built up to a sufficient level to justify further investment, it was clearly a limited assembly operation establishment. But as volume has increased and new market share has been captured, the high technology, high productivity nature of the operation has become a reality as well as a symbol. Local sourcing, but from overseas companies, has increased as Japanese suppliers have moved into the region (eg Sumitomo – tyres and Ikeda-Hoover – seats). Komatsu, makers of earth-moving equipment, acquired the factory abandoned by Caterpillar of the US and has sought to develop high quality suppliers in the region; however, even in traditional 'metal bashing' areas it is reported to have found difficulty in immediately obtaining the requisite quality. Notwithstanding the local adjustment lags in both cases, the scale of the investment and its continued

Reducing Regional Inequalities

upgrading in the context of the European market suggests that Japanese investment in the region may be more permanent than previous US activity.

The next section of the chapter considers the contribution of public policy, particularly regional industrial policy and local economic development policy to the North. This discussion of policy lays the basis for suggestions as to future priorities which can exploit some of the opportunities arising from the latest phases of structural change in the economy.

III Policies for the North

Economic development policies in the region

In a recent report Robinson (1990, p40) states that:

> The Northern Region has been variously described as a 'policy laboratory' and a 'state managed region'. It has tried and tested, sometimes to destruction, almost every kind of economic development policy option available within the context of a 'mixed economy'. There have been innumerable policy statements, economic plans and strategies and scores of policy initiatives. To the extent that the region is still disadvantaged relative to other parts of the country all this policy effort can hardly be judged to be a great success.

However, Robinson does go on to suggest that without the intervention the situation might have been worse. He also points to the direct role of government in the sense that much of the economic decline has occurred within nationalised industries and its indirect role in supporting high technology in the south through defence procurement and public R&D laboratories.

With the establishment of the Team Valley Estate in 1938 the North led the way in public action to promote economic development through the provision of advanced factories. Such public provision continues through the activities of English Estates, which still has its headquarters in Gateshead. But, according to the latest Audit Commission Report (1989), the Estates Corporation's expenditure now forms a small part of the patchwork quilt of government policies designed to assist area development. Moreover, traditional instruments of regional policy administered directly and indirectly through the DTI, now represent a relatively small share of total central government expenditure in area development. For example DOE expenditure on the Urban Programme and Urban Development Corporations at £560 million in 1988-9 is four times the expenditure on Regional Selective Assistance (£145 million).

No separate figures are available for the Northern Region. Robinson does provide some rough estimates with current annual figures of £80 million on Regional Selective Assistance, £25 million on the Urban Programme, £10-20 million on rate relief and tax concessions in the Enterprise Zones, about £50 million through two UDCs, £100 million through the training agency, £20 million support for the Gateshead Garden Festival and £50 million from the European Regional Development Fund and the European Social Fund (some of which substitutes for UK expenditure). This gives a total of over £300 million. Robinson *et al* (1987) provide more accurate figures for direct public financial assistance to industry on Tyneside from all sources drawing on a database for the 1974-84 period. This amounted to an annual expenditure of £40 million. A further £72 million in 1983-4 was allocated through the training budget and £35 million through the Urban Programme. To put this total of £147 million in perspective, Robinson *et al* suggest that the Exchequer costs of unemployment in the area in the same year amounted to £750 million.

Regional industrial policy

The shortcomings of regional industrial assistance are well known. The bulk of automatic Regional Development Grant went to support capital intensive projects in companies that were well established in the region. For example, between 1974 and 1985, three major companies - ICI, BNFL and British Steel accounted for 44 per cent of regional development grants paid to companies in the North.

What contribution has assistance policy made to addressing some of the shortcomings in the region's industrial structure identified earlier in the chapter? Has it contributed towards a reduction of industrial specialisation, stimulated industrial innovation, R&D and small businesses?

In this context, the North's poor record in job creation may be regarded as a *symptom* of these shortcomings. While there have been many evaluations of regional policy, these have concentrated solely on its cost-effectiveness in terms of job creation to the virtual exclusion of other aspects of industrial performance that lie behind poor employment growth. A notable exception to research of this genre has been the evaluation of all forms of financial assistance to firms in the Newcastle

Metropolitan region undertaken by Robinson *et al* (1987) covering the period 1974-84.

Robinson *et al* found that DTI assistance made a very limited contribution to industrial diversification. The largely responsive system of aid resulted in a concentration of Regional Development Grants (RDGs) into a very few firms and sectors - ten minimum list headings of the Standard Industrial Classification out of a total of 99 industries with more than 100 employees in the area received 54.8 per cent of the grant; these sectors accounted for 34.2 per cent of total manufacturing employment. The bulk of aid therefore went to the principal heavy engineering companies. However, a survey of assisted firms did reveal that 53.2 per cent of the projects receiving RDG and 75.8 per cent of those receiving regional Selective Financial Assistance (SFA) used the aid to introduce new products. And 45.5 per cent of the projects did involve additional research and development activity.

Notwithstanding the generally acknowledged need to support the small firm sector, the analysis did not reveal any bias in favour of small firms. In fact, 52.7 per cent of the assisted cases had under 25 employees, whereas 61.2 per cent of the manufacturing firms in the area as a whole were in this size range.

Survey research revealed that 57.4 per cent of the RDG and 75.0 per cent of the SFA projects had led to an increase of employment. Two-thirds of the jobs created were for skilled and semi-skilled manual workers and only 16.3 per cent were for technical and managerial grades, suggesting that the contribution of assistance to broadening the occupational structure of the area was limited.

An important additional role of financial assistance has been to preserve employment. A detailed comparison of assisted and non-assisted firms in the area over the period 1974-87 undertaken by Wren (1988, p107) has revealed that in the long run (ie after five to eight years), closure rates are significantly lower for RDG-assisted firms in the large and medium-size band. Wren suggests that RDG acted like bank finance, enabling some marginal firms to stay in business. RSA, which he likens to private sector risk capital, was in contrast less effective at reducing closure.

Further research by Wren (1989, p127) on Teesside suggests that prior to the revision of the RDG scheme in 1984, financial assistance made a negligible contribution to job creation. For example, ICI received £150 million in RDG between 1975-84 and is reported to have shed 10,000 jobs over the same period. One project, for example, received £150 million in RDG but created only ten jobs. In response to such criticisms RDG was modified to become what was in effect a marginal labour subsidy (15 per cent of eligible capital expenditure of £3,000 for each net new full-time job created). In a survey of firms assisted under the revised scheme in Cleveland, Wren estimated a cost per job figure of £7,500 (1989 prices) which compared very favourably with the figures for Tyneside of £17,650 (1985 prices) for the old scheme. Moreover, less than 2 per cent of awards and 11.5 per cent of expenditure went to firms with over 200 employees, suggesting that the revised scheme was more effectively reaching small firms where it was mainly taken in the form of a job grant.

In the vast majority of cases in Wren's 1989 survey the projects would not have gone ahead without assistance. He concludes that the scheme was highly cost-effective, creating between 1,500 and 2,000 jobs in Cleveland in a two-year period. Nevertheless, in the drive to avoid 'deadweight' public expenditure, the automatic RDG scheme was abandoned in 1988 in favour of a totally selective approach to regional industrial assistance.

The abolition of RDG marked the end of a long era of regional industrial policy. While selective assistance remains it is far from clear that the scheme is being used to bring forward the kinds of project which could address the many weaknesses in the region's economic structure identified in this chapter. While the academic and public accountability debate continues to rage about the cost-effectiveness of selectivity over automatic subsidies, political attention has very much switched to a wide range of local economic development initiatives which attempt to address the environment outside the firm in local areas – seeking to bring derelict land into productive use in Urban Development Corporation areas and Enterprise Zones, raising confidence through promotional activities, raising skill levels and addressing some of the bottlenecks to small firm birth and growth – all operating through what Robinson (1990) calls a 'multifarious range of initiatives and institutions which constitute the north's burgeoning economic development industry'. The next section of the chapter abstracted from Robinson, describes these initiatives.

Local economic development policies (Robinson (1990) pp42-5)

While regional policy declined in importance, both financially and politically, *urban* and *local* policy became much more prominent in the 1980s. This is partly because of a lack of regional level institutions, leaving local institutions, especially local authorities, to fill the gap. It stems, too, from a view that local initiatives can make a major impact; to

Reducing Regional Inequalities

a degree, central government subscribes to this view, and so, for example, has created Urban Development Corporations targeting resources just on parts of cities as opposed to whole regions.

From the early 1980s, local authorities in the North - as in other economically depressed regions - have wanted to be seen to be 'doing something' about local economic decline and unemployment. Nearly all of them have built up a variety of economic development policy initiatives - building small factory units to rent, providing industrial sites, giving loans and grants to industry and setting up advisory services to help small businesses. In the region's two conurbations, Tyne and Wear and Cleveland, such efforts are supported by central government through Department of Environment Urban Programme funding. Thus, Newcastle and Gateshead receive a total of around £12 million a year from the Urban Programme, more than half of which finances the economic development activities of local authorities and also some of the local development agencies.

The region also has a considerable number of enterprise agencies, most of them concerned with fostering the development of self-employment and small firms. Many of them are publicly funded - by central and local government, the EC, or by nationalised industries (the 'job creation' subsidiaries set up by British Steel, British Coal, British Shipbuilders). According to the Audit Commission the region has a higher number of enterprise agencies per head than any other region, but with fewer powerful programmes. Private sector sponsorship is rather limited - unsurprising perhaps, in a branch plant economy where local industrialists have little financial autonomy. Examples of local agencies and initiatives include the Tyne and Wear Enterprise Trust, Cleveland's CADCAM Centre and the West Cumbria Development Agency. At the regional level there are some similar, but more specialised, agencies: the Northern Region Co-operative Development Agency and the North East Media Development Trust, for example. Also at the regional level is the North's promotional agency, the Northern Development Company (NDC), which is responsible for 'selling' the region to attract inward investment (eg through 'The Great North' campaign). NDC was set up in 1986 and is supported by local and central government and also by business and the trade unions.

Central government also operates a remarkable array of local economic development policies. There is the Urban Programme, originally set up in the 1960s and revamped in the 1970s. In the early 1980s, government also introduced Enterprise Zones, quite small experimental areas in which businesses enjoy a rates holiday and generous tax concessions for physical development. The Northern Region has four Enterprise Zones (Tyneside, which includes the MetroCentre site; Hartlepool; Middlesbrough; and Workington) due to end their leases of life in the early 1990s. A further zone is now planned for Sunderland to counter the effects of the recent shipyard closures.

In the latter half of the 1980s, the major emphasis of the government's local economic development policy has been 'property-led regeneration' – the encouragement of physical development in the inner cities through subsidies to property developers. The Tyne and Wear and Teesside UDCs, designated in 1987 and with budgets of £150m or so over the next six years, are now major 'players' in the region's economic development industry. Both have announced several large-scale development schemes which are to be undertaken by private sector developers with substantial UDC grant subsidy. Retail and leisure schemes, housing and some office and business park developments are planned. 'Festival' shopping (eg at Newcastle Quayside) is proposed and also marinas (eg on the Tyne and at Hartlepool Docks). But there are also projects concerned more directly with attracting new employment to the region such as the Newcastle Business Park (primarily office development, currently under construction) and Teesside's Chemical Park proposal.

The full list of central government-funded economic development agencies and initiatives in the region is a lengthy one: 'a patchwork quilt of complexity and idiosyncrasy' according to the Audit Commission. The region has two inner city task forces, two City Action Teams, large-scale intervention in the labour market by the training agency and hosted the 1990 National Garden Festival in Gateshead. On top of this is the involvement of the EC which supports major infrastructure schemes and social fund projects in the North.

To all this has recently been added yet a further layer of activity and institutions: private sector-led regeneration agencies. 'The Newcastle Initiative' (TNI) was originally set up through the CBI 'Inner Cities Task Force'. TNI is supported by business and its main aim is essentially to encourage engagement in economic regeneration by the private sector. Working with Newcastle City Council and the Urban Development Corporation, TNI has launched a number of 'flagship projects', including the regeneration of the city's Victorian Grey Street and the development of a 'themed' 'Theatre Village'. Much of its efforts have been concerned primarily with promotion, image-building and generating confidence in the city centre, but TNI has also recently developed an innovative project to stimulate economic and social development on the Cruddas Park housing estate. Following TNI, Sunderland set up a similar organisation,

Reducing Regional Inequalities

'The Wearside Opportunity', and Teesside has now followed suit, establishing an organisation called 'Teesside Tomorrow'. Like TNI, these agencies are largely representative of private sector business and are heavily involved in the promotion of a 'positive' image of the region.

Opportunities for the future

This plethora of new initiatives reflects growing confidence in the region, associated with a new partnership between the public and the private sectors. Unlike previous periods of major economic change, the private sector is now an active rather than a passive agent in the region's economic development. But, while the private sector can be looked to for 'leadership', it also remains highly dependent on the public sector to implement various schemes. And although the private sector has become more fully engaged in the region than previously, it is only just beginning to provide the kind of strategic guidance for the public sector that might be derived from a business analysis of regional strengths, weaknesses, threats and opportunities. Such an analysis (which has been called for by the Audit Commission in its review of local economic development policies and which is equally applicable at the regional scale) could assist with the prioritising and focusing of the many initiatives that are in train.

The chapter concludes by describing one analysis being developed by a new task force of TNI which brings together industry, higher education and the public sector in the city. Although the focus is on Newcastle the framework is highly relevant to the development of the region as a whole. Newcastle is the focus of many of the institutions necessary to assist the region in its adaptation to the structural transformations outlined earlier in the chapter. It is the technological focus of the region, with two large higher education institutions, a number of R&D based companies, consultancy organisations and software firms and a wide range of support services for informational and technical activities. Newcastle is also a regional centre for business and financial services, retailing and cultural activities and a major node on national and international communication networks. In short, Newcastle provides industry in the region with many of the 'externalities' which are increasingly necessary in a changing economy and also with a gateway to a wider world; it is a location which can combine local resources with global opportunities.

Urban-based regional regeneration is not a new idea. The concept of 'technopoles' has been successfully applied in France to act as a counter-magnet to Paris: R&D has been decentralised from the Isle de France and linked with heavy investment in telecommunications services in provin-

Regional Economic Development: The Northern Region of England

cial centres. In Japan a 'technopolis' is a major science-based new town project developed around a mixture of public and private research laboratories and higher education facilities. This idea has been exported to Australia where a 'multi-function-polis' is under active discussion; this will be based on high technology industry, research and development, health and leisure services and will *inter alia* act as a focus for Japanese inward investment. While there are shortcomings with each of these schemes, a particular strength of the approach is that of focusing economic development initiatives around a few leading sectors or 'growth poles' with a view to generating change throughout the rest of the regional economy by highlighting excellence and linking other activities to it.

Four such poles are being considered by TNI Task Force. In each sector it is possible to identify research and production capacity in industry and in higher education, specific sites where new developments could take place and institutions which could bring together the various actors. By considering sites, firms and institutions the approach combines the best of urban policy and regional industrial policy.

The first example is in advanced engineering in which the North East has maintained excellence through successive recessions. Although much emphasis in economic development has been placed on new high technology industries like microelectronics, important opportunities remain in traditional engineering industries for new and improved products based on advanced engineering skills. Within the private sector in the North there is considerable R&D capacity: Northern Engineering Industries with its own R&D company; International Research and Development; British Gas Engineering whose R&D centre has won international renown for its on-line pipeline inspection systems; the offshoot from British Shipbuilders, British Maritime Technology, which exports ship design software worldwide. There are also many engineering consultancies such as R&P Appledore and Merz and MacLellan. In manufacturing, new engineering technologies are being developed by large firms like NEI, Vickers, Nissan and Press offshore and small firms like Tolag (disk drive mechanisms) and Whickham Engineering (ion implantation equipment). Within higher education there are strengths in engineering, some brought together in the Design Unit of Newcastle University's Department of Mechanical Engineering which undertakes consultancy work and houses a national gear metrology laboratory. Sites have been made available in an offshore technology park for new industrial development provided by Newcastle City Council and Tyne and Wear Development Corporation. Some new initiatives may be needed to pull these possibilities together – for example designating one of the available industrial sites as 'an advanced engineering campus'; the campus would need to be served by CADCAM facilities which could assist firms on the campus to participate in 'made to order' engineering products.

The second suggestion focuses on the application of advanced information and communication technologies in a wide range of manufacturing and service sectors. The North East is unlikely to be able to reproduce the conditions that have made the M4 Corridor the focus for state-of-the-art IT production. Information and communications technologies are being applied in the North, but there are dangers that the nature of its application will involve relatively low skills or back office operations. An offensive strategy based upon the research-intensive application areas in leading firms and available sites is being suggested. Particular stress is placed on a 'filiere' of potentially interconnected activities from hardware through software to value added information services that can be marketed over a telecommunications network. An important 'animateur' in this network would be the Micro Electronics Applications Research Institute, a spin-off from Newcastle University's Computing Department which now has a staff of over 200 engaged in a wide range of research and economic development initiatives in the region and outside.

The third pole highlights strengths in health care and biotechnology. Biotechnology is generally regarded as having potentially the same cross-sectoral applications as microelectronics. The North East has considerable strengths in research and production in these fields, particularly focusing on the health care side at the high value added end of the spectrum. Key components are the Medical School and Science Departments of Newcastle University and recent inward investment by multinational pharmaceutical companies. There are complementary strengths in the more basic end of the sector focused on ICI and an industrial park (Belasis Hall on Teesside).

The final pole is based in the service sector around the audio-visual and cultural industries. The arts and cultural industries have been identified in a number of cities as being able to make a major contribution to urban regeneration both as direct employers and through their role in improving the 'quality of life' in projecting the 'image' of an area through cultural 'products' and in creating a sense of local identity. Newcastle contains a significant concentration of cultural facilities (theatres, cinemas, galleries etc) and audio visual production activities (BBC and Tyne Tees Television, independent film production companies, media training workshops, public studio facilities, headquarters of a major cable company). Important opportunities to link the cultural and media activities on Tyneside are being pursued via a Northern Media Forum involving the private sector, Northern Arts, the Northern Development Company and the University of Newcastle's Cultural Industries Research Unit. The aim is to develop a critical mass of closely interrelated activities which enables a sector composed of many very specialised enterprises to thrive outside London.

Infrastructure for the future

If these growth poles are to develop in Newcastle and if they are to be effectively linked into the rest of the region the appropriate infrastructure will have to be maintained and enhanced. 'Themed' sites will have to be developed and linked together. Information networks will be needed to enable activities in the region to be plugged into global trends within each field. And the physical networks to facilitate this – road, rail, air and telecommunications – will need to be provided, often in advance of demand.

In past periods of rapid economic restructuring, the North has been assisted by the public provision of infrastructure ahead of immediately obvious need. In telecommunications, however, the main providers of much of the region's infrastructure are now in the private sector. Privatisation has had the positive effect of a greater direct involvement in economic development of the newly privatised utilities whose profitability is now closely linked to the economic strength of the local market. However, privatisation could create difficulties for an infrastructure led approach to regional economic development in the future. The region's utilities face the problem of a low population density region with much high-cost provision required outside of Tyneside and Teesside and with few opportunities remaining for inter-regional cross-subsidies. Private utilities are no longer eligible for support from the European Regional Development Fund (unlike their public counterparts elsewhere in Europe). It has to be recognised that public funding has contributed significantly to the region's presently outstanding infrastructure.

Whether or not the privatised industry will be able to maintain a high level of service provision is open to question. New competition from cable at the intra-urban level and from Mercury at the inter-urban level focusses on the most profitable areas of businesses, attacking profits which were traditionally ploughed back across the system to provide a national and uniform service. As a result of these market forces new investment will be concentrated in the most competitive areas of business leading to a relative neglect of the wider system.

These are all matters that cannot be resolved within the region. Central government still has an important role to play to ensure that the North and its infrastructure providers are not disadvantaged by the peripheral position of the region within England. Continued attention will still have to be given to the inter-regional allocation of public resources. The North's continued success in the context of a Europe post 1992 will depend on the maintenance of the quality of its infrastructure, much of which will probably never be economically 'viable'. Developments outside the region — for example a link to the motorway network and to the Channel Tunnel — will be a necessary if not sufficient condition for the future prosperity of the region. But these links will be one-way streets if they are not paralleled by initiatives of the type described which seek to create economic linkages within the region. Global integration and regional disintegration is not an ideal future for the North.

Bibliography

Alderman, N and Thwaites, AT (1987), 'R&D and Technical Change in Traditional Industries: A Regional Perspective', Paper presented at the 27th Regional Science Association European Congress, Athens.

Alderman, N, Davies, S and Thwaites, AT (1989) *Patterns of Innovation*, Training Agency, HMSO, London.

Audit Commission (1989) *Urban Regeneration and Economic Development: The Local Government Dimension*, HMSO, London.

Begg, I and Cameron, G (1989) 'High Technology Location and the Urban Areas of Britain', *Urban Studies*, 25, pp361-79.

Bianchi L, Gillespie, A, Nicklin, P, Wallstein, R (1989), *Telecommunications and Business in the Northern Region: Use, Awareness and Opportunities*. Microelectronics Applications Research Institute, Newcastle.

Coombes, M and Raybould, S (1989), 'Developing a Local Enterprise Activity Potential (LEAP) Index'. *Built Environment*, 14, pp107-17.

Coombes, M, Raybould, S and Wong, C (1990) 'Identifying the Indicators for Assessing the Potential of Inner City Areas for Physical and Economic Regeneration', CURDS report to the Department of Employment.

CURDS (1989), 'Computer networking and the spatial development of organisations', *Compendium of Research Findings*, Programme on Information and Communication Technologies, ESRC.

Dept of Employment (1985) 'Number of Businesses: Data on VAT Registrations', *Employment Gazette*, April.

Diamond, D and Spence, N (1989) *Infrastructure and Industrial Costs in British*

Industry, Department of Trade and Industry, HMSO, London.

Fothergill, S and Gudgin, G (1983), 'Trends in Regional Manufacturing Employment; The Main Influences', in Goddard, JB and Champion, AG (ed), *The Urban and Regional Transformation of Britain*, University —Press, Methuen.

Goddard, (1976), 'Positioning Older Industrial Regions in Relation to the Emerging Information Economy: The Case of North-east England', in Hebbert M (ed), *Unfamiliar Territories*, Avebury.

Harris, RID (1988), Technological Change and Regional Development in the UK. Evidence From the SPRU Database on Innovation, *Regional Studies*, 22, pp361-74.

Howells, J and Green, A (1986), 'Location Technology and Industrial Organisation in UK Services', *Progress in Planning*, 27, pp83-184.

Hudson, R (1985), 'Regional Development Policies and Female Employment', *Area* 12, pp229-34.

— (1986), 'Producing an Industrial Wasteland: Capital, Labour and the State in North-east England', in Martin, R and Rowthorn, B (eds), *The Geography of De-industrialisation*, Macmillan, London.

Ibbs, R (1988) *Civil Service Management Reform: The Next Steps*, HMSO, London.

Johnson, S (1987), 'Small Firms in the Local Economy: Some Evidence from Teesside and Reading', *Northern Economic Review*, 15, pp9-19.

Leyshon A and Thrift (1989) *South Goes North? The Rise of the British Provincial Financial Centre*. Working Papers on Producer Services, 10, Centre for the Study of Britain and the World Economy, Department of Geography, University of Bristol.

Marshall, JN (1990), 'Reorganising the British Civil Service: How are the Regions Being Served?', *Area 22.3*.

Massey, D (1984), *Spatial Division of Labour*, Macmillan, London.

National Audit Office (1989), 'Foreign investment in the North: Distinguishing fact from hype', *Northern Economic Review*, 18, pp50-61.

Northcott, J (1986), *Micro Electronics in British Industry*, Policy Studies Institute.

Owen, DW and Coombes, MG (1986), *An Index of Peripherality for Local Areas in the United Kingdom*, ESRC Research Papers, No6, Scottish Economic Planning Department.

Robinson, R, Goddard, J, Wren, C (1987), *Economic Development Policies: An Evaluative Study of the Newcastle Metropolitan Region*, Clarendon Press, Oxford.

Robinson, F (1990), *The Great North?: A Special Report for BBC North-east*, BBC, Newcastle.

Smith, JJ (1979), The Effects of External Takeover on Manufacturing Employment in the Northern Region, *Regional Studies*, 13, pp425-37.

Smith, I and Stone, I (1989) 'Foreign Investment in the North: Distinguishing Fact from Hype', *Northern Economic Review*, 18.

Storey, DJ and Johnson, S (1987), *Job Generation and Labour Market Change*, Macmillan, London.

Thwaites, A, Oakey, R, Nash, P (1981) 'Industrial Innovation and Regional Development', *Report to the Dept of Employment by CURDS*.

Thwaites, AT *et al* (1986), 'The Regional Dimension to Technological Change in Great Britain', in Amin, A and Goddard, JB (eds), *Technological Change, Industrial Restructuring in Regional Development*, Allen and Unwin.

Williams, H and Charles, (1986), 'The Electronics Industry in the North East: Growth or Decline?', *Northern Economic Review* 13, pp29-38.

Williams, H, Carr, T, Taylor, S (1988) 'The Use and Impact of Computer Networks in John Carr PLC', *CURDS PICT Working paper on Networks No 1*.

Wren, C (1988), 'Closure Rates Among Assisted and Non-assisted Establishments', *Regional Studies*, 22, pp107-20.

— (1989), 'The Revised Regional Development Grant Scheme: A Case Study in Cleveland County of a Marginal Employment Subsidy', *Regional Studies*, 23, pp127-28.

8
Transport Infrastructure and Regional Policy

Alan Armitage and Derek Palmer, of the CBI

I Transport in the context of regional issues

Differences in growth rates have led to a variation in the speed with which different industries and therefore regions adjust to changes in the pattern of demand.

Regional disparities are felt by some to have become too wide and are considered to be a considerable constraint on balanced economic growth. Resources remain under-utilised in some areas, generating unemployment, vacant land and under-used facilities such as schools. At the same time, in other locations strong demand has led to shortages, particularly of labour and of land for new development, as well as congestion on both the road and rail networks. A high quality integrated transport network can help level out such regional disparities.

Historically, regional policy has attempted to encourage business location into less-favoured areas by subsidising capital, and sometimes operating costs. This has attracted criticism since successful companies are taxed to help finance developments in the poorer regions and often to subsidise industries created there. Direct attempts at stimulating economic growth have not always been successful in generating self-sustaining prosperity. Thus, throughout the last decade, the government has withdrawn financial assistance and targeted it on specific locations such as inner cities.

Nevertheless, substantial regional differences remain. Companies based in the north and west of Britain face severe difficulties in transporting goods. Business faces problems not only in distributing goods to the South East – the largest and most prosperous market in the UK – but also in getting their products around London through to the ports and to the Continent. They must compete from the geographic periphery both of the UK and the European Community. Regions in the south and east of Britain are important transit routes. A first-class transport network is vital if firms in the North and West of Britain, as well as those in the South East, are to be able to compete effectively within Europe. Good access is required, not only to the central markets of Continental Europe, but also to raw materials.

Good transport facilities are fundamental to regional development. Research by the DTI in the early 1970s identified three factors that determined business investment in development areas – labour supply, especially skills, government financial assistance and transport infrastructure. Early in 1988, when the 13 CBI regional chairmen met to discuss what factors would inhibit continued economic growth in their area following the completion of the Single European Market, it was clear that transport inadequacies, especially the road network, was the number one priority for action.

Relocation of business is a key element of regional policy. The introduction of the uniform business rate, which favours companies operating in the north compared to those in the south, is one key element of the government's approach. It also hopes that congestion in the South East will encourage relocation, but the success of this depends on the provision of a good transport infrastructure in other regions. To be an effective regional policy, transport infrastructure must be integrated nationally.

II Pressures on the UK transport network

Business needs no convincing of the need for more and better roads. Britain's current motorway network reflects our past economic history, when trading activities were dominated by importing goods from across the Atlantic. These were then processed in the Midlands and sold in the South East. The nation's economic geography reflects this, focused on London (the centre of power) and extending north and west. As a result, our links with Europe are poor. But Europe now accounts for over half of our exports, so it is vital that the links to the Continent are improved.

Distribution costs to British business account for some 15–20 per cent of output costs. Efficient freight distribution is therefore vital for business, especially for companies in the manufacturing and retailing sectors. The introduction of just-in-time management methods requires reliable and predictable distribution systems in order that 'delivery windows' of 15 minutes can regularly be achieved. Failure to achieve such specifications imposes severe cost penalties on business.

Table 8.1 *Land-borne traffic, 1988 (percentage)*

	Passengers (billion kilometres)	Freight (million tonne kilometres)
Road	93.3	87.9
(Motorways	14.0	35.0)
Rail	6.7	12.1
Total	100	100

Source: Transport Statistics Great Britain, 1978–88
Department of Transport

Roads account for the vast bulk of goods and passengers transported – 88 per cent of land-borne freight and 93 per cent of passenger miles – and rail accounts for the remainder (see Table 8.1). Because of its flexibility, road distribution will remain the most important delivery mode in urban areas. But two-thirds of loaded road freight journeys are less than 50 miles. Thus, the benefits to industry of a better road network are obvious. The current road network is suffering from 'thrombosis', with inadequate arteries clogged by congestion.

Some long-distance freight will be shifted from road to rail when the Channel Tunnel opens. But, nationally, a 50 per cent increase in rail freight would only lead to a 5 per cent reduction in road transport and it is unlikely that the present rail network could cope with such an increase.

Yet investment in new and improved roads has not kept pace with demand. The CBI has estimated that congestion is costing the nation £15 billion a year – well over £10 per household per week. This is damaging business, especially in London and the South East, and has an impact on companies elsewhere. The waste is a competitive handicap which needs to be removed.

III The continental challenge

When the Single European Market has taken shape in 1992, it will be severely lacking in at least one important respect: the transport infrastructure will be inadequate to cope with the growing demands placed on it. The distances between many areas of the EC, especially between those on the periphery of the UK and the centre of the Continent, will remain large. Firms in all parts of Britain will be placed at a competitive

Reducing Regional Inequalities

disadvantage: poor infrastructure can be just as effective a barrier to trade as high tariffs.

The economic geography of Britain needs to be changed if the whole country is to benefit from the Single European Market, and not just the already overcongested South East corner. Britain has twice the traffic density of West Germany (as was) and France, and 3.6 times that of the US. Yet other nations have invested heavily in preparation for growth in future traffic levels: between 1987 and 1997 France will double its motorway network.

Britain's existing motorway network is inadequate. New motorways are needed along the south and east coasts linking the ports and leading directly to the Channel Tunnel. Some 50 years ago a Royal Commission called for better links to our ports. Even when the proposals in the government's White Paper, 'Roads to Prosperity', are implemented, it is doubtful whether access to the ports will be adequate. A failure to invest heavily would mean that companies based in the North will be further away from the prosperous South East than their competitors based in France using the Channel Tunnel.

Furthermore, overseas nations are investing heavily in rail. A £20 billion investment programme has been announced for the high-speed Train à Grande Vitesse (TGV) network which will eventually be 7,000km long, including 2,300km of new line, allowing running speeds of up to 300kph. The TGV Sud-Est has been operating since 1981 and the TGV Atlantique has recently opened. In addition, the TGV Nord Européen will link to Brussels, Amsterdam and Cologne, with a branch to the Channel Tunnel from Lille. Studies are under way for linking Paris with Strasbourg and with Marseilles/Cannes by TGV. Such railway investment is deliberately designed to stimulate economic development in the regions, especially the Nord/Pas-de-Calais. Charles de Gaulle airport, which is to have a third runway and has the capacity to be expanded to five runways to cater for 90 million passengers per annum, over twice the current size of Heathrow, will be fully integrated into the TGV network via a station at Roissy.

The Federal Republic of Germany is considering proposals for a 'Transrapid' electromagnetic hovertrain costing DM30 billion to link seven major airports from Hamburg to Munich along a 640-mile corridor. The project is designed to reduce road traffic by 8 per cent and air traffic by 35 per cent by the year 2000. Although Germany and the UK are of comparable area and population, the more heavily susidised German rail network (30,600km) is already almost double the length of that in the UK (17,000km), while its motorway network is almost three times longer – 8,400km in the Federal Republic compared to 3,000km in the UK.

The transport ministers of Germany, Austria, Italy and Switzerland are also considering the construction of two new rail tunnels through the Alps, one of which would relieve pressure on the Brenner Pass. These would link in with the proposals for a European high-speed train network to connect the industrial and commercial centres of the Continent: Paris, Amsterdam, Brussels, Hamburg, Cologne, Frankfurt, Munich, Basle, Milan, Lyons, Barcelona and Madrid. London and other British cities will be linked into the network, but will not reap the full rewards of high-speed operation.

Continental companies also have the benefit of an extensive network of rivers and canals which offer cheap transport. To this will be added another advantage: the Rhine–Danube canal, due to open in 1992, which will create a waterway corridor allowing vessels up to 1,350 tonnes dead weight and barges up to 185m long carrying 3,300 tonnes to cross Europe from the North Sea to the Black Sea. Thus, North Sea ports will

Table 8.2 *Infrastructure investment, 1984–7*

	As Percentage of GDP, Average		
	Rail Rolling Stock	Rail Infrastructure	Road Infrastructure
Belgium	0.12	0.21	0.49
France	0.05	0.12	0.61*
Federal Republic of Germany	0.07	0.27	0.75
Italy	0.14**	0.31**	0.54
Netherlands	0.04	0.12	N/A
Spain	0.07	0.18	0.38
UK	0.02	0.08	0.47

*   For 1984 only
** 1984–6

Source: Statistical Trends in Transport 1966–87, European Conference of Ministers of Transport

Reducing Regional Inequalities

be open for direct traffic to Hungary, Yugoslavia, Rumania and Bulgaria, and ultimately the Near and Middle East.

The UK has invested less in the infrastructure than our principal European competitors (see Table 8.2) and our plans for the future are markedly less ambitious. This is an inevitable consequence of the lower priority given by the UK to transport infrastructure investment and damages the potential for economic growth in the regions. A good transport infrastructure is a necessary, but not sufficient, condition to encourage business investment.

IV New business opportunities

Since the existing infrastructure was planned, several important developments have taken place which could bring considerable economic benefits to the regions of the UK – provided always that the appropriate infrastructure is put in place in time.

It is difficult to overstate the impact that the completion of the Single European Market and the opening of the Channel Tunnel one year later will have on the British economy. The changes associated with 1992 will lead to a further competitive stimulus to British business. The UK represents only a small part of the EC (18 per cent of the population) and is distant from the centre. The Channel Tunnel could transform the economics of rail freight, particularly if container-handling facilities are improved at BR stations so that containers can be readily transhipped anywhere in the UK and delivered the following day in central Europe. At present, freight by rail is uneconomic at distances much under 250 miles, except for some bulk materials such as coal, coke and aggregates. But within the UK some 58 per cent of road freight is moved over 100km. Whereas in France and West Germany (as was) some 35 per cent of such long-distance freight is moved by rail, in Britain only 15 per cent is carried by this mode. However, traffic projections for the Channel Tunnel suggest that rail freight tonnages will increase by at least 8 per cent by 2003, reversing the steady decline of the last decade.

To capitalise on the opportunities offered by the Channel Tunnel, British Rail is planning to construct a network of 12 'freight villages', each of which would provide a variety of services. Major regional terminals are planned for Strathclyde, Teesside, West Yorkshire, Merseyside, Greater Manchester, East Midlands, West Midlands, South Midlands, South Wales, Avon, Willesden and Temple Mills. More might be needed if the full benefits of the Channel Tunnel are to be reaped. The freight facilities grant must be reformed to encourage the development of more private yards.

Compared to the approximately 1,500 rail freight yards in the UK, there are about 15,000 in West Germany (as was) and 16,000 in France.

Completion of the Single European Market will bring greatly increased demand for local air connections for business people seeking to improve their links with customers on the Continent, or to manage their Continental operations more effectively. An efficient network of regional airports will be a vital corollary of UK commercial leadership in the new Europe.

Since shipping carries over 90% of Britain's trade, good access to the ports is vital. The competitive position of south coast UK ports in particular could be enhanced as possible entry ports for Europe, competing with Rotterdam, Hamburg, Antwerp or Le Havre, for instance. The savings in sailing time up the Channel would be attractive to shipowners when a 50,000DWT container vessel costs at least £4,000 a day to operate. The possibilities for Plymouth, Poole, Southampton and Portsmouth are considerable, always provided that the necessary links to the Channel Tunnel by road and rail are in place. Similarly, the concept of a 'Landbridge', though currently considered uneconomic, might be an opportunity in the longer term.

Another opportunity lies in increasing the UK share of international tourism. This comes at a time when the domestic tourist industry has been investing substantially to upgrade the quality and variety of its 'product'. Tourism within Britain is inhibited by poor infrastructure. Neither the roads to our national parks, nor those to our major tourist centres in the South West can cope with peak demand. This hampers the growth of both domestic tourism and international visits to the UK.

The major infrastructure requirement for the tourist industry is sufficient international air gateways - combined with a competitive domestic road and rail infrastructure - which spreads the benefit of tourism beyond the overcrowded South East and encourages UK tourists to take more short breaks. The development of international airports outside the South East - in particular Manchester and Glasgow - is thus of more than local economic significance. They are of national strategic importance, given the key role of tourism in the economy. Income from overseas visitors amounted to £6.2 billion in 1987, some 29 per cent of invisible earnings and equal to 16 per cent of manufactured exports. Tourists account for 59 per cent of airport arrivals and by 1993, expenditure in the UK by overseas tourists is forecast by the British

Reducing Regional Inequalities

Tourist Authority to have grown by 69 per cent, making the sector more important than domestic tourism.

V Developing the nation's infrastructure

Thus, business needs good access for freight, by both road and rail, to Continental Europe. This requires a reduction in congestion and improvements to the routes around London and through the South East. It is vital that urban congestion is also reduced both to ease distribution and journey to work for employees.

Detailed proposals for infrastructure investment have been outlined in the CBI's 'Trade Routes to the Future'. Four strategic priorities were identified.

Roads

The motorway network must be extended to provide fast and reliable links to the Channel Tunnel down the east coast, along the south coast and from the Midlands and North West without the need to use the M25 or go through London; similarly, links to the east coast ports must be improved. These should provide an 'arrowhead' at Europe.

Rail

Investment in the railways must be sufficient to capitalise on the potential that the Channel Tunnel affords for a railway renaissance. Britain must be fully linked to the high-speed lines being built on the Continent. Rail will continue to provide three services effectively: moving freight over long distances and short-distance movement of bulk commodities; providing mass passenger transit into cities; and rapid inter-city travel. The Channel Tunnel could call forth a railway renaissance by providing effective long-haul freight services competitive with roads. Diversion of some freight from road to rail will help relieve road congestion on inter-urban routes and yield environmental benefits. The Channel Tunnel will also provide rapid city centre to city centre links between London and such important centres as Paris and Brussels. On these routes rail will compete for business traffic with air travel.

Air

In order of priority, improvements to air traffic control (ATC), expansion of terminal capacity with suitable surface access and additional runway capacity are urgently required in the South East. Regional

Transport Infrastructure and Regional Policy

airports need to be developed to take a larger share of air transport away from the London area. Heathrow Airport, with its two parallel runways, is the UK's only airport capable of competing internationally for major hub status and is vital to maintain the UK as a centre for Europe and worldwide business HQ locations. Some 75 per cent of the demand for international air travel originates in the South East and although 15 UK regional airports have long-established, relatively frequent services to Europe, the cost of lack of capacity at Heathrow and Gatwick is considerable. The most urgent need now is for approval of the expansion and modernisation of London Heathrow by building a fifth terminal as soon as possible. Terminal 2 needs to be replaced and the central area redeveloped. Road and rail infrastructure, including the provision of the Heathrow to Paddington link, needs to be improved and extended to match the increased passenger demand.

So as to provide more capacity in the South East, parliamentary approval to lift the air transport movement limit at Stansted to enable the terminal to be expanded from 8 to 15 million passengers per annum should be given. The rapid development of Stansted is needed as a new major international airport for London and the South East and its further expansion will make an important contribution to meeting demand. Furthermore, the proposed extension of the runway at London City Airport to cater for larger aircraft must be built.

Progress in developing regional airports must also be maintained in order that a larger share of air traffic can be served away from the London area. The recent development plans for Manchester and Birmingham airports are most welcome and will act as a considerable stimulus to the regional economies. The review of the Scottish Lowlands airports policy currently being undertaken by the government will, it is hoped, advocate Glasgow being given international gateway status.

But, in the longer term, more runway capacity will eventually be needed in the South East. Capacity should be provided where the need arises. Interlinking, on which the viability of many services to regional airports depends, can only effectively be offered through London. Thus, further consideration needs to be given to providing a third runway at Heathrow or to adding a second main runway at Gatwick, to cater for interlining.

Urban areas

Congestion in the South East and London must receive urgent attention. It is a national as well as a regional problem. This requires improvements

Reducing Regional Inequalities

to the existing road network as well as enhancing public transport. Road congestion in the capital is rapidly nearing crisis proportions. Since there is less scope for new road building in urban areas, especially London, public transport must be strengthened. The pre-eminence of the City as a financial centre cannot be ensured without improvements in public transport in the capital.

The implementation of the CBI's proposals (costing £20 billion over a ten year period) would not be easy. In addition to continued increases in public expenditure on transport, a more integrated view of planning is required. Furthermore, planning procedures need to be speeded up and levels of compensation for those adversely affected increased.

VI Business and the regions

The completion of the Single European market in 1992 will open up new competitive pressures for business in the regions. Their geographic peripherality is a disadvantage that can be overcome by greatly improving the transport network. But other disadvantages will need to be addressed – lack of local ownership, skill shortages, housing, etc.

Inward investment by companies from North America and the Pacific Basin wishing to build a manufacturing base in Britain frequently locate outside the South East, notably in South Wales, the North East and Scotland. Many wish to supply the EC via the Tunnel, and this will generate futher infrastructure demands. Unless there is much greater investment in our transport infrastructure, the considerable opportunities presented by the Channel Tunnel and the Single European Market could well be lost. In the 1980s, one quarter of the 200 companies that have located in the Nord/Pas-de-Calais have been British. And the UK companies acquiring businesses on the European mainland at the rate of around 200–300 acquisitions a year will find it necessary to locate their headquarters on the Continent. This trend could all too easily accelerate.

Inward investment from overseas is vital for the continued economic prosperity of the regions. The UK is the base for half of Japanese investment in Europe. Yet, continued investment cannot be relied on in the future without the provision of a good transport infrastructure. It is noticeable that inward overseas investment in the UK has located near to good transport facilities – motorways and airports – since these facilities provide accessibility to European and worldwide markets.

The CBI regions have identified the key infrastructure investment necessary to ensure both that their areas are attractive to inward investment and that indigenous companies need to remain competitive in the 1990s (outlined in the Annex to 'Trade Routes to the Future'). Various themes can be identified: better links to the ports, especially on the south and east coasts; improved connections, both road and rail, to the Channel Tunnel; the expansion of regional airports and more scheduled flights; better roads within their region; improved transit through the South East; and better rail freight facilities.

It is clear that difficult decisions need to be taken. Higher investment is needed in all forms of transport infrastructure. But it must also be co-ordinated to ensure that the system is integrated. Poor planning in the past means that our transport system is not well co-ordinated. Furthermore, infrastructure takes a long time to plan and build: new roads can easily take 15 years from conception to completion and other facilities can take just as long to provide. Public opposition – the NIMBY (not in my back yard) syndrome – is understandably strong. If all the regions of the UK are to be fully integrated into the Single European Market, decisions to invest heavily must be taken now.

Bibliography

Armstrong, H and Taylor, J (1987), 'Regional Policy: The Way Forward', Employment Institute.
Breheny, M and Congdon, P (eds) (1989), 'Growth and Change in a Core Region', *London Papers in Regional Science*, No 20 (Pion).
British Railways Board (1989), 'International Rail Services for the UK', December.
Business Strategies Ltd (1989), 'Chunnel or Transmanche: Business Location and Regional Opportunities'.
Chisholm, M (1986), 'The Impact of the Channel Tunnel', *The Geographical Journal*, November.
Civil Aviation Authority (1989), 'Traffic Distribution Policy for Airports Serving the London Area – Advice to the Secretary of State for Transport', *CAP 559*, July.
— (1989), 'Traffic Distribution Policy for the London Area and Strategic Options for the Long Term', *CAP 548*, January.
Community of European Railways (1989), 'Proposals for a European High-Speed Network'.
Confederation of British Industry (1988), 'Transport Plan', CBI Eastern Region.
— (1989), 'The Capital at Risk – Transport in London', *Task Force Report*, CBI London Region.

— (1989), 'Into the 1990s – Future Success of the Northern Ireland Economy', CBI Northern Ireland.
— (1989), 'Prosperity for the 1990s – the South-west's Needs', CBI South West Region.
— (1989), 'Roads to Growth – a Competitive Transport System for Southern England', CBI Southern Region.
— (1989), 'Towards 2000 – Transport and the North-west', CBI North West Region.
— (1989), 'Trade Routes to the Future', November.
— (1989), 'Transport Infrastructure in the South East – The Gateway to Europe', CBI South Eastern Region, November.
Department of the Environment for Northern Ireland (1989), 'Transportation Programme 1989-93'.
Department of Transport (1989), 'Roads for Prosperity', May.
— (1989), 'Transport Statistics Great Britain 1978-88', September.
— (1990), 'Trunk Roads, England: Into the 1990s', February.
European Conference of Ministers of Transport (1989), 'Statistical Trends in Transport 1965-86'.
House of Lords Select Committee on the European Communities (1988), 'Inter-Regional Air Services', December.
— (1989), 'Transport Infrastructure', November.
Institution of Civil Engineers (1989), 'Congestion' (Thomas Telford).
Lock, D (1989), 'Riding the Tiger: Planning the South of England Town and Country Planning Association'.
OECD (1988), 'Cities and Transport'.
Palmer, D (1989), 'Improving Britain's Transport Links with Europe', *National Westminster Bank Quarterly Review*, August.
PIEDA (1989), 'The Implications of the Single European Market and the Channel Tunnel for Rail Links to the North West Region', July.
Rees Jeffreys Road Fund (1989-90), 'Transport and Society Discussion Papers', Transport Studies Unit, Oxford University.
SERPLAN, (1989), 'The Channel Tunnel: Implications for the South East Region', The London and South East Regional Planning Conference, June.
Vickerman, R (1989), 'Regional Development Implications of the Channel Tunnel', Paper presented to the Infotrans European Planning and Transportation Conference, April.

9
Regional Development in the 1990s: A Trade Union Perspective

Trades Union Congress

This chapter is concerned with the policy issues likely to dominate the debate on regional policy and regional development in the 1990s. Although regional policy has fallen out of fashion, the TUC believes it is re-emerging as a key element in the future conduct of economic and industrial policy and is set to assume a new importance in the 1990s.

The chapter is divided into four sections. Section I is a brief review of policy developments in the 1980s and the outlook on unchanged policies for the 1990s. Section II discusses the issues likely to dominate debate in the coming years – the local dimension, inner city policy, regional planning, the European Community, inward investment, and devolution and decentralisation. Section III considers the elements of a regional development framework for the 1990s – the role of infrastructure investment, research and development, and policies directed to labour and to capital. Section IV looks at possible institutional frameworks – the role of local authorities, the European dimension, and new regional structures.

I Policy change and outlook

Policy in the 1980s

Regional policy made a major contribution to job creation in the assisted areas before the 1980s. One widely accepted estimate is that by 1981 employment in the assisted areas was half a million higher than it would have been had regional policies not been in force since 1960. The progressive weakening of regional policy in the 1980s and its virtual abandonment as a major industrial policy after 1984 has to be seen in the light of this relatively good performance.

Reducing Regional Inequalities

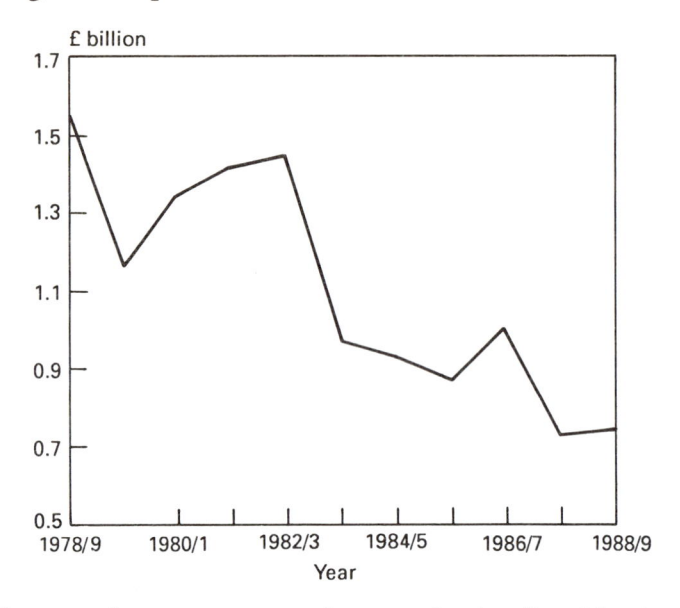

Note: All figures are for government spending on regional preferential assistance to industry, 1988/9 prices

Source: Hansard, 20 November 1989

Figure 9.1 *Government spending on regional policy in the 1980s*

The traditional basis of regional policy has been to redistribute growth and jobs within a national economy close to full employment. After the 1973 oil shock, a redistributive regional policy became less effective. The massive decline in manufacturing and record unemployment levels of the early 1980s further undermined the policy's effectiveness. By 1981 the government saw regional policy in social rather than economic terms, and also as a means of offsetting Britain's financial contribution to the EC.

A further factor was the higher priority being given to urban policy after the 1981 inner city riots, and as part of a wider recognition that regional problems were linked to the decline of older towns and cities. The importance of inner city policy was reaffirmed by the then Prime Minister shortly after the 1987 general election.

Regional policy also suffered as part of the government's move away from an interventionist industrial policy and, in particular, the provision of direct support to an approach which emphasised the role of market forces. This was a strong theme in the 1984 White Paper, and has been taken up in other areas, notably national pay bargaining.

A weaker regional policy and greater reliance on market forces have seen regional imbalances grow in the 1980s. Whether measured by GDP growth, unemployment rates, or the creation of new employee jobs it has been the South East, East Anglia, and the South West which have gained relative to the national average, while most of the traditional assisted areas have fallen behind. Indeed, the decline in manufacturing saw unemployment rise in the West Midlands to levels previously associated with the north, a blow from which the region has only recently started to recover.

Perhaps of even greater concern has been the tendency for the gap in prosperity between the south and the Midlands and the rest of Britain to increase over the past six years of sustained economic growth. Between 1983 and 1988 the South East experienced GDP growth of 23 per cent, and East Anglia growth of 32 per cent: in contrast the North experienced an increase of barely 14 per cent. The contrast is especially marked in the distribution of employment growth: between 1983 and 1989 the South East, East Anglia, and South West accounted for 80 per cent of the national increase in employee employment. The view that the full benefits of the recovery in the national economy since 1983 would automatically spread to the higher unemployment regions does not appear to be correct.

Prospects for the 1990s

The prospects for the 1990s on unchanged policies suggest that the pattern of relative regional disadvantage will persist. Forecasts by Cambridge Econometrics and the Northern Ireland Economic Research Centre suggest population and employment growth will be concentrated in the south and parts of the Midlands, with increasing congestion and labour shortages bringing some limited dispersion in economic activity to the South West and Wales. The forecasts to the year 2000 are summarised below in Figure 9.2.

These trends are already apparent in parts of the south. Registered unemployment was around $3\frac{1}{2}$ per cent in the South East and East Anglia, but many travel to work areas were recording unemployment rates of 2 per cent or less. Congestion in and around Greater London is severe, labour shortages are reported on an increasing scale, and development pressures are growing. As well as concern at the economic cost of these developments there is concern with the impact on the environment.

The immediate economic prospects are not encouraging, and some increase in registered unemployment must be anticipated in 1990, unless interest rates come down. For the longer term, however, labour markets

Reducing Regional Inequalities

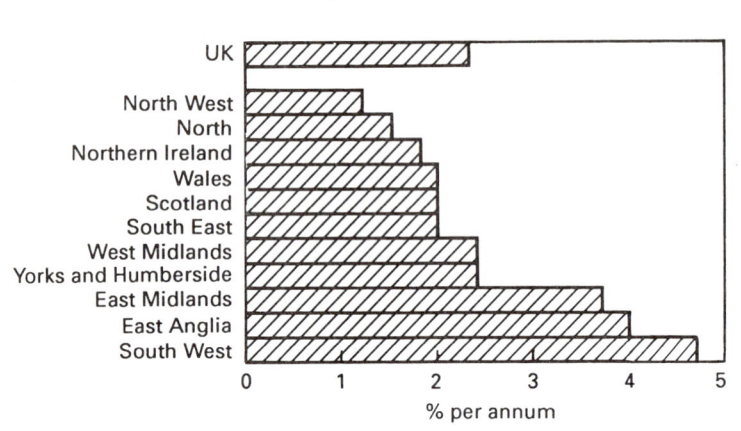

GDP change 1988-2000 (% per annum)

UK	~2.3
North West	~1.4
North	~1.5
Northern Ireland	~1.8
Wales	~1.9
Scotland	~2.0
South East	~2.1
West Midlands	~2.3
Yorks and Humberside	~2.6
East Midlands	~3.4
East Anglia	~4.2
South West	~4.7

% per annum

Source: Cambridge Econometrics and Northern Ireland Economics Research Centre: Regional Economic Prospects, December 1988

Figure 9.2 *Regional growth prospects to 2000*

in the 1990s may be tighter than in the 1980s, especially after 1992. Regional imbalances will persist and will retard growth prospects in the South East and parts of East Anglia – and this in turn will restrain national growth rates as well as adding to underlying inflationary pressures.

In the 1990s the impact of the 1992 process and moves towards economic and monetary union within Europe will tend to accentuate regional imbalances unless counter-balanced by stronger regional policies at the European level. The regions outside the 'Golden Triangle' will tend to fall behind – and this includes much of the UK. The Channel Tunnel's opening will further increase regional differences unless more active steps are taken to distribute the economic benefits more widely.

Despite these trends and their anticipated continuation into the 1990s, regional policy continues to occupy a low priority within government. How far it will continue to do so given the economic and industrial pressures of the next decade is a matter for debate, but the continued failure to develop coherent long-term regional development policies is likely to exact a growing economic, industrial and social price.

II Policy issues for the 1990s

The local dimension

Although regional policy achieved much in the 1960s and 1970s, the circumstances of the 1990s require a different approach. Although a 'north–south' divide clearly exists, it is a simplification of a much more complex picture. There exist considerable differences between the economic performance of parts of Kent and the rest of the South East, between the coastal areas of East Anglia and growth centres such as Cambridge, and between West Cornwall and areas such as Bristol. Similarly, in the North West, GDP per head in Cheshire is 16 per cent above the national average. Regional policies in the 1990s will need a strong local dimension.

Inner cities

There is growing recognition that the regional problem is often associated with the problems of older industrial towns and cities. Growth has favoured smaller towns and semi-rural areas. Some commentators have suggested that a major reason is that modern industry needs more floorspace because of the impact of new technology, leading to relocation and siting of new investment on 'greenfield' sites, rather than redevelopment of costly and difficult inner city locations. Whatever the reason, the 1990s will need to see a more effective integration between inner city policies and regional development policies. The TUC's 1988 policy statement, 'Trade Unions in the Cities', sets out detailed proposals on the development of inner city policy in the 1990s (TUC, 1988c).

Regional economic planning

Above all, however, there will need to be a recognition that a successful long-term approach to regional development must be underpinned by a commitment to regional economic planning. The formal abolition of the regional economic planning system in the early 1980s has left a strategic vacuum which can only partially be filled by local authority bodies and regional agencies. Regional policy in the 1990s must be concerned with the infrastructure, the provision of education and training, and the encouragement of research and development, as well as the distribution of grants and subsidies. This can only be successfully tackled by a planned, strategic and long-term approach at the regional level. The TUC's 1982 policy statement, 'Regional Development and Planning', set out proposals for how this might be implemented in practice (TUC, 1982).

European regional industrial policy

A related point is the impact of European industrial policy. The European Commission is likely to rule out the large-scale direct industrial subsidisation which underpinned the Nissan investment in Sunderland, and regional policy in the 1990s is likely to be more concerned with improving the economic potential (the supply side) by means of transport infrastructure investments, training programmes and developments such as science parks to encourage research and development. At the same time, the Commission itself is seeking to take a more active role through (in the case of Britain) the development of Community Structural Funds based on the EC Social and Regional Development Funds. The TUC's Committee on European Strategies has been actively encouraging trade unionists in the regions to make themselves fully aware of the opportunities offered by the Funds, as part of the Delors initiative on social dialogue and the social dimension.

Inward investment

Regional policy in the UK has always reflected the importance of inward investment, especially when new investment by UK based companies has slumped. Foreign investment could well increase even further in importance in the 1990s. According to a recent UN report, technological and organisational pressures are likely to divert investment by transnational corporations to the industrialised world, as the need for skilled workforces to operate advanced technology effectively becomes paramount in order to maintain a competitive edge. For the UK, Japanese inward investment – which so far has been heavily skewed towards the assisted areas – will be of growing importance. A TUC memorandum to the National Economic Development Council in November 1988 looked at the issue of international investment, and later in this chapter the industrial relations implications are considered (TUC, 1988a).

The 1992 process

The 1992 process and further moves towards economic and monetary union within Europe are likely to have beneficial impacts on European growth, employment and investment levels – but their exact magnitude and even more their distribution are highly uncertain. For the UK, investment by non-EC countries worried by protectionist barriers after 1992 is an important factor, but perhaps even more important will be merger and acquisition activity by European based multinationals. The recent takeover of Rowntree-Mackintosh not only has implications at national level: in Calderdale, West Yorkshire, Rowntree is the biggest single private sector employer. Multinational investment has been criticised because of the vulnerability of plants to strategic decisions taken in Detroit or Tokyo, but in the 1990s decisions taken in Frankfurt, Paris or Milan, by European multinationals will be of increasing importance. The TUC has issued two reports on the implications of 1992 (TUC, 1988b, 1989a), and is undertaking further work for trade union negotiators and in developing the 'social dimension'.

Environment

A new concern for regional policy in the 1990s will be concern for the environment. Evidence from the British Social Attitudes Survey (BSAS) suggests that public concern and hostility to new development in areas of intense economic pressure, notably in the South East, is strong and growing. This is not to suggest that people living in areas of higher unemployment do not feel just as strongly about the environment, or that more polluting industries should be relocated in these areas in anticipation of less adverse public reaction. What it does mean is that an environmental framework for the 1990s needs to embrace the idea of balanced regional development. The TUC is developing work on this issue through the TUC's Environment Action Group.

Devolution

Finally, there is again growing interest in the question of regional government and devolution for the English regions, Wales and Scotland. However this debate develops in the future, there is an urgent need for decentralisation to the regional and local level of many industrial development functions and powers. This would also help meet a growing concern that the genuine dialogue between the 'social partners' at regional and local level as part of the Delors programme is being constrained by central government hostility.

III Towards a regional development framework

A regional development framework for the 1990s will need to go beyond the traditional policy instruments of regional grant and subsidy, and

Reducing Regional Inequalities

consider the role of infrastructure investment, of research and development, and training. More traditional policy instruments aimed at labour or attracting capital investment will still have a role, but will need to be part of a wider planned approach to regional development. The following paragraphs discuss how each of these elements would be developed in the 1990s.

Infrastructure investment

There has been widespread concern at Britain's long-term investment performance. The boom period of 1987-8 cannot disguise the fact that as a proportion of GDP Britain is still investing less than in the 1970s, and less than our major European comeptitors. According to figures published recently by the European Commission (1989), gross fixed capital formation averaged just over 19 per cent of GDP between 1971 and 1980, but fell to an average of just under $17\frac{1}{2}$ per cent between 1981-90 (including the Commission's forecasts for 1989 and 1990).

Moreover, within this total two sectors have underperformed, namely investment by manufacturing industry and public investment in the infrastructure. Manufacturing investment has barely recovered to its 1979 level, and has averaged well below the investment levels recorded in the 1960s and 1970s. Investment by government fell sharply in the mid 1970s and has not recovered in the 1980s - indeed, in recent years it has fallen in real terms. According to European Commission estimates, general government gross fixed capital formation as a percentage of GDP averaged nearly $3\frac{1}{2}$ per cent between 1974 and 1981, but this has fallen throughout the 1980s and in recent years has been averaging $1\frac{1}{2}$ per cent. This is far lower than any other member of the EC; and for most EC members general government investment has either been maintained or increased in the 1980s.

Increasing national levels of investment, both in the crucial manufacturing sector and in the infrastructure is therefore critical, but it will be equally important to boost investment in the higher unemployment regions and inner cities. Providing an infrastructure in the higher unemployment regions which is suitable for modern industry not only means suitable sites with adequate housing for the workforce, but access to good communication links by air, road and rail. It is notable that many growth centres in the faster growing regions - such as Cambridge - have benefited from large-scale infrastructure investment projects, including the electrification of railway lines to London and the completion of the M11 motorway and dualling of the A45, and development of Stansted

Regional Development in the 1990s: A Trade Union Perspective

Airport. These facilities exist in some areas in lower growth regions, but many others are lacking one or more elements. However, even if a suitable infrastructure is provided within a particular region, communication links also need to be made between the high and low unemployment regions if firms are to retain access to national and international markets. The development of east coast ports such as Felixstowe depended in part on the dualling of the A604, which links the port to the north and the Midlands. This is becoming all the more important as the British economy is increasingly integrated with that of the EC, a process due to receive a further boost from 1992 and the Channel Tunnel.

The 1980s have seen a remarkable concentration of major transport and other public sector investment projects in the South East, not least the completion of the M25, electrification of key rail links into London, airport expansions at Heathrow and Gatwick, London Docklands, Milton Keynes and telecommunications investment centred in the City. These are to be followed in the 1990s by the Channel Tunnel, further crossings at Dartford, expansion at Stansted and Luton airports, and further major work on the M25 and the M1. This is not, of course, to argue that such investments should not take place. The concern is that similar commitments of public (and private) resources are not being made in the higher unemployment regions. This lack of balance is perhaps best seen in infrastructure links to one of the biggest projects of the 1990s, the Channel Tunnel.

The Channel Tunnel

Of the three main routes linking northern regions to the tunnel, only two of them, the east coast main line (ECML) and the west coast main line (WCML), are currently being electrified or upgraded. The other route, the Midlands main line (MML), is only electrified to Bedford, and major centres such as Leicester, Derby, Nottingham and Sheffield will be disadvantaged. The TUC considers that the Midlands main line should be electrified as a priority. A study commissioned by Midlands local authorities has shown that the earlier an electrification scheme is begun, the better the rate of return of investment would be and they estimate that the financial return would be close to the 7 per cent rate required for Intercity investment. Other electrification schemes being proposed by local authorities include West Drayton-Bristol/Swansea, Manchester-Blackpool, Crewe-Holyhead, Edinburgh-Glasgow and Aberdeen and Northallerton-Stockton-Ferryhill. The Cambridge-Kings Lynn scheme should be brought forward for early electrification. The TUC would also wish to see Channel Tunnel services extended to Ireland via Holyhead.

Research and innovation

As well as good communications, an important factor behind the growth of cities such as Cambridge and Bristol and the development of the M11 and M4 'corridors' has been the attraction of higher tech and other companies by access to key business and educational services. A number of local authorities and higher education institutions have promoted 'science parks' and similar initiatives to build links between research and development work and innovation. The provision of such facilities will help attract newer industries and newer processes, but also help overcome any bias towards locating 'head-office' functions (including research functions) in the South East.

These developments have necessarily occurred in an *ad hoc* and localised way, given the lack of national industrial policy direction and the absence of regional planning. An interesting contrast is with Japan, where MITI in 1980 drew up a programme for the regional dispersal of high technology research and development and production through the 'Technopolis' plan (a similar concept was pioneered in France in the Alpes-Maritimes region in the 1970s). According to an OECD assessment published in 1986, 14 such 'technopoles' have been designated with the principal universities in the selected areas targeting specific industrial sectors and these are generally regarded as a success. It is important to note that MITI regarded the provision of good infrastructure and communications as an essential prerequisite for the success of the scheme.

The TUC believes an essential element in future regional development should be the promotion of research and innovation centres through a co-ordinated government strategy, including additional resourcing for local authorities and other local public bodies, and a recognition of the need to see improved infrastructure go hand in hand with measures to encourage research and development location in the regions. An important element will be efforts to attract research activities associated with inward investment, as well as the more standard forms of production to the assisted areas.

Labour mobility

In the 1980s we have seen a migration towards the lower unemployment regions, and some higher unemployment regions face a decline in their populations in the 1990s on current trends. Yet, traditional policies to encourage labour migration have proved costly and of mixed effectiveness, and are more likely to lead to widening regional imbalances. Moreover, increased labour mobility between regions may have limited effects on national problems, such as unemployment. In a memorandum to the National Economic Development Council in 1986 the then Secretary of State for employment, Lord Young, concluded it would be wrong to suppose that a higher level of mobility would lead to a major reduction in unemployment. Since then, of course, unemployment has fallen and regional imbalances worsened, but the Secretary of State's conclusion than within a region most vacancies could be filled by improving local mobility and additional training is still valid.

At the same NEDC meeting the then Chancellor of the Exchequer, Nigel Lawson, criticised national pay bargaining. He said:

> A bricklayer or a bank clerk or a bureaucrat is paid much the same on Merseyside as in Maidenhead. Yet local conditions are obviously different.

He went on to state:

> If we had greater variations, workers in higher unemployment areas would be encouraged to move to areas where there was more demand for labour. And companies would have more of an incentive to set up or expand in higher unemployment areas where labour costs would be lower.

This conclusion appears in part to be in direct contradiction of the views of the Secretary of State for employment. Moreover, it ignores the workings of the housing market. There is a considerable house price differential between the expanding regions and localities and higher unemployment areas, and with the decline in both public and private rented accommodation this has been identified as a major barrier to mobility. Work by Bover, Muellbauer and Murphy has indicated a close link between wages and house prices and wider regional pay differentials would imply wider regional house price differentials. Workers in the north could not afford to move to jobs in the south, while workers in the south would be reluctant to relinquish appreciating capital assets. This has already been identified as a problem by the CBI for firms moving managerial and other grade staff between different regions.

The direct evidence to suggest the sort of link between regional earnings and employment growth suggested by the former Chancellor (and since repeated by other ministers and commentators) appears thin. Broadly speaking, those regions with above average growth in employment have experienced above average growth in earnings, while the opposite has held true. However, analyses based on simplistic assertions should be treated with caution: the relative earnings level in a particular locality will depend on the industrial and occupational structure. A DE article published in 1983 eventually concluded that once these factors were taken into account, there was a small fall in the relative earnings of manual men in the special development areas between 1976 and 1982 (Pike, 1983).

The idea that bigger pay differentials would allow local labour markets to clear by attracting firms to low wage regions is a persistent one. There is little evidence, however, beyond theoretical assertion to suggest that this would be very effective. There appears to be little up-to-date research on the factors which influence firms to move, at both the national and international level. A review of regional policy published by the DTI in 1983 quoted a survey carried out in 1976 which found labour availability, regional incentives, local authority aid and transport facilities featured as the major reasons quoted by firms in their decision to move (DTI, 1983). A more recent survey of multinational location decisions in 1970 and 1983 also found that relative labour costs had little influence – access to markets was by far the most important. This is shown in Table 9.1, facing.

The evidence also suggests that most firms are very reluctant to consider moving across regions simply in response to congestion and labour shortage. There is probably some decentralisation occurring from London, and some movement out of the South East to East Anglia and the South West, but there is no evidence to suggest large-scale movement of industry in response to 'overheating' pressures. A survey of employers in Croydon by Incomes Data Services in 1989 (with registered unemployment of $1\frac{1}{2}$ per cent) found few would consider moving as a response to labour shortage – most preferred to raise pay.

It has been claimed that one barrier to bigger pay differentials on a regional scale is national pay bargaining, which is claimed to set the same pay levels, regardless of where a person works. National pay scales, however, do not set basic pay or total earnings, but establish a pay floor. There exists considerable flexibility within national agreements. Moreover, while important, the extent of national bargaining should not be exaggerated. As a result, there are considerable variations in regional earnings: taking the Chancellor's 1986 examples, managerial grades received 23 per cent less in the North West than in Greater London, non-

Table 9.1 *Main influences on foreign direct investment decisions in industrialised countries*

Reason	Percentage of respondents	
	1970	1983
Access to host domestic market	89	67
Access to host regional market	41	37
Avoid tariff trade barriers	24	16
Avoid non-tariff trade barriers	13	8
Integrate with existing investment	26	37
Change in industrial structure	20	22
Slow growth at home	17	18
Access to raw materials	13	10
Inducements by host country	11	12
Integrate other companies' investment	4	8
Comparative labour costs	4	6

Notes: Figures are percentage of respondents mentioning a factor in their 'top three'. Other factors put to respondents were comparative material costs, shifts in political and social stability, tax advantages, market presence, distribution of risk, return on R&D, development of local market, acquisition opportunities, exchange rate shifts on siting investment.

Source: International Investment and Multinational Enterprises – Recent Trends in International Direct Investment, OECD, Paris 1987.

manual women in the banking sector receive 23 per cent less, and full-time manual men in construction received 18 per cent less (all figures refer to 1986 earning levels).

In a recent article in the *National Institute Economic Review* (August 1990), Brown and Wadhwani suggest that much of the pressure to decentralise pay bargaining has come from the unprecedented expansion in pay differentials between 1979 and 1989 in the South East, when male average earnings rose from 10 per cent to 23 per cent above the national average. However, setting aside the South East, Brown and Wadhwani find average earnings have continued to converge despite policies of labour market deregulation aimed at undermining national collective bargaining and national determination of pay scales. They conclude that this is because decentralisation in both the public and private sector has followed a model in which 'there appear to be strong reasons for firms internal bargaining structures to reflect less where employees work than what they work at'. Decentralisation of pay bargaining along these lines appears to make pay less responsive to external labour market pressures, such as unemployment.

A policy of regional development on the basis of widening pay differentials, even if it could be implemented, would have a potentially devastating impact on regional economies. This would come both from the loss of effective demand and the knock-on effects on local business, and in the longer term the encouragement of low pay–low productivity employment. Those regions within the EC which have prospered have not been associated with either low pay or low productivity employment. Measures which promote higher labour productivity, notably training, can, in contrast, have a highly beneficial impact on regional economies.

Training

The TUC has recently issued a statement, 'Skills 2000', which sets out the key objectives and policies the TUC believes will be needed to meet Britain's training challenges for the 1990s (TUC, 1989b). There is widespread evidence – from NIESR and NEDO studies – that much of Britain's poor productivity performance stems from an undertrained and under-educated workforce in comparison with our major competitors. The provision of training and education and retraining is especially critical in the regions to cope with the run-down in traditional sources of employment, and the development of new industries. A widespread criticism of the London Docklands Development Corporation was the failure to either plan for or directly provide training programmes for local people as redevelopment proceeded and new jobs were created. A major concern with the proposed system of local Training and Enterprise Councils (TECs) is that the coverage and quality is likely to be highly variable, so that within a particular region the amount and quality of training delivered may be unrelated to the wider industrial and economic needs for successful future regional development. The TUC has suggested that a national training agency be established to provide a national strategy for training and overseeing the work of TECs at local level within each region.

Regional incentives

The 1984 White Paper reflected a move away from the traditional automatic subsidy to large-scale capital-intensive industries towards a more selective use of incentives aimed at smaller and medium-sized companies and more labour-intensive sectors, and this was reinforced by the 1988 industrial policy White Paper which abolished regional development grants. The TUC welcomed some of these changes in principle, provided they were clearly part of an overall expansion in regional provision. Unfortunately, it appears that the reverse has occurred.

Regional incentives have been criticised on a number of grounds, but research by Moore *et al* in 1986 and an assessment published by the DTI in 1980 rejected the view that regional policy had led to the inefficient allocation of resources between firms (DTI, 1980). The DTI report said: 'we cannot conclude that the diversionary effects of regional policy have led in general to any inefficient use of real resources'. The Moore *et al* study concluded that regional policy had not subsidised inefficient firms:

> there appears to be no evidence for the view that it has been the least profitable industries which have responded most to regional policy (the so-called lame ducks). If anything the balance of evidence supports the opposite view.

The Moore *et al* study estimated regional policy created (net) 630,000 jobs in the assisted areas between 1961 and 1980, with 280,000 created between 1971 and 1981 when economic conditions were much less favourable. Of course, the effectiveness of different policy instruments varied particularly on the criteria of cost per job. One of the most effective was the use of planning controls (the Industrial Development Certificate (IDC)) on industry in the south and Midlands. Such a control might well prove highly effective again in the 1990s, with two important provisos – that national growth continues to bring unemployment down and that regional investment in the infrastructure means UK firms can relocate without significant competitive disadvantage in relation to Europe.

The traditional policy instrument of direct capital subsidy such as regional development grant was criticised as being very costly, and benefiting highly capital-intensive industries such as chemicals. The cost-effective argument is important, but can be taken to extremes. For example, it could be argued that low quality training programmes were vastly more effective because the cost per person was so low – yet such programmes make little or no contribution to regional development. Although job creation is one of the key objectives of regional policy, the regional economy also needs capital-intensive industries and these are as essential to long-term economic stability and prosperity as is a thriving service sector. An equally important concern should be about the quality of jobs. The Welsh Affairs Committee, in their 1988-9 session report on

Reducing Regional Inequalities

inward investment in Wales, highlighted the fact that unemployment and poverty could still increase when new job creation was over-dependent on low-paid, relatively unskilled assembly work traditionally filled by women, young people, and part-time workers.

However, the return to the use of such subsidies on a large scale in the 1990s may be increasingly hindered by the EC Commission's active discouragement of such aids, especially for inward investment projects. The Commission was clearly concerned at the competition between EC member states to attract inward investment in the 1980s, and it is unlikely that the Nissan factory in Sunderland would be allowed to receive large capital investment subsidy today. This does not, of course, stop the development of different forms of incentives – for example, the new Toyota factory in Derby has been underpinned by the provision of cheap development sites and infrastructure investment. For the 1990s, incentives for industry will need to take non-traditional forms – with particular emphasis on the provision of infrastructure and through training policies a skilled workforce. This would also help make maximum effective use of the EC's new structural funds (based on the European Regional Development and Social Funds).

Community Structural Funds

The reform of the EC's Structural Funds was seen as a key part of the 1992 process, in particular to cope with the enlargement of the EC caused by the accession of Spain, Portugal and Greece which doubled the population of 'least favoured' regions. The three Structural Funds involved are to have total resources doubled in real terms by 1993. Most of Britain benefits under 'objective two', which is to rise by 50 per cent by 1993. The total amount available is less than clear as yet: Commission figures have been incomplete and indicative. A stronger EC regional policy would, however, benefit Britain and the TUC has argued that it should be an essential element in any further moves towards monetary union, following Britain's entry into the exchange rate mechanism (ERM). As well as increasing the resources available under all the objectives, it should be made clear that all money received is additional to the existing national regional policy programmes: the suspicion is that in Britain additionality has not been as transparent as it might have been. The reforms will mean a much greater level of co-ordination between the existing Structural Funds and a greater emphasis on an integrated programme approach developed in partnership with local and regional authorities and the social partners (employers and trade unions).

Unfortunately, the government has resisted the spirit of these reforms by attempting to exclude the social partners and severely restricting the input of local authorities. Such an approach is likely to hinder the fund's effectiveness in the future.

Inward investment

Inward investment is of particular importance and concern to Britain, where inward investment has formed an important part of the industrial base for many years. It is estimated that foreign owned companies account for 14 per cent of UK manufacturing employment, 19 per cent of net output and 22 per cent of capital investment. In 1987 it was estimated there were 300 new inward investment projects creating or safeguarding just over 30,000 jobs. About 35 per cent of total UK employment in foreign owned establishments in manufacturing was in Wales, Scotland, the North West and the north.

The TUC has taken a balanced view on inward investment. It can undoubtedly help bring badly needed jobs and investment to high unemployment areas, and by drawing on worldwide resources and markets in areas such as research and development, it can strengthen the UK industrial base. The TUC broadly welcomes inward investment into the UK, but it would be wrong to suggest that inward investment is a solution for all of Britain's economic ills. Suggestions that inward investment – particularly from Japan – will remove the balance of payments problem in the 1990s are, in the TUC's view, exaggerated. There are also obvious concerns about inward investment projects – the impact on domestic industry and employment; the extent of domestic sourcing of components; the level of technology and the transfer of higher value added functions alongside so-called screwdriver assembly jobs. The EC requirements on local sourcing content are important safeguards in this respect.

For trade unions, the industrial relations question is clearly an important one. Trade unions would be concerned if new projects to the UK were being delayed by a misleading, atypical or negative image of industrial relations in the UK. On the contrary, trade union recognition and organisation will provide the inward investor with the basis for long-term, stable industrial relations. The TUC is aware that Japanese companies in particular pay careful consideration to this and other labour market issues in deciding on the final location, but typically trade unions themselves are not involved in the inward investment process until very late in the day. The TUC believes that in general earlier contact between inward investors and trade unions could be mutually beneficial.

All the evidence suggest that this has been the experience of inward investors into the UK. The vast majority have recognised trade unions and enjoy stable and harmonious industrial relations. The TUC's Special Review Body published changes to the TUC's guidance to unions specifically to help with inward investment and investment in new sites. The TUC's code of practice requires that all affiliated unions co-operate with the procedures laid down by the TUC, the Scottish TUC, the Wales TUC and TUC regional councils in regard to inward investment agencies. On single union deals, the code sets out a procedure by which the TUC will offer advice and, if necessary, arbitration to resolve possible difficulties. On new sites it has proved possible to accommodate the desire for single union recognition, and the TUC has recently introduced new procedures at the national level to assist this process.

Small firms and local economic initiatives

The change in focus towards smaller and medium-sized firms was in part a reflection of the government's obsession with the enterprise culture. Small firms are important and will make a considerable contribution to employment growth, but their role has been exaggerated – especially that of the start-up enterprise. It has been claimed that particularly low levels of new firm formation in some regions has hampered their development, but given the scale of run-down in traditional industries it would be totally unrealistic to expect new and small firms to provide more than a small proportion of the new jobs required. Nonetheless, it is true that the modest rise in net new businesses registered for VAT in the 1980s has favoured the south and Midlands.

There is therefore increased scope for the use of local economic initiatives focusing on smaller enterprises – including local authority enterprise boards, local enterprise agencies and the work of organisations such as Business in the Community, and the urban development corporations. These have an especially important role in the inner cities, especially for the problem companies which have reached around 25 employees. In Britain, at least, such companies appear to run out of steam, and recent research by Gallagher *et al* (1990) shows companies in the 20–49 employment size category made a negative net contribution to employment growth between 1985 and 1987. The Local Government and Housing Act 1989 provides some opportunities for expanding the positive role of local authorities in promoting such activities, albeit there are also restrictions on more direct investment activity. Local authorities will need to be properly resourced, however, and the TUC would like to see their powers to hold direct investments restored, and their general development powers backed by index-linked provision.

IV Regional development institutions

Government policy has been to reduce the active role of local government, despite an impressive track record of local initiatives ranging from training to innovation centres and the provision of direct capital investments, as well as more traditional provision of industrial estates, industrial rented property and infrastructure. The reliance on non-governmental agencies, such as UDCs, has brought considerable criticisms of their insensitivity to local community needs, the lack of planning, and the failure to work alongside local authorities. These weaknesses are being corrected, especially by the 'new' generation of UDCs set up from 1987 onwards, but UDCs can still be placed under pressure from central government to concentrate on land assembly and infrastructure investment, and the LDDC in particular may scale down its own social programmes on training, housing and community development. In 1988, the TUC issued a statement and note of guidance for trade unions on UDCs noting the criticisms, but also seeking to exert constructive influence on future policies, especially in areas like consultation and provision of social programmes such as training and affordable rented housing.

Local authorities

The TUC believes that the positive contribution of local authorities needs to be recognised in the future development of regional policy, both by restoration of their powers of direct intervention and by additional resources for economic and industrial regeneration. Local authorities should also be able to directly approach the European Commission for assistance for particular projects. This would help to ensure the money from the EC is genuinely additional to that provided by national government. However, increasing the role of local authorities by itself could also lead to fiercer competition between authorities and the development of *ad hoc* and fragmented policies from the regional perspective.

Regional planning and development

The TUC's 1982 policy statement, 'Regional Development and Planning', (TUC, 1982) suggested that the development of a local dimension to regional policy must take place within a regional economic planning framework. The aim would not be to impose a view of regional development on local authorities, but to provide a common framework within which policies could be developed on a coherent and co-operative basis. The TUC has therefore suggested setting up tripartite regional industrial development boards in all English regions, to plan at the regional level, to co-ordinate the resources of central government, and to assist local authorities' own development programmes. Arrangements in Wales would reflect the development of distinctive tripartite institutions and the role of the Welsh Office.

New regional institutions for economic and industrial development could be quickly set up on the basis of existing institutions (eg the regional offices of central government and regional development agencies and companies). This would help provide the strategic approach to regional development which the Scottish Office and the highly successful Scottish Development Agency and Highlands and Islands Development Board have already shown can work. It would also ensure that industry had an effective voice working in partnership with local government within a tripartite forum – a model which has worked well in bodies such as the former Training Commission, NEDO, and the Health and Safety Commission. The relative improvement in the economic fortunes of Wales in the second half of the 1980s also owes much to development of tripartite structures backed by government intervention and resources and local economic initiatives, in all of which the Wales TUC has played an active role. The Welsh Affairs Committee's recent report was highly favourable to the co-operation between employers, trade unions, and development agencies and the positive impact this had had on industrial relations and the attractiveness of Wales to inward investment.

Regional government

However, the debate on regional institutions has in recent years seen a growing interest in regional government. There has long been support for a separate elected assembly in Scotland, reflecting a strong national identity, and the 1989 congress reaffirmed the TUC's support. There have been calls for regional government in the English regions (and for Wales), perhaps linked to further local government reform. This in part reflects the legitimate concerns of northern regions in particular that Scotland with a regional assembly would have an unfair advantage in terms of attracting new investment (as well as more freedom in developing new initiatives). However, it has also been noted that in countries with a federal structure, such as the US or Germany, the scope for active regional policies by regional or State government is much greater. For example, the EC Commission has recently made unfavourable comparisons between resources available to Yorkshire and Humberside from central and local government and resources available for the Westphalia region in Germany from the State government. The issue of accountability is also important, and it is often argued that agencies charged with regional development should in some sense be accountable to the regional electorate, either by direct election or through elected local government representatives.

Decentralisation

Regardless of what side of the debate is taken, the fact remains that decentralisation of resources and functions and the creation of development bodies with strategic roles at the regional level would be important immediate steps in addressing regional problems in the 1990s. An equally important element of decentralisation will be the provision of regional accounts showing the distribution and levels of central government spending. At present, separate figures are regularly published only for Scotland and Wales, but in future similar statistics should be available for all the English regions. As well as public spending totals, the regional impact of fiscal and other national policy measures should also be assessed. This would allow national government to take account of regional needs in developing policy and to be aware of particularly adverse or beneficial impacts of national policy changes.

V Conclusions

A regional development framework will be required which is based on regional economic planning and the integration of inner city and regional policies. This will have a central role in improving the supply side of regional economies through investment in the infrastructure, in research and development, and through education and training.

With national policies committed to growth and continuing falls in unemployment, a redistributive regional policy based on the reintroduction of some form of planning control would prove highly effective in the 1990s. Direct capital subsidies would be less important and incentives would increasingly focus on providing attractive centres of development for business and industry.

Local government's positive role in promoting local economic initiatives and developing inner city policies should be recognised and adequately resourced. This could be assisted by the development of tripartite regional development institutions and agencies in all English regions and in Wales. Scotland would have an elected regional assembly. The issue of regional government elsewhere will need to be examined on the basis of wide consultation.

Bibliography

Bover, O, Muellbauer, J and Murphy, A (1989) 'Housing, Wages, and UK Labour Markets', *Oxford Bulletin of Economics and Statistics*, vol 51, March pp97-162.

Brown, W, and Wadhwani, S, (1989) 'The Economic Effects of Industrial Relations Legislation since 1979', *National Institute Economic Review*, vol 131, pp57-70.

Gallagher, C, Daly, M, and Thomason, J (1989), 'The Growth of UK Companies 1985-7 and their Contribution to Job Generation', *Department of Employment Gazette*, February, 1990.

Moore, B, Rhodes, J, and Tyler, P, (1986) *The Effects of Government Regional Policy*, Department of Trade and Industry.

OECD (1987) *International Investment and Multinational Enterprises: Recent Trends, in International Direct Investment*, OECD, Paris.

Pike, T (1983) 'Relative Earnings in Counties with SDAs', *Department of Employment Gazette*.

TUC (1982) *Regional Development and Planning*.

TUC (1988a) *Inward Investment*, Memorandum to NEDC.

TUC (1988b) *Maximise the Benefits, Minimise the Costs*.

TUC (1988c) *Trade Unions in the Cities*.

TUC (1989a) *Europe 1992: Progress Report on Trade Union Objectives*.

TUC (1989b) *Skills 2000*.

10
*Regional Economic Disparities: Some Public Policy Issues**

Alex Bowen and Ken Mayhew

The second National Economic Development Office Policy Seminar was concerned with the economic disparities among the regions of the UK. These disparities include differences in productivity, the composition of output, the components of the labour force, standards of living and levels of unemployment. The range of unemployment rates among regions is perhaps the most obvious indicator that resources in some areas are not being utilised fully. Other indicators like house prices and the degree of traffic congestion suggest that other locations may be suffering overheating. Thus the question arises, can the supply side performance of the British economy as a whole be improved by changing the spatial pattern of output and employment? If so, how can economic agents – business people, employees, the unemployed, consumers – be encouraged to bring about the desired changes? This final chapter attempts to draw attention to some of the answers prompted by the contributors to the seminar.

I What are the disparities?

The chapters presented marshal evidence that there are considerable economic disparities among regions. Jim Taylor, in particular, demonstrates that, according to many criteria, the concept of a 'north–south divide' in the economy makes sense. He looks in some detail at spatial variations in unemployment, GDP per capita, employment growth, income and expenditure, house prices and efficiency wages (real

---

* This paper expresses the views and judgements of its authors; it does not represent any official view of NEDO.

wages divided by labour productivity). There is, of course, room for debate about where precisely the divide is located, particularly given the arbitrariness of the standard planning regions (Goddard and Thwaites draw attention to the problem of defining regions). Variations within regions are often substantially greater than those between them and there are pockets of economic deprivation within all regions which merit particular concern. However, Taylor notes some clear-cut differences on average between his 'north' - Scotland, Wales, Northern Ireland, the North of England, the North West and Yorkshire and Humberside - and his 'south' - the South East, East Anglia, the South West and the East Midlands (with the West Midlands usually being part of the 'south', except in the early 1980s). The difference in rates of employment growth is particularly noticeable in the 1980s (high in the south, negative in the north), and regional unemployment rates diverged substantially in absolute terms at the beginning of that decade. Rankings of standard regions by different variables are not identical and some of the regions have changed position over the years; for instance, the West Midlands has suffered relatively over the last ten years or so, while East Anglia has prospered. But the degree of stability is striking; McCormick (Chapter 5) observes:

> only in the UK and Japan do relative regional unemployment levels of two decades ago readily predict recent relative positions - a fact which will comfort only those who had given up hope of finding anything that the UK and Japan labour markets share in common.

Jim Taylor explains how disparities in regional unemployment rates can be regarded as evidence of the regional mismatch of labour demand and labour supply. If the mismatch could be reduced, the country as a whole could have a lower unemployment rate at any given level of aggregate demand, and inflationary pressures arising in labour markets might fall. However, the reasons for the persistence of mismatch have to be understood; otherwise, attempts to correct it may be counterproductive (for instance, a fiscal stimulus applied in high unemployment regions might be inflationary if local wages do not become more sensitive to local unemployment levels). Several of the chapters in this book pursue the issue of how to make regional labour markets function better for the benefit of the economy as a whole. Taylor also reminds us that society's 'social overhead' capital - such as the transport infrastructure and the telecommunications network - would be better used in these circumstances (because in many respects it is overburdened in the South East and under-utilised in 'the north'), a theme picked up by a number of contributors at the seminar.

II Labour market problems

If it is accepted that the documented economic disparities reflect in part an inefficient allocation of resources, the question arises as to why market forces do not put matters right. The chapters identify a number of obstacles, some of them the consequence of government policy, others of market failures.

Jim Taylor (Chapter 2) writes:

> perhaps of greatest significance ... in explaining differences in performance between the north and the south is the apparent inability of northern labour markets to improve their competitive position through lower unit labour costs.

This is a theme of most of the chapters (though the TUC express some scepticism in Chapter 9). Patrick Minford and Peter Stoney, for instance, construct a model of regional economic interactions in which the downward inflexibility of real wages – particularly for manual workers – plays a crucial role in generating the high unemployment rates of 'the north'. They adapt the traditional economic theory usually applied to international trade (so-called 'Heckscher – Ohlin' trade theory, discussed along with other theoretical perspectives in Chapter 1) to explain where the sources of the problem lie. This theory stresses the importance of a region's endowments of different types of land, labour and capital and the potential for mutual gains to be had from inter-regional trade. For instance, a region with a disproportionate number of unskilled workers should concentrate on producing goods and services which do not require a high skill level. One factor of production – land – is inevitably immobile, but Minford and Stoney argue that manual labour is also effectively immobile, largely because of public intervention. Excessive reservation wages (the wages below which workers will not accept job offers) induced by social security benefits maintain excessive wage rates and, combined with subsidies to council house tenants, remove the incentive to migrate. Neither are nearly as important for non-manual workers.

However, the functioning of the labour market is not the only culprit in their view. The reason why absolute unemployment differentials for manual workers increased sharply in the early 1980s was a decline in the relative price of the output of manufacturing, which uses manual labour to a greater degree than do other industrial sectors. This terms of trade effect has been exacerbated by unnecessarily high transport costs for manufactured goods (pushed up by monopoly and excessive regulations) and, implicitly, less competition in the market for services. Planning regulations and procedures also inhibit the proper pricing of land in its alternative uses. The authors believe that the prospects for business migration to 'the north' narrowing regional disparities are good, thanks to various government reforms, such as the abolition of the dock labour scheme, the abolition of household and *local* business rates, and the sale of council houses.

Minford and Stoney's chapter is particularly interesting as an attempt to organise an analysis of regional problems by using an explicit theoretical framework. As Chapter 1 argues, their 'trade-theoretic' perspective is not the only one available, but the articulation of some theoretical framework is desirable if the reasons for persistent economic disparities are to be understood. Some participants at the seminar were unhappy with the dichotomies between manufacturing and non-manufacturing and between manual and non-manual workers. For instance, a possible alternative to spatial mobility is occupational mobility, in this case the conversion of manual into non-manual workers, and not all the participants were as confident as Minford and Stoney that current retraining programmes for the unemployed will achieve a sufficient occupational transformation. In the absence of econometric estimates of the crucial parameters in the model, the quantitative importance of the various hypothesised causes of regional unemployment is uncertain, but their paper reports some preliminary estimates for Merseyside and North Wales.

Barry McCormick, in Chapter 5, also points the finger at problems in the markets for housing and labour. He is perhaps less sanguine about the effectiveness of market forces to bring about the demise of dysfunctional regional disparities, writing:

> we need a keen sense of why, even if markets in labour and capital were perfectly mobile, the *laissez-faire* competitive allocation of labour between UK regions in unlikely to be that which a wise government might arrange.

Some of the problems he perceives echo the concerns of Minford and Stoney. He believes that land for housing development is probably undersupplied in the south, because there is no market mechanism in the planning system to allow the potential gainers to compensate the losers (one could envisage, for instance, local auctions of planning permits, the receipts being used to compensate those whose amenities are reduced). He also notes that problems can arise from the substantial variation of housing subsidies (including mortgage tax relief) across housing tenure types. But other problems, such as the costs imposed by congestion and how to decide on appropriate levels of investment in social infrastructure, can in principle be alleviated by sensible government intervention. Using taxes or other charges to make those who contribute to congestion internalise the costs they are generating is an example. Some of McCormick's specific proposals will be mentioned later.

III Migration

McCormick, in Chapter 5, focuses on whether migration reduces inequalities between geographic areas. This depends on how responsive people are to market signals (for instance, do workers tend to move towards areas with higher real wages?) and on whether the market signals are pointing in the desired direction (for instance, do differential housing costs provide an incentive to people to leave labour surplus areas?). McCormick's review of the evidence suggests that people do respond as expected to real wage differentials (although these are difficult to measure), and that if an individual is unemployed he or she is more likely to migrate, other things being equal. But empirical investigation shows that a high regional unemployment rate tends to reduce the propensity to migrate of both the employed and the unemployed in that region. Also, the net out-migration from depressed regions tends to consist of non-manual workers, whereas the regional unemployment problem is largely one afflicting manual workers (a point also stressed by Minford and Stoney). The mobility of British manual workers is very much lower than in the US. McCormick points out that making housing markets work better may not improve matters greatly if other markets – particularly labour markets for different occupational groups – do not. He draws attention to the danger that the ratio of manual to non-manual workers in 'the north' may increase excessively in these circumstances because of the inflexibility of 'north–south' manual wage differentials. Overall, McCormick's work does not suggest that migration is likely to be a major equilibrating factor in the medium term. For most regions, the net flows are not big enough, even when they are moving in the right direction, which is not always the case. However, there are measures that could be taken to ensure that migration helps rather than hinders the removal of unnecessary regional disparities; improving the supply of low cost housing to rent in the South East is one example. Net migration flows may not be big, but they can make a considerable contribution to increasing labour market flexibility at the margin. At a more disaggregated level, increasing mobility could be much more important, given the very limited geographical scope of job search by unskilled workers living in council accommodation. This is one area where focusing on broad regions may distract from more localised problems, particularly in the inner cities.

IV The locus of pay bargaining

Janet Walsh and William Brown, in Chapter 4, also express doubts about the virtues of untrammelled *laissez-faire*. They investigate the consequences of the decentralisation of wage bargaining, which has been encouraged by the present government (see, for instance, the 1988 White Paper on employment policy). They argue that the shift from multi- to single-employer bargaining has led to more wage-setting at the level of profit-related product centres. This makes sense from the company point of view, and fits in with modern economic theory of wage determination, which stresses the limitations to competition in labour markets which are imposed by long-term employment relationships. As far as regional earnings are concerned, this decentralisation has so far failed to achieve the responsiveness of pay to *local* labour market conditions advocated by the government (and by most, but not all, participants at the seminar). The South East region is a possible exception, but Walsh and Brown's discussion of the evolution of 'London allowance' payments expresses scepticism on this point. If greater *local* wage flexibility is desirable, their evidence suggests that the current developments in wage-setting are insufficient as yet to bring it about.

V Housing market problems

At the seminar, Anthony Murphy gave voice to some misgivings about the way in which housing markets worked, at least in the short to medium term. He and John Muellbauer elaborate these misgivings and present a summary of their work in this area in Chapter 6. In particular, they remind us that the immobility induced by the council house system is not the only problem: owner-occupiers have low mobility by international standards, too. Their concerns are twofold; first, the ways in which house price movements may inhibit migration and, second, the encouragement they may give to wage pressure, particularly in the South East. Both factors work against the better functioning of regional labour markets. The authors do not deny that a regional house price differential in favour of the South East is an appropriate price signal, to builders and developers to supply more housing, and to families that real housing costs are higher. But they think that differentials are increased by shocks (eg financial liberalisation) other than regional demand shocks, and that there is evidence of price overshooting which discourages desired inward migration, also pushing up wage demands and increasing labour market mismatch. Muellbauer and Murphy's suggestions as to how overshooting in house prices can be avoided deserve attention. They propose increasing the price responsiveness of the supply of land for housing, which would entail a relaxation of planning restrictions in the South East. On the demand side, they want a reform of mortgage tax relief (removing its net cost to the Exchequer) and the introduction of a residential property tax, related to residential land prices, but levied at a national rate.

VI A 'worst case'

John Goddard and Alfred Thwaites investigate, in Chapter 7, the economy of the north of England as a case study in regional problems. They note (at p272) that:

> in considering the smallest and most peripheral region within England and one which has had the longest history of public intervention to arrest economic decline, the chapter examines a 'worst case' situation for public policy.

They make a detailed study of characteristics of the region in which the seminar took place, such as industry structure and the size distribution of firms, drawing attention in the process to a large range of regionally differentiated factors which can influence industrial location decisions (this is in contrast to the more schematic approach of Minford and Stoney, in Chapter 3, who emphasise the importance of just two such factors – manual wages and land prices). These authors stress the long-run changes in the structure of industry and commerce which are being brought about by such factors as the functional division of labour between conception and execution of tasks, the globalisation of markets and the increasing importance of innovation. Several of the changes which Goddard and Thwaites seek to map concern the allocation of resources within corporations, instead of via the market place: examples include foreign inward investment, the location of R&D and headquarters functions, training, and the use of communications and information technology to alter intra-firm location decisions (eg the relocation of large parts of the *London* directory enquiries service to the Newcastle area cited by the authors. In these circumstances, when location decisions have to be taken, current market signals to potential investors compete with internal bureaucratic procedures and, more important, expectations about these future developments in technologies and markets. Goddard and Thwaites are concerned that the tendency of peripheral regions to be 'branch plant economies' may mean that the information about future developments, and the developments themselves, are slow to diffuse through them. There certainly seems to be scope in the North for public and private organisations to improve information flows and to stimulate local entrepreneurship.

VII Improving supply side performance

The chapter by the TUC (Chapter 9) represents a departure from traditional regional policies of input subsidies and control of industrial location, in the sense that it stresses the importance of improving the structural performance of the regions. Though the TUC believes there are arguments for the reintroduction of some form of planning controls, traditional capital and labour subsidies are seen as less important. It is concerned with providing the conditions in which local entrepreneurship can flourish. The TUC paper considers a number of supply side factors. Two areas in which it is generally recognised that the government can help to overcome 'market failures' in private provision are R&D and training. Citing the Japanese initiative through the 'Technopoles' plan, the TUC advocates the promotion of regional research and innovation centres. This, the TUC believes, will help to ensure that newer industries and processes are attracted to the regions. In the same spirit, a national training agency is advocated to ensure a reasonably uniform coverage and quality of training and enterprise councils; more detailed proposals are contained in the TUC's document 'Skills 2000'. Thus, their focus is on providing attractive centres of development for business and industry. An issue which arises in this context is the appropriate level of official decision-making. The TUC's position is the most unequivocal. A more positive, better resourced role for local government is advocated, as is the establishment of regional planning bodies. Several other participants placed more emphasis on the role of urban development corporations and other business-led bodies, although there was general recognition that their activities need to be monitored and evaluated by a public body at some level (Minford and Stoney, for instance, opt for supervision by central government). The TUC emphasises the importance of the continuing economic integration of the EC, which will stimulate intra-European trade and make investment more footloose. They recommend measures to avoid the UK becoming more peripheral. Making use of EC funds is one of them (Chapter 1 reviews the scope of EC programmes of regional aid). Another is the improvement of the country's infrastructure, in particular in the transport sector.

The transport infrastructure takes centre stage in the contribution from Alan Armitage and Derek Palmer of the CBI in Chapter 8. They argue for a high quality integrated transport network, citing the example of the large investments which are being made by our European competitors. Like the TUC, they attach particular importance to the consequences of the Channel Tunnel. Links to the tunnel and the ports facing the Continent must be improved, particularly given the advent of 'just in time' or 'right on time' production methods (the TUC appears to differ on the extent to which investment should be directed to the South East). Specific and detailed proposals, drawing on the CBI publication, 'Trade Routes to the Future', are made for improvements of road, rail, air and public transport. At the seminar there was much discussion of the dangers of peripherality and whether businesses selling to the UK would tend to locate in northern France in future if the transport system alone were improved. Our labour markets, technologies and entrepreneurship will have to be more competitive *internationally* if this is to be avoided.

VIII Conclusions

There was universal agreement at the seminar that the scale of economic disparities across regions is evidence of inefficient use of resources. There is considerable scope for increases in output, employment and living standards in the more disadvantaged regions, without consequent losses in regions where overheating of the economy is apparent. This could be achieved without increasing overall inflationary pressures in the economy if regional labour markets and housing markets in particular were more responsive to local economic conditions. It is important to find

Reducing Regional Inequalities

appropriate supply side policies to improve the structure of incentives facing economic agents, so that these changes can come about. Some of the main areas in which the seminar stimulated proposals are noted below.

Local wage flexibility

Greater responsiveness of pay to local economic conditions could help to reduce unemployment, by improving the competitiveness of firms in high unemployment regions, encouraging business formation and relocation, and perhaps making migration more of an equilibrating force in the labour market. Current developments in pay bargaining have not yet had much effect in this respect. Better information to feed into local wage-setting processes is needed, in particular, reliable evidence on regional variations in the cost of living, including housing costs. But it would be wrong to expect greater wage variations alone to be the solution to the regional unemployment problem. The factors which appear to prevent some regions from sustaining full employment at average wage levels must be tackled. Reforms on other fronts, particularly the housing market and the fostering of entrepreneurship, must be sought at the same time.

Migration

The net flow of migrants can only be expected to have a second-order effect on unemployment differentials directly, but the structure of incentives could be improved. Higher migration rates could help labour market flexibility at the margin considerably. Barry McCormick makes specific proposals to:

- put the tax treatment of the expenses of migrants moved by their employers on the same basis as that of other migrants who do not get tax relief;
- earmark a greater proportion of council housing for migrants;
- encourage retired people to move out of the area of housing shortage in the South East, so that economic activity rates there can increase with the same housing stock.

Reforms to housing markets

Housing markets are a major culprit in discouraging migration and there is evidence that they contribute to aggregate wage pressures, worsening the consequences of regional labour market mismatch. The inflexibility of the council housing system needs to be remedied. Council house sales are one way of achieving this (McCormick enters a caveat here), but there are also problems in the owner-occupier sector, particularly the different rates of capital gain expected in different regions. The differential response of regional house prices to macroeconomic shocks may have adverse effects on migration and wage bargaining (especially in the South East). Some participants advocated moves towards equivalent tax treatment of all tenure types, so that the private rented sector could provide greater mobility at the margin.

Education and training

The wasted resources measured by unemployment are concentrated among manual workers. The trend in the occupational requirements of the economy works against them, too. Upgrading the labour force further (especially the unemployed) will encourage individual mobility and regional development. A more highly educated labour force also provides a more fertile source of entrepreneurship and innovation, to which we turn next.

Entrepreneurship

The north loses out when it comes to entrepreneurial effort. This is reflected in the birth-rate of new firms, the supply of venture capital and the propensity to innovate. Business led efforts to revitalise urban areas show how public policy can help here; at the seminar, we had the opportunity to see the excellent example set by 'The Newcastle Initiative'. The regulatory framework and the role of public authorities need to be sorted out. Disseminating information about the economic environment and business opportunities in disadvantaged regions is a continuing task for public policy.

Infrastructure

Further development of our infrastructure is needed, particularly in transport. The South East is under particular strain in this respect. Even if other measures helped to reduce overheating in the South East, improvements in the transport network would still be necessary if other regions are to participate fully in the growth of intra-European trade. Measuring exactly how much investment is needed and where is difficult, and the CBI's efforts in this respect are to be commended. More explicit discussion of the pricing of private and social costs and benefits would help. It would also help people use what we have more efficiently; the under-utilised resources of 'the north' should be remembered here. A better infrastructure helps foreign exporters to the UK, too, so this goal should be seen as complementing measures to improve the competitiveness of British regions.

This list has not included the more traditional regional policies of subsidies to labour and capital, planning controls and a fiscal policy that discriminates among regions (Chapter 1 reviews briefly their history). It is fair to state that participants were divided in their evaluations of the success of such policies in Britain in the past. In the future, however, the scope for such policies will be limited by the guidelines set down and the money disbursed by the EC. Hence, we have concentrated on areas where sustainable supply side improvements can be encouraged. Such improvements are feasible and desirable both for enhancing the performance of the economy as a whole and for improving the economic prospects of people in currently disadvantaged regions.

Index

accountability 44, 311, 353
acquisition 93, 307
Adams, C D 36
adjustment
　delayed 125
　house price 81
　investment 281
　labour market 35, 234, 242
　mechanisms 30
　problem of 35
　process 248
　wage 53, 229, 260
aerospace industry 52
African trade 128
agglomeration economies 33, 52
aggregates 40, 41, 107
agreements
　multi-employer 191
　non-union 144
　wages 209
agriculture 46, 121, 193
　areas dependent on 48
　land prices 110, 112, 119, 168
　policy 50
Ahlstrand, B 198
aid, see assistance
air 317, 327, 328-9
airports 147, 324, 329, 340-1
Alderman, N 287, 288, 289, 290
A-levels 87
Allen, K 42
Alliance and Leicester Building Society 208
Alnwick 291
Amsterdam 324, 325

Antwerp 131 n., 172, 173, 327
Aoki, M 197
Armitage, A 363
Armstrong, H 29, 37, 72, 85 n., 101
arts 316-17
Ashcroft, B 95
Ashton, P 54
Asian trade 128
assistance 43, 281, 305, 309, 310
　DAs and 39
　discretionary 41, 42
　fall in value of 51
　linked to national policies 49
　major reduction in 276
　need for 37
　use of 44
　see also grants; loans; regional issues (aid); subsidies
Atlantic 128
auctions 148
Audit Commission 308, 312
Australia 315
Austria 325
autonomy 49
Avonmouth 174

BAA (British Airports Authority) 147
backward countries/regions 31-2
Bailey, R 210, 211 n.
Bain, Professor 134
balance of payments 31, 259
Balchin, P N 72
Baltic 171
Bank of England 191

367

Index

banking 95
bargaining 134, 185, 209, 211, 276, 360
  collective 31, 56, 60, 190, 345
  decentralisation of 98, 196-202, 210, 212, 213, 346
  level of 191, 193-6
  multi-employer 205
  national 97, 334
  selective 36
barriers 133, 134, 324, 339
  entrepreneurship 281-2
  migration 205, 239-40
  removal of 99-100
  wage 53, 36-7
Baum, A G 36
Beaumont, P 197
Begg, I 284, 286
Belgium 49, 325
benefits 44, 202
  costs and 60, 99, 100, 144, 148, 365
  see also unemployment benefits
Berkshire 291
BES (Business Expansion Scheme) 45, 91, 92
Beveridge principle/Worktest 141, 145
Bianchi, L 297
biotechnology 316
Birmingham 173, 175, 329
Blackaby, D H 97, 191
Blanchflower, D 200
BNFL (British Nuclear Fuels Ltd) 309
Board of Trade 38
boom 38, 255, 259, 260, 340
Booth, A 197
Bover, O 54, 55, 57, 98, 212, 233, 234, 247, 252, 256, 260, 343
Bowen, A 109 n.
Bowers, J 38
Breheny, M 51, 52
Bristol 34, 207, 303-4

British Airways 207
British Gas 290, 291, 315
British Rail 210, 326
British Steel 210, 291, 309, 312
British Technology Group 289
British Telecom 303
Brixton 73
Brown, W 97, 192, 199, 201, 345, 360
Brussels 324, 325, 328
BSAS (British Social Attitudes Survey) 339
budget 47, 51, 52
Building Societies Act (1986) 265
Bulgaria 326
Burridge, P 229
business rates 111
  high 136, 137
  uniform (UBR) 138, 139, 322

Cahill, J 200
Cambridge Econometrics 205, 212
Cameron, G 284, 286
canals 325
capacity utilisation 101-2, 103
capital 30, 41, 122-4, 357
  access to 89, 90, 99
  allocation of 239-40
  equity 91
  flows of 31
  losses 99
  mobility of 30
  movements 170
  social overhead 101, 104
  stock of 143
  venture 45, 92, 106, 365
capital equipment 35
capital goods 32
cargoes 146, 147
  bulk 130, 131, 133, 172-3, 174
cartel practices 146
case law 145
CBI (Confederation of British

368

Index

Industry) 273, 322, 328, 330, 343, 365
  and bargaining 194
  and capacity utilisation 103
  and lost production due to congestion 101, 323
  and pay 200, 202-3
'Celtic Fringe' 220
centralisation 274, 275, 276, 300
'centre-periphery' effects 113
Champion, A 221
Channel Tunnel 133, 318, 326-7, 336, 341-2
  accessibility to Europe for North 296
  freight shift and 323
  landbridge idea 146-7
  links needed to 324, 328, 331, 363
  opportunities presented by 330
Cheshire 180
childcare 43, 212
Chisholm, M 40
CIPFA (Chartered Institute of Public Finance and Accountancy) 136
cities 33, 35, 105
  see also inner cities
Civil Aviation Authority 205
Cleveland 43, 273, 294, 311, 312
Clydeside 34, 35, 36, 43
Coats Viyella 199
collectivism 276
colleges 45, 88 n.
Cologne 324, 325
communications 164, 275, 282, 296, 297, 303
community charge, see poll tax
competition 110, 198, 205, 302, 317, 358
  foreign, protection from 133
  imperfect 257
  internal 133-4
  international 180

protected sector and 111, 143, 145, 146
rail 147
regional disparities 30-1
shipping 133, 135, 152
within EC 100
competitiveness 75, 83-4, 297
  improving 364, 366
  inability to improve 106
  innovations necessary for 94
  international 258
  loss of 70
computer network 299, 300
computerisation 305
congestion 100, 236, 237-8, 239, 243, 329-30
  Channel 146
  costs 101
  encouraging relocation 322
  London 105, 107, 328, 335
  road networks clogged by 323
  taxes on 138, 242, 268
Conservative government 39
Consett 276, 289
containers 130, 131, 133, 171-2, 173, 174
  costs 327
contracting out 145
Control of Office and Industrial Development Act (1965) 38
conurbations 34, 35, 43, 192, 312
Coombes, M 292, 293, 294, 296
corporation tax 240
costs 36, 84, 109, 133, 170
  executive 147
  factor 53
  haulage 171, 172, 173, 175; see also transport costs
  land 113
  living 53, 55, 258
  materials 33, 110
  psychic 99
  transaction 33
  unskilled labour 136

369

Index

urban squalor 43
wage 41
see also benefits (costs and)
council housing 59, 139, 211, 218, 360, 364
  estates 46, 73
  huge investment in 230
  immobility caused by 56-7, 60, 232-3
  privatisation 240-1
  relets 241-2
  rents 118, 137
  sales 145, 266, 365
  share of 251
  stocks of 58, 241
  tenants 232-3, 241, 252
Courtaulds 199, 277
co-operative ventures 44
Craig, C 197
Crawley 27, 148
credit 54, 55, 268, 269
Creedy, J 221 n.
cultural industries 316-17
Cumbria 273, 274
CURDS 298, 299

Damesick, P 45
DAs (development areas) 38-9, 42, 43
Da Vanzo, J 226, 227
Davies, S 288
DD curves 168
Deaton, D 197
decentralisation 35, 43, 212, 344
  bargaining 60, 97, 98, 196-202, 210, 346
  operational decisions 198
  pay structures 191
  population 44
  research and development 291
  resources 353
  routine office functions 34
decision-making 49, 93, 94, 115, 217

defence 52, 60, 121, 308
deflation 71
deindustrialisation 40
demand 134, 168, 246, 258, 259
  aggregate 29, 107, 168, 170, 266; excess 53, 54
  elasticity of 32, 181, 182, 184
  goods 31
  housing 266-7
  labour 183, 260
  land 119
  management 38, 40
  migration 241
  relocation of 105
  shocks 53, 54, 55, 225, 249
  and supply 101, 102, 107
Denmark 49, 50
densely populated regions 242-3
Department of Economic Affairs 38
Department of Employment 141, 190, 192, 204, 297, 344
Department of the Environment 43, 105, 309, 312
Department of Health and Social Security 276
Department of Social Security 304, 305
Department of Trade and Industry 296, 305, 308, 322, 344, 347
  grants and aid by 143-4, 310
  and shipping conferences' practices 147
Department of Transport 152
dependence 34, 35, 93
depreciation allowances 39
depressed regions 219-21, 242
depression 377, 71, 274
deprivation 43, 356
deregulation 143, 145, 190, 205, 213, 300
Derelict Land Act (1982) 36
dereliction 44, 112, 136, 138, 144
destination choice 228
devaluation 29

Index

developing economies 31
development 137, 143, 148, 149, 272–318
  economics of 31
  institutions 351–3
  prospects 335–6
  promotion of 50, 180
devolution 101, 339
Diamond, D 296, 297
diseconomies of scale 32–4
disequilibrium 218
disincentives 35, 39, 42, 99
disinvestment 307
dislocation 249
disparities 246–69, 321, 355–66
  causes and consequences 70–107
Distribution of Industry Act (1958) 37, 238
diversification 286, 287, 289
dividends 170
divisions of labour 273, 282
Dixon, R J 96
dock labour scheme 130, 131, 133, 135, 146, 152
Docklands 44, 114, 341, 346
dole 141, 145
dominant employers 273–4
Doncaster 299
Donovan Commission (1968) 192
Dover 131 n., 133, 172, 173
Durham 273, 274
Dusseldorf 173, 175

earnings, *see* wages
East Anglia, *see* regional issues
East Midlands, *see* regional issues
EC (European Community) 321, 346, 351, 353, 363, 366
  Britain's financial contribution to 334
  CAP (Common Agriculture Policy) 46, 50, 110
  Channel Tunnel and 341

developing role in regional policy 28
  distance between areas 323
  Guidance Section 46, 48, 49
  Regional Development Fund 273, 309
  regional economic disparities 98–9, 100
  regional industrial policy 338
  Social Fund 309
  structural policies/funds 46–51, 100, 348–9
Economic Planning Councils 38
economies of scale 32–4, 100, 110, 147, 197
ECU (European Currency Unit) 46, 48
Edinburgh 301, 302–4
education 121–2, 143, 315, 337, 365
  attainment/qualifications 87, 89, 90
  levels 229, 230–1
  migration and 230–1, 232–3
efficiency 60, 101, 148
  and densely populated regions 242–3
  improving 27, 97, 139
Egginton, D M 52
electrification schemes 341–2
electronics industry 52, 284, 285
Elias, P 221 n.
Eltis, W 109 n.
emerging national companies 274
employment 34, 37, 52, 90, 167, 170
  aggregate 59
  creation 38, 39, 41, 310, 347
  decentralisation 43, 44
  decline 85
  distribution 94, 105
  growth 36, 77–8, 84, 88
    differences/disparities 77–8, 356

371

Index

negative 86
highest proportion of 35
illegal 145
level 40, 50
opportunities/prospects 45, 70, 72, 75, 145
  self- 278, 312
  trends 121
Employment for the 1990s (White Paper) 185, 190, 192
Employment Protection Act 1979 196 n.
engineering 315-16
English Estates 275, 308
enterprise agencies 312
enterprise allowance scheme 45-6
enterprise culture 351
Enterprise Zones 44-5, 309, 311
entrepreneurship 100, 138, 364, 365
  activity 88-93, 106
  barriers to 281-2
  firms lacking in 35
  incentives 45
  local 283, 301, 362
  survival 70, 292
environment 105, 151, 339
equalisation 30, 53
equilibria 54, 60, 98, 167, 170, 181-2
  distribution of labour 239
  locational 32
  long-run 30
ERM (exchange rate mechanism) 348
ESRC (Economic and Social Research Council) Inner Cities Research Programme 34-5
ET (employment training scheme) 137, 141, 150
ethnic minorities 43
Europe 128, 146, 171, 282, 327, 328
  economic and monetary union 336
  Heathrow vital to UK in 329
  see also EC (European Community)
European Industrial Relations Review (1989) 202, 207
European Investment Bank 47, 49
Evans, A 267
exchange rate 29; *see also* ERM
expansion 36, 151, 180, 275, 302, 305
expenditure 27, 28-34, 39-41, 74, 78-81
  EC 48, 49, 51
  public 277, 311
external control 93-5
externalisation 294, 301
externalities 36, 37

factors of production 30, 42, 357
Family Expenditure Surveys 248, 262
Far East 100, 171
federal structure 353
Federation of London Clearing Bank Employers 199
Felixstowe 131, 133, 135, 146, 172, 341
ferry companies 174
Fields, G 228
financial control 95
financial services 205, 206, 207, 301, 302
Financial Times 73, 100, 105
first-time buyers 267
fiscal policy 28-34, 38
Fothergill, S 283, 284
Foxley Wood New Town 105
France 49, 314, 324, 325, 326, 342
Frankfurt 325
fraud 145, 151
freight 171, 172, 323, 326, 328
  distribution 322
fringe benefits 208, 211
fund managers 92
funds 143

372

Index

EC 46–51, 100, 338, 348–9
venture investment 45

Gallagher, C 351
Gateshead 274, 308, 312, 313
Gatwick 329, 341
GDP (gross domestic product) 114, 122–7, 278, 302, 337, 340
  EC 47, 48
  growth 81
  manufacturing percentage of 123, 153–63
  per capita 50, 76–7, 99, 355
Germany (West) 49, 219–20, 324–5, 326, 327, 353
Glasgow 34, 175, 303–4, 327, 329
Goddard, J 275, 356, 361, 362
Gordon, I 216 n., 221 n., 222 n., 225, 229
government 137, 185, 190–2, 275, 318, 322
  agencies 144, 152
  decentralisation 210
  expenditure 51–2, 308–9
  inner city problems and 43
  intervention 37
  investment 340
  policy 43–6, 93, 312–13, 334, 351, 357
  pressure on employers 209
  regulation 133, 134
  specialised parts of 164
  UDCs and 144
  see also protected sector
graduates 87, 88, 291
grants 36, 38, 41, 44
  capital 275
  discretionary 143
  EC 48, 49
  freight facilities 327
  rate support 137, 276
  regional development 42, 310–11, 347
  shift from 39

urban development 45
Greater London
  BSE funds 45
  congestion problems 105, 107, 335
  earnings differential 189
  expenditure per capita 27
  house price differential 254
  migration: from 83, 121, 225; into 236, 243
Greece 48, 49, 100, 348
Green, A E 235 n., 290, 291
Green Belt 105
Greenock 152
Greenwood, M J 227
Gregory, M 200
growth 93, 105, 123, 217, 230, 281
  basis for 289
  earnings ratio 187
  employment 36, 85, 190, 262
  excessive 101
  GDP 81
  high-tech 'jobless' 41
  income and liquidity 259
  in situ 305
  levels of 38
  manufacturing 124
  national 347
  northern, prospects for 145
  OECD 150
  output 103
  potential 39
  profitability 299
  services 125–6
  slow 70
  technological change, a consequence of 96
  undermining 239
  urban 33
Gudgin, G 283, 284

Hailsham, Lord 37
Hamburg 324, 325, 327
Hamnett, C 205

373

Index

Hampshire 106
Hansen, B 256
Harris, R D 93, 94, 95, 96 n.
Harris, R I D 287
Harrison, R 45, 72, 91
Hartlepool 276, 313
Hart, R A 221 n.
Harwich 172
haulage costs 171, 172, 173, 175
health 121, 122, 143, 239, 316
Heathrow Airport 147, 329, 341
Heath, Edward 37
Heckscher-Ohlin trade theory 30, 31, 109, 167, 170, 357
Hendry, D 264, 265
Highlands and Islands Development Board 352
HM Customs & Excise 164
Holland, *see* Netherlands
Holmans, A 246, 253, 266
Home Office 43
home ownership 151
Hotelling, H 32
house prices 56, 205, 256-62, 268, 355, 365
  convergence 278
  determinants 246, 264-6
  differentials 54-5, 60, 79-81, 233, 247, 263;
    determinants 248-9
  disparities 212
  distortions 59
  inflation 103-4, 204
  owner-occupied 247
  pattern 253-6
  rates as a percentage of 135-6
  structure 250-3
housing 28, 202-13, 229
  action trusts 46
  adequate 340
  associations 139, 241
  costs 53, 151, 164
  land for 113, 120, 236
  private ownership 276
  proposals for 137, 143
  reforms 139-40, 141
  rented 137, 267, 351; private 251, 252, 266, 269
  role of 246-69
  subsidies 99
  see also council housing; house prices; housing markets; owner-occupation; tenure
housing markets 57-8, 231-2, 242, 359, 363
  distortions 59, 267, 269
  imperfect 43
  influence on migration 226
  instability 265
  liberalising 145
  owner-occupied 56, 212, 256
  problems 360-1
  reforms 364-5
  rented 112, 119
  role of 218
  speculative frenzy and 262
Howe, Sir Geoffrey 71
Howells, J 290, 291
Hudson, R 275
Hughes, G 54-9, 99, 221-2, 225, 226, 228, 231, 232, 234, 243, 246, 252, 261, 263
Hull 131, 133, 173
Hungary 326

Ibbs, Sir Robin 304
IBM 152, 205, 289
ICI 205, 273, 290, 309, 310, 316
ICTs (information and communication technologies) 299, 300, 316
IDS (Incomes Data Service) 199, 204, 207-8, 209, 211-12, 344
IMF (International Monetary Fund) 276
Immingham 174
immobility 56, 57, 118, 144, 252

Index

IMP (Integrated Micro Products) 289
incentives 139, 241, 346-8, 353, 364
  ad hoc expansion of 208
  automatic 38
  government reaction to 135
  investment 43, 92
  labour, arguments against 41-2
  migration 57, 243, 357
  recognition of need to provide 44-5
  relocation 205
  tax 212, 246
income 41, 78-81, 123, 279
  distribution 31, 46, 280
  low 70, 72
  per capita, Denmark/Ireland 50
income support 50, 99
incubator firms 89, 90
Industrial Development Act (1966) 38
Industrial Development Certificates 38, 40, 42-3, 60, 238, 347
  controls reduced 39
  introduction of 37
industrial development executive 39
industrial revolution 273
industrial structure 34, 35, 90, 121-2
Industry Act (1972) 39
industry mix 84-6, 89, 100
inefficiencies 27, 35
inflation 207, 230, 249, 336
  build-up of 107
  house price 103, 104
  severe 101
  short-run 59
  wage 30, 54
information technology 362, *see also* ICTs
infrastructure 52, 144, 275, 337, 348, 351
  dilapidated/poor 70, 72

EC money for 48, 50
  for the future 317-18
  high quality 33
  investment in 281, 325, 328, 340-1, 351, 353
  investment decisions for 60
  transport 115, 363, 365-6
  UDCs and 44
  under-utilised resources for 59
Ingram, P 200
injustice 101
Inland Revenue 164
inner cities 43-6, 337, 340, 351
  contraction of 34
  development schemes for 313
  disadvantaged people in 37
  physical environment 36
  riots 334
Inner Urban Areas Act (1978) 43-4
innovation 39, 45, 351, 365
  diversification by 286, 287, 288, 289
  landbridge 147
  policy 40
  poor (product) 35
  potential for 100
  pressure for 152
  propensity 95-7, 106
  research and 342
  technical, in transport and communications 275
inputs 107, 133
insider-outsider theory 258
Institute of Personnel Management 202
interest rates 103, 259
intervention 27, 41, 43, 45, 125, 239
  capital market 238
  EC 46
  break with/move away from 39, 334
  fragmentation of 282
  inappropriate 143
  local authority 351

375

Index

responsibility for 37
to boost entrepreneurship 281
investment
  ancillary 144
  appraisal 42
  attracting and supporting 44
  capital 29, 238
  decisions 60, 345
  diversion of 282
  EC 48, 50
  infrastructure 281, 325, 328, 340-1, 351, 353
  manufacturing industry 280-1
  new 41, 318
  opportunities 92
  premium scheme 42
  rates: falling 40; low 35
  subsidies to 37
  tax relief on 91
  telecommunications 314
  see also inward investment
inward investment 307, 315, 330, 342, 349-50, 362
  Enterprise Zones and 45
  Japanese 94, 305, 307-8, 330, 338, 349
  non-EC countries 339
Ipswich 147
Ireland, Northern, *see* regional issues
Ireland, Republic of 48, 49, 50, 100
IRS (Industrial Relations Services) Employment Trends 198, 200
Italy 49, 219, 220, 325

Jackman, R 227, 260, 264
Japan 219, 220, 289, 315, 342, 356
  inward investment into UK 94, 305, 307-8, 330, 338, 349
Jarrow Marches 274
job creation/prospects, *see* employment
Johnson, G E 220
Johnson, S 88, 90, 292, 293, 294
joint ventures 289

Kan, B 227
Kings Cross 73
Kleinman, M 237
Knowsley 175 n.

labour 43, 101, 102, 110, 150, 357
  allocation 239-40
  costs 35, 106, 131, 146, 183, 199, 344
  demand 356
  excess supply 190
  intensity 130
  relations 35
  shortages 79, 103, 213
  skilled 114, 135, 136, 138, 190; costs 110, 111; shortages 202; *see also* skill shortages
  supply 137, 167, 184, 190
  turnover 53
  unskilled 112, 114, 115, 133, 136, 137
  see also labour markets; manual labour
Labour Force Surveys 116, 142, 226, 228, 261, 263-4
labour-intensiveness 33, 35, 130, 282, 305, 346
labour markets 33-4, 41, 52-4, 257, 263, 363-4
  adjustment 35, 234, 242
  circumstances 226-9
  competitive conditions 185
  deregulation 213
  differentiated 198
  distressed/depressed 230, 243
  dominated by government employer 304-5
  external 201
  functioning 60, 361
  influence on inflation 104
  labour shortages 107
  liberalising 145
  migration and 218-26
  nominal wages in 192

376

owner-occupation and 258-60
performance 59, 106
pressures 36, 191, 209, 346
problems 357-9
rigidities 190
segmentation 56, 246
threatening 38
tightening 196, 205
young people entering 203
Labour Party 276
government 38, 40
laissez-faire 358, 360
land 114, 119, 357
acquisition 44
derelict 36, 311
housing 113, 361
industrial 138, 139
market for 36, 43
rentals 136, 168
shortages 151
supplies 112, 137, 149, 152, 361
under-utilised 281
land prices 118, 119, 129, 139, 150, 358
agricultural 110, 112, 119, 168
differentials 141, 151
increased 111, 125, 137, 167
with planning permission 110
Landbridge 146, 147, 327
landlords 145
large employers 206
Lawson, Nigel 255, 343
Layard, R 142, 220, 257
Lee, K 185, 187, 189, 191
Leeds 301, 302-4
lending practices 268
less developed countries/regions 31
Lewis, J 72
Leyshon, A 301, 302
linkages 42, 95, 223-5
liquidity 56, 259
Liverpool 121, 149, 303, 304
and shipping 128, 131, 133, 146, 147, 172

living costs 53, 55, 258
loans 38, 44, 47, 49
Lobban, P 200
local authorities 111, 113, 208, 267, 299
agencies supported by 273
behaviour 135-7
housing 99, 276; vacant 59
initiatives 312
property rights and 148
reform 352-3
resources 350-1, 353, 363
UBR and 139
urban policy 44, 45
local development agencies 312
local government, *see* local authorities
Local Government and Housing Act (1989) 350
Local Government, Planning and Land Act (1980) 44
locations 35, 110, 217, 361-2
allowances 209
choice of 55, 164, 226-9
determined by profit maximisation 33
optimal 28
rigidity of 241
theory 32
transport differential 151
urban 36
lolo (lift-on, lift-off) 131, 133, 173
London 91-2, 211, 237, 274-6, 325
computer networks 298
congestion 328
council estates 73
Docklands 44, 114, 341, 346
earnings 188-9, 209, 304, 344
fund managers 91-2
moves: into 55; out of 212, 301, 302, 344
new airport for 329
rates 36
shift of producer services from 33

377

Index

transport differentials 131, 173, 175
see also Greater London
London allowance 185, 203-8
long-term unemployed 138
Love, J H 95
low-income families 267
Lowry, I 228
Luxembourg 49

McCormick, B 54, 56, 57-9, 99, 216-43, 246, 252, 261, 263, 269, 356, 358-9, 364-5
McGregor, B E 36
McIntosh, J 31
McMaster, I 221 n.
Macmillan, Harold (Lord Stockton) 38
Makower, H 223
management 134, 146, 152, 322
Manchester 36, 114, 173, 303, 304
  air transport 147, 327, 329
  shipping 131, 147, 172
  terminals planned for 326
Mann, K 202
manual labour
  allocation 242
  firms with high ratio of 95, 357-8
  market 112, 145-6
  migration 57, 58, 139, 141; effects 140; from South East 59, 260; low 225-6, 229-35
  mobility 144, 151
  percentage as council tenants 139
  skilled 227, 290
  supply 114, 141, 150, 152, 180-1
  technological shift away from 128
  unemployment 223, 359
  unused 123
  wages 52, 114, 125, 129, 167, 237; decline 152; equalising 151; rising prosperity and 279-80

manufacturing 49, 90, 93-7, 284, 294, 358
  concentration on 106
  creation of jobs in 42, 94
  decline 41, 111, 122-3, 335
  expenditure of value added 133
  foreign-owned companies 306, 349
  GDP and 123, 153-63, 302
  growth 124
  investment 180-1, 280, 305-8, 340
  long-established 293
  reliance on 85, 121
  revival 147
  and services 34, 110, 113, 114, 152; decline of pay agreements 193; labour supply 184; wages 183
  small companies 43, 310
  terms of trade 124-5, 129
  traditional areas of 40, 48
Marginson, P 198, 199
market failure 97-9, 235, 357
market forces 201, 317, 357
  power of 115
  protected sector and 143, 146-7
  reliance on 334-5
  regional policy and 109-84
markets 41, 53, 57, 70
  access to 344
  capital 47, 92, 238
  demand 33, 89, 90
  distortions 143, 144
  financial 205
  globalisation of 361
  goods 143
  imperfect 43, 281
  internal 99
  international 301
  land 113
  output 238
  price 144, 148
  property 37

378

regulation of 282
world 110
see also housing market; labour market; market failure; market forces
Marseilles 324
Marshall, J N 305
Martin, R 45, 92, 204, 277
Mason, C 45, 72, 91, 92
Massey, D 275
Mayhew, K 109 n.
mechanisation 110
Merscyside 43, 44, 109, 119, 149, 176-84
MetroCentre 313
microprocessors 287, 288, 289
Middle East 326
Middlesbrough 313
Midland Bank 199
Midlands, *see* regional issues
migration 36-7, 54-60, 70, 81-3, 98-9, 119-21
  barriers to 205, 239-40
  business 152, 358
  housing and 226, 359-60, 361
  incentives 57, 243, 357
  not always required 30
  predicted rates of 252
  and regional policy 216-43
  to and from the South East 259-64
  unemployment and 183, 249, 342-3, 364
  variations 248
  see also manual labour; Greater London
Millward, N 193
Minford, P 41, 54, 56, 99, 135, 137, 139, 143, 166, 252, 253, 259, 357, 358, 359, 361
Ministry of Defence 51
MMC (Monopolies and Mergers Commission) 147

mobility 118, 120, 212, 218, 260, 342-6
  barriers to 239-40
  degree of 217
  as equilibrating mechanism 54
  factor contributing to decline in 305
  housing tenure and 56, 246-7, 252-3, 257, 262-3, 266
  improving 59
  incentives 139
  index 57, 252
  low 41, 229-35, 360
  occupational 358
  perfect 238
  privatisation of council housing and 241-2
  restricted 258
modernisation 42
Molho, I 221 n.
monetary policy 29, 103
monetary union 336, 338
monopoly 111, 133, 146, 147, 358
Moore, B 36, 42, 43, 185, 189, 347
Morgan, K 45, 51
mortgage interest tax relief 56, 234, 259, 266, 267, 359
motorway networks 275, 318, 322, 324-5, 328
Moyes, A 89
Muellbauer, J 54, 55, 99, 216 n., 260, 263, 269, 343, 361
multinationals 94, 277, 288, 305, 316, 339
multiplier effects 29, 30, 41, 42, 260
Munich 324, 325
Murphy, A 54, 55, 99, 260, 263, 269, 343, 360, 361

National Audit Office 305
National Board of Prices and Incomes 208
National Economic Development Council 338, 343

379

Index

NDC (Northern Development Company) 273, 312
Neary, J P 31
Netherlands 42, 49, 226, 227, 268, 325
New Community Instrument (1978) 46-7
New Earnings Survey (1978/1988) 142, 187
new firms 88-93, 106, 291-4
New Towns 37, 44, 275
Newcastle University 315, 316
Newcastle upon Tyne 273, 290, 309, 362, 365
  cultural industries 316-17
  development proposals 313, 314, 315
  downgrading as office centre 274
  local authority 312
  service industries 276, 300, 301, 302-4, 305
NHS (National Health Service) 60, 209, 211, 261, 264
Nickell, S 53, 257, 258
Nissan 138, 152, 307, 315, 338, 348
NJICs (national joint industrial councils) 192, 195, 199
Nolan, P 212
Nord/Pas-de-Calais 324, 330
North/North East/North West, *see* 'Celtic Fringe'; Cleveland; Liverpool; Manchester; Merseyside; Newcastle upon Tyne; Northumberland; regional issues; Sunderland; Teesside; Tyneside; Tyne and Wear; Wearside
Northcott, J 287
Northern Development Company 297
north-south divide 28, 31, 32, 70-107, 337, 335
  and UK trade orientation 18
Northumberland 273, 274, 291

occupations 54, 78, 116, 232, 233
  structure 89, 90, 289-91
OECD (Organisation for Economic Cooperation and Development) 150, 202, 219, 236, 343
OFT (Office of Fair Trading) 131, 147
oil 174, 205, 206
open sector 152, 167, 168, 169, 170
Open University 289
opportunities 100, 144, 238, 314-17, 326-8
output 29, 122-4, 167, 184
  aggregate 59, 168-70
  growth of 103
  manufacturing 97
  national 101
  non-manufacturing 182
  protected sector 168
  services 183
  subsidies and 42
overheads 291
Owen, D W 296
owner-occupation 90, 249, 256, 269
  distortions and 56, 212
  labour markets and 258-60
  mobility and 252-3
  national employment and 116
  price of 218, 226, 266
  proportion in UK 57-8, 251
  tax relief to encourage 232
Oxford Economic Papers 223
Oxford group 233

Palmer, D 363
Paris 175, 314, 324, 325, 328
Parsons, W 38, 39
Patten, Christopher 105
pay, *see* wages
Peel, M 54
performance 27, 86, 299
  corporate 202, 293
  differences in 106

growth 85, 105
investment 340
poor 287
regional disparities and 74–84
supply side 362–3
peripherality 294–300
Phillips curve 30, 38, 53
Pianelli, M 27 n.
pilotage 131, 171
Pissarides, C A 53, 98, 221 n., 227, 232, 264
pit closures 277
planning 115, 143, 330, 337–8, 353, 356
  controls 37, 148, 180, 267, 362; distortions on 56
  decisions 144
  and development 351–2
  land restrictions 139
  national 38
  permissions/permits 36, 44, 110, 358–9
  poor 331
  regulations 358
  system 137
  zones 45
policy 190–1, 308
  analysis 235–42
  change 333–6
  intra-regional 143–4
  options 266–9
  reform 239–40
political unity 100, 101
poll tax 138, 237, 238, 242, 267, 268
polytechnics 88 n.
ports 130–3, 146, 324, 325, 327
  charges 171, 172
  costs 174
Portsmouth 131, 147
Portugal 48, 49, 50, 100, 348
Post Office 209, 210
poverty 73
prices 53, 113, 133

differentials 174
elasticity 183
factor 29, 30, 31
relative factors/trends 115, 125
stability 40
see also house prices; land prices
privatisation 143, 145, 147, 210
  council housing 230, 240–1
production 29, 34, 35, 53, 101, 170
  mass 275
productivity 55, 141, 146, 150–1, 346
  backward countries/regions 31
  competitiveness and 84
  concessions 200
  educational attainment and 87
  growth 184
  indicators 200, 202
  unions' effect on 142
  variations 27
profitability 35, 97, 123, 142, 200
property developers 313
property rights 148
prosperity 138, 274, 278, 279, 335
protected sector 167–70, 206, 208–11, 301
  industrial structure 121–2, 123
  relocation 113–14, 138, 164
  union power and 111, 142–3, 133–5, 145
public administration 121, 122
public sector, *see* protected sector
Purcell, J 198
'push' factors 89, 90

QSP (Quality Software Products) 289
Queen's Awards 291
quotas 48, 111

race relations 43
rail 317, 324, 325, 326, 327, 328
  competition 147
  freight rates 175

Index

rate support grant 137, 276
rates
  business 36, 111, 135, 136, 137; uniform (UBR) 138, 139, 322
  freight rates 171, 172, 175
  personal 135
  revaluation 137, 138
rationalisation 274
raw materials 110, 114
Raybould, S 292, 293, 296
RDGs (regional development grants) 42, 310-11, 347
Reading 293, 294
reallocation 31, 43
rearmament 274
recessions 50, 76, 94, 274, 277, 315
recovery 277-81
recruitment 209, 211, 291
Redcar 291
redevelopment 44
redundancy 89
reflation 38
reforms 50, 51, 60, 139-41, 268-9, 364-5
regional issues
  aid 36, 46, 48, 143-4
  development 333-54
  development grant 309
  earnings 185-213
  economic analysis 28-34
  economic disparities 70-107; spatial 28-34
  employment premium 29, 39, 40, 43
  government 352-3
  industrial policy 309-11
  industrial structure 283-9
  pattern of house prices 253-6
  Planning Councils 38
  policy 37-43, 109-84, 216-43, 321-31
  problems 281-308
  profiles 153-63
  unemployment 115-19

regulations 133, 134, 252, 358
  market 282
relocation 40, 105, 212, 302, 322, 362
  defence establishments 52
  economic sense of 100
  incentives 205
  protected sector 113-14, 138, 164
remuneration packages 205, 207
Rent Act
  (1965) 56, 252, 253
  (1974) 252, 253
rents 33, 56, 118, 129, 137, 241
  savings in 164
  subsidy reforms 241
  tribunals 145
resale price maintenance 134
research and development (R&D) 51, 53, 95, 164, 308, 309
  in France 314-15
  importance 289-91
  initiatives 316, 338
  location 362
  role 340
resources 48, 168, 170, 341, 351
  allocation 148, 347, 357, 362
  decentralisation 353
  inefficient use of 363
  reallocation 43
  under-utilised 366
Restart 138, 141-2, 150
restructuring 39, 42, 97, 300, 317
retail distribution 206
retraining 46, 138, 152
Reward Survey (1989) 207
Rhodes, J 36, 42, 43, 185, 189
Ridley, Nicholas 105, 237, 241
Rifkind, Malcolm 100
rivers 325
roads 275, 317, 322-3, 327, 328, 331
  distribution 323
  network 330
Roads to Prosperity (White Paper) 324

Index

Robertson, D 192
Robinson, F 308, 309, 310–14
Rogaly, J 72–3
Roper 260
roro (roll-on roll-off) 131, 133, 173
Rotterdam 131 n., 147, 172, 173, 175, 327
Rowntree-Mackintosh 339
Rowthorn, B 277
rural areas 50

satellite links 305
Savouri, S 264
Sayer, A 45, 51
school leavers 87, 90
schools 45, 239
Scotland, *see* 'Celtic Fringe'; regional issues
Scottish Development Agency 352
Scottish Office 352
segmentation 56, 246
self-determination 101
semi-bulks 130, 131, 133, 172–3, 174
services 133, 167, 300–5
  financial 205, 206, 207, 301, 302
  goods and 33, 357; non-traded 111
  growth of 124–5
  manufacturing and 34, 110, 113, 114, 152; decline of pay agreements 193; labour supply 184; wages 183
  obsolescent 48
  percentage of employees in 90
  producer 35
  terms of trade 129
  tradability of 29, 30
SFA (selective financial assistance) 310
shareholders 134
Shell Chemicals 164
shift-share analysis 34, 85–6, 284, 287
shipping 147, 327
  deep sea 146, 130, 171–2, 174
shocks 31, 218, 220, 266, 365
  demand 53, 54, 55, 225, 249
  product market 234
Silicon Valley 289
Simmons, P 109 n.
Single European Market 322, 323–4, 326–7, 330, 331
  peripherality and 296
  reinforcement of regional disparities 99–100
skill shortages 202–12
small firms 88, 289, 291–4, 311, 312, 350–1
  policy 90–1
Smith, D 72, 105
Smith, I 307
Smith, I J 275
social attitudes 115
South/South East/South West, *see* regional issues
Southampton 173, 207
Spain 48, 49, 50, 100, 325, 348
spatial economices 32–4
  disparities 28–34, 89, 107
Special Areas Act (1934) 37, 71
special social need 43
speculation 36
Spence, N 396, 297
sponsorship 312
SPRU (Science Policy Research Unit) 96 n., 97, 286
SS curve 167, 168
standard of living 27, 46, 53, 55, 78–81
Stanlow refinery 174
Stansted Airport 340–1
staple industries 75
start-ups 35, 106
Steedman, H 212
Stevens, M 193
Stone, I 307
Stoney, P 357, 358, 359, 361

383

Index

Storey, D J 88, 90, 292
Strabane 27
Strasbourg 324
Stuttgart 34
subsidies 29, 37, 41, 313, 347-8
   automatic 311
   migrant 243
   mortgage 209
   rent 118
   resulting in spin-off 42
   shipping 131, 146
   tenure 237
substitution effect 39
Sunderland 313, 338, 348
supply 32, 362-3
   housing 54, 248, 267
   labour 137, 184, 167, 190, 246; manual 112, 114; mismatch of demand and 356
   land 119, 137, 150; for development 235-7
Sussex University 96 n., 97
Swales, K 42
Swan Hunter 291
Sweden 220
Switzerland 145, 325

takeovers 94, 106
tariffs 31, 100, 173, 324
task forces 44, 314
taxation 51, 139, 230, 240, 269, 361
   allowances 44, 242
   concessions 312-13
   domestic rates as 135; *see also* poll tax
   effort to reduce 40
   incentives 212, 249
   local government 111
   property 268
   regional pattern 29
   relief 91, 232; mortgage interest 56, 234, 259, 266, 267, 359
Taylor, J 29, 31, 34, 37, 72, 85 n., 101, 216 n., 355, 356, 357

Team Valley 274, 308
technological change 95-7, 128, 130, 289
technology 110, 288, 338, 362
   capital intensive 33
   communications 282
   high 284, 286, 289, 308, 315; microchip-based 96
TECs (Training and Enterprise Councils) 273, 346
Teesport 174
Teesside 273, 290, 313, 314, 316, 326
   new firms 293, 294
telecommunications 296, 314-15, 316, 317, 341, 356
   importance of development 48
tenure 55, 246-7, 262, 365
   migration and 56, 58
   national employment by 116
   patterns 220, 231-2
   structure of 249-53, 266
   subsidies and 237, 359
   wage pressure and 55, 257
terminals 326
terms of trade 124-5, 129, 152, 358
Terry, M 192
TGV train 324
TGWU (Transport and General Workers Union) 146
Thirlwall, A P 96
Thomson, A 200
Thwaites, A T 288, 289, 290, 356, 361, 362
Tory Reform Group 101
tourism 48, 180, 327, 328
Town and Country Planning Act (1947) 37
Townsend, A 772
Toyota 348
trade theory 30-1, 52
trade unions 27, 138, 197, 200, 231, 276
   civil service 210

384

Index

monopoly 111
power 133–5, 142–3, 145, 183, 258; and prices 53; and wage pressure
role 201
strong 97
traded sector 113, 121, 136, 143, 180
trade-offs 32, 55, 59, 256–60
traditional industries 273, 274
training 145, 281, 337, 340, 362, 365
  courses/schemes/programmes 48, 338; *see also* ET; Restart; retraining; TECs; youth training scheme
  reorganisation of strategies 212
  vocational 50
transport 104, 114, 152
  infrastructure 115, 321–31, 338, 341, 363, 365
  see also air; airports; motorway networks; rail; roads; shipping; transport costs
transport costs 29, 32–3, 133, 135–6, 296–7
  disadvantage 128–33
  export/import 171–5
  high 358
  negligible 30
  reduction in importance 34
  rising 111
Treaty of Rome 46
TUC (Trades Union Congress) 273, 333–54, 362–3
  see also trade unions
turnover 209, 228, 231, 291, 292, 298
Tyler, P. 36, 42, 43, 204
Tyne and Wear 138, 152, 273, 291, 312, 313
Tyneside 34, 35, 43, 273, 276, 313
Tyneside Metro 275

UBR (uniform business rate), *see* business rates
UDCs (urban development corporations) 44, 138, 143–4, 152, 273, 309, 351
Ulph, Alistair 216 n.
unemployment 99, 113, 137, 144, 355, 365
  aggregate 216, 223, 247
  differentials 36, 59, 115, 218, 262; equilibrium 220
  disparities 72, 75–6, 97
  excessive 238
  fall in 142, 145
  high 37, 46, 58, 190, 357; changing circumstances and 70; ESF support 48
  immobility and 56, 57
  in Japan 219
  land 112, 136
  long-term 50
  low 79, 90–1, 98
  lowering/reducing 138, 143
  migration and 183, 221, 226–7, 249, 342–3, 364
  by occupation and tenure 116
  perpetuating 141
  producing 110–11
  of resources 170
  rising 40, 71, 118, 125, 228–9, 277; decline of manufacturing and 335
  structural 101, 102
  union power and 135
  vacancies and 53, 54, 248, 249, 256–60
  variations in rate 80, 103, 104, 107, 274
  wages and 30, 52, 73, 117, 119, 139
unemployment benefits 145, 257
  wages relative to 110–16, 119, 150, 168
United States 228, 274, 289

Index

closure of UK manufacturing plants 307
education 230-1
housing markets/prices 57, 268
migration 58, 225, 226, 240
mobility 229, 359
unemployment 219
urban areas 34-7, 329-30
  dereliction 111
  policy 43-6
  regeneration teams 273

vacancies 53, 54, 248, 249, 256-60
value-added activities 35, 124, 133, 135
Van Dijk, J 226, 227
venture capital 45, 92, 106, 365
Vines, D 31

Wadhwani, S 201, 345
Wadsworth, J 98, 227, 232, 264
wages 54-9, 110-20, 183, 246, 357
  adjustment 53, 229, 260
  aggregate 234, 247, 256, 262, 263
  agreements 209
  and benefits 150, 168
  cost 41
  cut in 29
  differentials/disparities 80, 98, 138, 196, 202-13, 343-6,. 359; demand variations and 53; encouraging 60
  disparities 80
  distribution 219
  efficiency 53, 83, 84, 106, 355
  flexibility 97, 98, 185-213, 241, 364
  levels 79, 98, 280
  non-union 142
  pressures 103, 248, 257, 258, 263
  public sector 276
  reduction in, and migration 228-9
  restricting 192
  rigidity 31
  skilled 135, 137
  unemployment and 30, 52, 73, 117, 119, 139
  see also bargaining; manual labour (wages)
Wagner, K 212
Wales, *see* 'Celtic Fringe'; regional issues
Walker, Peter 101
Walsh, J 30, 97, 199, 360
Walsh, T 27 n.
Warrington-Runcorn 180
Warwick Survey (1978) 192
wealth 73, 276
Wearside 273, 314
Welsh Affairs Committee 347
Welsh Office 352
West Cumbria Development Agency 312
West Midlands, *see* regional issues
Westhead, P 89
White Papers
  (1944) 37
  (1984) 334, 346
  (1988) 185, 190, 192, 347, 360
  (1989) 324
Whitehead, C 237
Williams, H 284, 299
Williamson, J 31
women 278, 345
Wood, A 27 n, 45
Workington 276, 313
Workplace Industrial Relations Survey (1984) 95, 192, 193, 194, 197, 202
Wren, C 43, 310

Yorkshire and Humberside, *see* regional issues
young people 50, 203
youth training scheme 45, 138
Yugoslavia 326

Zeebrugge 173
zoning 36, 56